T0303858

Vanishing Boundaries

How Integrating Manufacturing and Services Creates Customer Value

Second Edition

RECENT TITLES

Vanishing Boundaries: How Integrating Manufacturing and Services Creates Customer Value, Second Edition
by Richard E. Crandall and William R. Crandall
ISBN: 978-1-4665-0590-2

Food Safety Regulatory Compliance: Catalyst for a Lean and Sustainable Food Supply Chain
by Preston W. Blevins
ISBN: 978-1-4398-4956-9

Driving Strategy to Execution Using Lean Six Sigma: A Framework for Creating High Performance Organizations
by Gerhard Plenert and Tom Cluley
ISBN: 978-1-4398-6713-6

Building Network Capabilities in Turbulent Competitive Environments: Practices of Global Firms from Korea and Japan
by Young Won Park and Paul Hong
ISBN: 978-1-4398-5068-8

Integral Logistics Management: Operations and Supply Chain Management Within and Across Companies, Fourth Edition
by Paul Schönsleben
ISBN: 978-1-4398-7823-1

Lean Management Principles for Information Technology
by Gerhard J. Plenert
ISBN: 978-1-4200-7860-2

**Supply Chain Project Management:
A Structured Collaborative and Measurable Approach, Second Edition**
by James B. Ayers
ISBN: 978-1-4200-8392-7

Modeling and Benchmarking Supply Chain Leadership: Setting the Conditions for Excellence
by Joseph L. Walden
ISBN: 978-1-4200-8397-2

New Methods of Competing in the Global Marketplace: Critical Success Factors from Service and Manufacturing
by William R. Crandall and Richard E. Crandall
ISBN: 978-1-4200-5126-1

Supply Chain Risk Management: Minimizing Disruptions in Global Sourcing
by Robert Handfield and Kevin P. McCormack
ISBN: 978-0-8493-6642-0

Rightsizing Inventory
by Joseph L. Aiello

Vanishing Boundaries

How Integrating Manufacturing and Services Creates Customer Value

Second Edition

Richard E. Crandall
William R. Crandall

CRC Press
Taylor & Francis Group
Boca Raton   London   New York

CRC Press is an imprint of the
Taylor & Francis Group, an **informa** business

CRC Press
Taylor & Francis Group
6000 Broken Sound Parkway NW, Suite 300
Boca Raton, FL 33487-2742

© 2014 by Taylor & Francis Group, LLC
CRC Press is an imprint of Taylor & Francis Group, an Informa business

No claim to original U.S. Government works

Printed on acid-free paper
Version Date: 20130710

International Standard Book Number-13: 978-1-4665-0590-2 (Hardback)

Library of Congress Cataloging-in-Publication Data

Crandall, Richard E., 1930-
[New methods of competing in the global marketplace]
Vanishing boundaries : how integrating manufacturing and services creates customer value / Richard E. Crandall, William R. Crandall. -- Second edition.
pages cm. -- (Series on resource management)
Includes bibliographical references and index.
ISBN 978-1-4665-0590-2 (hardback)
1. Management. 2. Manufacturing industries--Management. 3. Service industries--Management. I. Crandall, William, 1956- II. Title.

HD31.C687 2014
658--dc23   2013026188

Visit the Taylor & Francis Web site at
http://www.taylorandfrancis.com

and the CRC Press Web site at
http://www.crcpress.com

This book is dedicated to Jean and Sue for their encouragement and patience not only during its preparation but also throughout all the other years we have been together. They are each "one of a kind."

Contents

Foreword xxiii

Introduction xxv
    Organization of the Book xxv
    How to Use This Book.. xxvii
    Who Can Use This Book xxviii

The Authors xxix

Acknowledgments xxxi

1   The Vanishing Manufacturing/Services Boundary 1
    Differences between Manufacturing and Service 1
    Forces That Are Eliminating the Boundary 3
        The Economy Is Going through a Natural Evolution 3
        There Is a Need to Identify Critical Success Factors.................... 4
        Supply Chains Are Expanding to Integrate Manufacturing and
        Service 4
        It Is Easier to Blend Tasks, Resources, and Techniques into
        Programs 4
        Service Businesses Are Looking for Ways to Operate More Efficiently ... 5
            Health Care Providers Are Looking for Ways to Cut Costs
            and Improve Quality 5
            Banks Are Facing Competition from Lower-Cost Banks........... 5
            Investment Bankers Are Facing Competition
            from Commercial Banks and Discount Brokers 6
            The Retail Industry Is Moving in Two Strategic
            Directions—Low Cost and Focus 6
        Manufacturing Is Adding Services to Be More Customer Focused..... 6
            Post-Sales Services 7
            Financing of Purchases... 8
            Online Purchasing.. 8
            Customer Input to Product Design... 8
            Enhanced Customer Relationship Management........................ 8

Technology Development 9
A Movement toward a Process Perspective 10
Outsourcing Focuses More Attention on the Need for Good
Service 10
Changing Personal and Organizational Relationships 11
The Need to Integrate Companies into Supply Chains 12
Services Become More Relevant to All Types of Customers..........12
Continuous Improvements Continue to Blur the Boundary
between Manufacturing and Services 12
    There Is a Movement toward an Industry Focus and away
    from a Functional Focus..12
    The Mass Customization Concept Requires a Blending of
    Manufacturing and Service 12
The Vanishing Manufacturing/Services Boundary......................14
    Phase 1: Separated Disciplines 14
    Phase 2: Internal Improvements in Costs and Quality..........14
    Phase 3: Customer Service Improvements.......................16
    Phase 4: Integrated Product and Service Functions..........16
Summary 17
References 17

2 Critical Success Factors and Strategic Planning......................19
What Are Critical Success Factors? 19
The Evolution of CSFs in the United States 21
Other Changes during a Country's Economic Life Cycle 25
The Need to Be Effective 27
A Hierarchy of the Planning Process....................................29
    What Is Strategic Planning?......................................30
    The Strategic Planning Process 31
    The Business Planning Process....................................33
    Project and Program Plans 34
    CSFs Plan 34
A Hierarchy of Critical Success Factors.................................34
    Background of CSFs in Strategic Planning 35
    Characteristics of Strategic CSFs 37
    The Temporal Nature of CSFs 38
        Startup 39
        Growth...39
        Maturity...40
        Decline..40

The Role of CSFs in Operational Planning 40
    Positioning Strategy 42
    Manufacturing Flows 43
    Manufacturing Strategies 43
Role of CSFs in Selecting Management Programs 44
Performance Measurement and CSFs 45
Summary 46
References 46

3   The ITO Model 49
   The ITO Model 49
   Introduction to Models 50
     Types of Models 50
     Why Are Models Used? 52
     Benefits of Using Models 52
   The Basic ITO Model—Inputs, Transformation, and Outputs 53
     The General Systems Model 54
     The Basic Model Extended 56
     Components of the Model 57
        Customers 58
        Outputs 58
        Transformation Process 59
        Inputs 60
   Extending the Basic ITO Model into Supply Chain Configurations 62
     A Look at the Supply Chain 62
     The Closed and Open Systems Model 62
        Closed Systems 62
        Open Systems 63
        Closed-System Strategy 65
        Open-System Strategy 66
        Feedback 67
     The Fine Art of Building Relationships 68
   The Concept of Reverse Logistics 70
     Why the Interest in Reverse Logistics? 71
     Benefits of Reverse Logistics 72
     Barriers to Reverse Logistics Implementation 73
     System Design and Implementation 73
   Summary 75
   References 75

4 The Role of Management Programs in Continuous Improvement.......77
What Are Management Programs? 78
   Normal or Sustaining Day-to-Day Operations 81
   Problem-Solving Activities...81
   Improvement Programs 82
Management Program Life Cycles 85
   Life-Cycle Stages 85
   At the End of the Life Cycle 87
   Implications of Program Life Cycles for Management 88
Why Are Management Programs Important? 89
Where Do Management Programs Come From? 90
   Overview of Employee Management Theories 90
   Scientific Management 91
   Human Relations Management..91
   Administrative Management 92
   Systems Theory...95
   Contingency Theory...95
Why Are Some Programs Successful and Some Not?...........................98
   Failure to Match Program with Need 98
   Implementing the Program Correctly 102
      Planning and Preparation 102
      Execution and Evaluation 104
Future of Management Programs 105
   Management for the Twenty-First Century 105
Summary...107
References..107

5 Adapting Manufacturing Techniques to Services 109
Introduction 109
Description of Manufacturing Process Types.................................... 110
Product–Process Relationship 113
Service Industry Classifications... 113
   The Chase Model.. 114
   Schmenner Model 114
Comparison of Manufacturing and Services...................................... 114
Manufacturing Objectives 116
   Reduce Product Costs... 116
   Reduce Inventories... 117
   Increase Resource Utilization.. 118
   Improve Quality 119
   Reduce Response Time..120
   Reduce Product Development Time 121
Service Objectives...123

Programs That Work in Services 123
    JIT and Lean 124
    TQM and Six Sigma...127
Programs More Difficult to Adapt to Service Operations.......................130
    Product Costing 131
    Activity-Based Costing 131
    Materials Requirements Planning...131
    Enterprise Resources Planning...132
    Performance Measures..132
    Automation..132
    Resource Utilization 133
Keys to Extending Manufacturing Techniques to Services 133
Conclusions 134
References..135
Appendix 5A: Amazon...138
    Overview 138
    Retailing—Physical Products 140
    Retailing—Electronic Products 142
        Physical Products Downloads (Books)......................................142
        Amazon Web Services 143
    Distribution (Fulfillment Centers) 143
    Manufacturing (Kindle) 145
    Origination (Book Authors)...146
        Amazon Publishing..146
    Summary...146
References..149
Appendix 5B: United Parcel Service (UPS)..150
    A Brief History of the Company...150
    The Crusade for Continual Improvement 152
        Long History of Embedded Culture of Efficiency....................152
        Heavy Use of Industrial Engineering Techniques 152
        Progressive Use of Technology...153
    Other Keys to Success..154
        Stability at the CEO Level (Promote from Within) 154
        Emphasis on Drivers as the Key Customer Interface.............154
        Concern for Employee Welfare...155
        Relationship with Unions (Teamsters) 155
        Sustainability..155
    Evolution along the Supply Chain 156
        Heavy Focus on Customer..156
        Movement Upstream in Supply Chain 156
        Movement Downstream in Supply Chain—Retail Stores......157
        Provides Integrated Logistics Services—Domestic and Global ...157

Business Segments 157
    Global Small Package 157
    Supply Chain Solutions 160
    Supply Chain Capital 162
Key Financial Results 162
Summary 162
References 169

6 Extending Service Techniques to Manufacturing 171
Introduction 171
    The Rise of Services as a Part of the Economy 172
    The Swing of Power from Manufacturing to Retail 172
    Maturity of the Customer as a Shopper 173
    Increasing Complexity of the Marketplace 173
    Need for Manufacturing Companies to Add Services 173
    Move from Product-Centric to Customer-Centric 174
    Services as a Separate New Business Segment 177
    Economic Growth and the Need for Added Services 178
    The Movement from Make-to-Stock to Make-to-Order .. 178
    The Movement toward Mass Customization 178
What Are Services? 179
    Distributive Services 182
    Personal Services 183
    Self-Service 183
    Producer (Business) Services 183
    Social Services 183
Knowledge Transfer from Services to Manufacturing 184
    Areas of Manufacturing Expertise 184
    Areas of Service Expertise 184
Examples of Programs Developed in Services 185
    Customer Relationship Management 185
        Definitions 187
        Background 188
        What Does It Do? 188
        Benefits 189
        Problems 190
        Relation to the Supply Chain 190
        CRM's Future 191
    Response Time Reduction 191
        Quick Response Systems 191
        Continuous Replenishment Programs 192
        Efficient Consumer Response 192

Vendor-Managed Inventory 193
Sales and Operations Planning 193
Collaborative Planning, Forecasting, and Replenishment 193
Supply Chain Management 194
Present Status 194
Future 196
Flexibility 196
Product and Service Flexibility: Mass Customization 196
Horizontal Communication and Organization Structure 197
Demand Management 200
Location near the Market versus Lowest Cost 201
Evolution from Job Specialization to Self-Directed Teams 202
Replacement of Inventory with Information 202
Inter-Organizational Communications 202
Traditional EDI 202
Internet EDI 203
Other Service Developments 205
Working in an Open-System Environment 205
Product Development as a Result of Customer Inputs 205
Quality as Customers' Perceptions, Not Just Conformance
to Specification 206
Managing the Customer Encounter 206
Nonquantitative Performance Measurement 206
The Use of Business Intelligence 206
Summary 207
Differences between Manufacturing and Services 208
From Services to Manufacturing 208
Collaborative Efforts 210
Conclusion 210
References 211
Appendix 6A: GE—An Example of How to Blend Services into a
Manufacturing Company 217
Background 217
The Early Years 217
Why Are Services Important to a Manufacturing Company? 218
Ways for Manufacturers to Add Services 218
Develop a Market for the Product 218
Extend Known Technology into a New Product Area 219
Provide Post-Sales Support 220
Help Sell Products by Financing Sales 220
Provide Infrastructure in Emerging Countries 220
Enter a More Profitable Business 220

GE Progress over Time 221
    Growth during World War II 221
    Growth Council Product Proposals (1970s).........................224
    Growth Council Service Proposals (1970s) 225
More Recent Product Profiles 227
    The Welch Years 227
    The Immelt Years...234
    Recent Changes 243
A Model to Illustrate GE's Strategy 244
Conclusions 249
References..250
Appendix 6B: Hewlett-Packard—From Scientific Instrumentation to
Business and Consumer Products and Services 251
    Introduction 251
    Transformation One (1939–1959).......................................253
    Transformation Two (1959–1968) 254
    Transformation Three (1968–1976) 254
    Transformation Four (1976–1986) 255
    Transformation Five (1986–1999).......................................256
    Transformation Six (1999–Present).....................................260
        Fiorina as CEO (July 19, 1999, to February 9, 2005) 263
        Hurd as CEO (April 1, 2005, to August 6, 2010)................. 264
        Apotheker as CEO (September 30, 2010, to September
        22, 2011) 265
        Whitman as CEO (September 22, 2011, to present)..............265
        Financial Results during 1999–2011....................................267
    Services since 2000..267
    Summary..273
References..276

7 The Role of Technology in Continuous Improvement.......................277
Definitions... 280
The Role of Technology in Continuous Improvement....................281
    Technology and the Infrastructure 281
    Technology and Organizational Culture 283
    Technology Transfer 286
Technology for Process Improvement 286
    Agriculture, Mining, Construction, and Manufacturing: Goods
    Producers...287
    Services..290
    Self-Service 292
    E-Business 293
Technology for Resource Enhancement...296

Human Resources 296
   As an Aid to the Employee 296
   As a Substitute for the Employee 297
   As an Integral Part of the Process 298
Equipment..300
   Enhance the Performance of Equipment 300
   Provide a Source of Performance Information 300
Facilities...300
   Design 301
   Location 302
   Layout 303
Information Technology...303
Integrated Systems 304
   Enterprise Resource Planning Systems...............................304
   Inter-Organizational Systems 304
   Service-Oriented Architecture 306
   Structure, Trust, and Collaboration 306
Criteria Used in Decision Making 306
   Strategic Needs versus Short-Term Needs 306
   Behavioral versus Scientific Management Issues 307
   Costs versus Benefits of Added Technology 308
Steps in Adding Technology to the Process...............................309
   Step 1: Communicate, Communicate, Communicate 309
   Step 2: Identify Needs and Opportunities 309
   Step 3: Evaluate Alternatives and Select the Optimum Alternative . 310
   Step 4: Educate and Orient..310
   Step 5: Develop the Implementation Plan..........................311
   Step 6: Implement the Technological Changes 311
   Step 7: Evaluate Results, Redefine Needs, and Redefine
   Additional Increments 312
Future Considerations for Technology......................................312
   Customer Acceptance 313
   Workforce Acceptance 314
   Economic Feasibility..314
   Technical Feasibility 315
Summary...316
References..318

8 The Role of Infrastructure in Continuous Improvement 321
What Is Infrastructure?..321
Strategies..322
   Corporate Strategy...322
   Business Strategy 323

The Four Classical Management Functions 324
    Planning..325
    Organizing 325
    Directing...326
    Controlling..327
Organization Structure...327
    Functional 327
    Product..328
    Geographic 329
Alternate Organizational Structures 329
    The Matrix Organization...330
    The Horizontal Organization..330
    The Virtual Organization 331
Trends in Organizational Structures...333
    Moving from Centralization to Decentralization...........................333
    Moving from Vertical Structures to Horizontal Structures.............334
    Moving from Autocratic Managers to More Empowered
    Employees..334
    Moving from Job Specialization to Higher Skill Variety.................335
    Moving from Line Managers to Self-Directed Work Teams 336
    Moving from Specialized Departments to Cross-Functional
    Teams 337
    Moving from Top-Down to Multidirectional Communications.....337
    Moving from Rigid Policies and Procedures to More Flexibility.....338
    Moving from Mechanistic Structures to Organic Structures 339
The Role of the Internet in Changing Organizational Structure 340
    Transaction Costs..341
    Information Symmetry and Asymmetry... 342
The Integration of Knowledge Management into Organizational
Structure.. 342
    Data, Information, Knowledge, Wisdom....................................... 342
    Why Knowledge Is Not Transferred 345
Does Your Business Need a Change in Its Infrastructure? 347
Notes 348

9 Understanding Organizational Culture—The Elusive Key to
   Change 351
Introduction 351
What Is Organizational Culture?...352
Why Is Organizational Culture So Important?......................................353
    Organizational Culture Gives the Company an Identity 354
    Organizational Culture Helps Employees Make Sense of Things ...355

Organizational Culture Enables Employees to Be Committed to the Company 356
Organizational Culture Helps Add Stability to the Company 358
What Are the Components of Organizational Culture? 359
The Components of Culture 359
Values 360
Artifacts: The Display of Organizational Culture 360
Stories 361
Language 361
Symbols, Ceremonies, and Rituals 362
Identifiable Value Systems and Behavioral Norms 363
The Physical Surroundings Characterizing a Culture 364
Organizational Rewards and Reward Systems 365
What Types of Organizational Culture Are There? 367
Types of Cultures 367
Aligning Culture and Strategy 368
Changing Organizational Culture 368
State What You Want the Culture to Be 369
Promote an Ethical Culture 370
Hire the Kind of People You Want to See Perpetuate Your Desired Culture 371
Take Care of Your Employees, and They Will Take Care of Your Customers 373
Implement the Strategy Proposed 373
Change the Physical Cultural Artifacts 375
Tell Stories 375
Recognize Employees Formally 376
Conclusion 377
Notes 377

10 Integrated Supply Chains—From Dream to Reality 381
Introduction 381
Present Status of Supply Chains 381
Background of Supply Chains 382
Material Flows 382
Organization Hierarchy 384
Systems Theory 384
Role of the Supply Chain 386
Setting the Stage 386
Supply Chain Models 391
Evolution of Supply Chain Models 391
A Comprehensive Supply Chain Model 393

The World of Lean Production 396
  Product..397
  Purchasing Process 397
  Production Process 397
  Delivery Process 398
  Demand Variation..398
Steps to Achieve a Lean and Agile Supply Chain 398
  Commitment..399
  Concept...400
  Configuration 400
    Marketing..401
    Purchasing...401
    Manufacturing...402
    Distribution...402
    Finance and Accounting..403
    Top Management 403
  Communication 404
  Culture..405
  Customization 406
  The Integrated Supply Chain..406
  Coordination 407
  Collaboration...407
Steps in the Change Process...408
  Investigate..408
  Initiate 408
  Invigorate 409
  Implement 409
  Integrate 409
  Institutionalize 409
  Innovate...409
A Look Ahead..410
  Complexity...410
  Clairvoyance...411
Conclusions 413
References...414

11 The Role of Services to Complement the Supply Chain 419
Introduction 419
  Producer Services versus Consumer Services.............................. 420
What Are Producer Services? 424
  Why Are Producer Services Important? 425
  The Role of Outsourcing... 426
  The Role of Producer Services...427

Classes of Producer Services...427
Reasons to Acquire Producer Services...................................427
    Gain a Cost Advantage...427
    Remove Noncore Types of Activities 428
    Supplement Internal Staffs with Added Expertise...............428
    Provide Flexible Capacity or Avoid Overload on Key
    Departments.. 430
    Acquire Expertise Not Available Internally...........................431
    Assist in Strategic Planning 432
    Comply with Regulatory or Otherwise Mandatory
    Requirements 432
    Manage Major Programs 433
Steps to Interfacing Business with Producer Services 434
    Recognize the Role of Producer Services in the Company.... 434
    Develop a Strategy for Each Producer Services Area............. 434
    Select Suppliers to Provide a Continuing Relationship 435
    Prepare a Plan That Interfaces with the Supply Chain Plan ...435
    Implement the Service—Fit It to the Situation....................435
    Measure the Progress 435
    Revise and Assimilate 436
The Future of Producer Services 436
Producer Services Firms Buy Producer Services 437
What Are Social Services?..437
Why Are Social Services Important?.......................................437
The Role of Social Services 437
Classes of Social Services 438
    Government 438
    Nonprofit Organizations 438
    Quasi-Private Firms...439
Specific Social Services 439
    Information 439
    Communications Systems...439
    Standards...439
    Education and Training.. 440
    Location Incentives.. 440
    Support Infrastructure... 440
    Protection 440
    Health Care.. 440
Steps to Interfacing Business with Social Services........................441
The Future of Social Services 442
    Level of Regulation.. 442
    Level of Privatization 442

Level of Effectiveness and Efficiency.................................... 443
Emphasis on Business-Related Topics................................ 443
What Are Consumer Services?... 443
Roles of Consumer Services... 443
Future of Consumer Services... 444
Integrated Service Package.. 444
The Need for an Integrated Service Package 445
Steps in Developing the Integrated Service Package....................445
Recognize the Need to Change 445
Identify the Service Package Components............................445
Develop Objectives for the Service Package...................... 446
Develop Strategies for the Service Package 446
Develop Objectives for Each Component............................ 446
Develop Strategies for Each Component 447
Implement the Strategic Plan..447
Evaluate the Results...447
Revise as Needed 447
Summary..447
References.. 448

12 The Future of Improvement Programs 449
Introduction 449
The Background to Improvement Programs...............................449
The Vanishing Manufacturing/Services Boundary 449
The Foundation Topics 450
Knowledge Transfer across the Manufacturing/Services Boundary...450
Agents of Change 451
Integration of Related Entities 452
Future Areas of Emphasis 452
Services Will Continue to Increase as a Critical Success Factor
in Business..453
Continuing Need to Integrate the Product and Service Bundle......453
Continuing Increase in Globalization.................................453
Outsourcing Will Become a More Focused Activity......................453
Increased Need for Project Management Competencies 453
Decision Making Will Deal with a Blend of Hard and Soft
Variables...454
Decisions Will Become More Complex 454
Future of Improvement Programs 454
The Drivers of Change..455
Technology 455
Structure...455
Culture 456

Most Likely Future Methodologies 456
    Integrated Supply Chains 456
    Outsourcing...457
    Total Cost of Ownership 457
    Performance Measurement 458
    Project Management...459
    Mass Customization 459
    Virtual Organizations..460
    Information Technology 460
    Environmental Design...460
    Service Sciences 461
    Chaos and Complexity 461
Most Likely Improvement Programs..462
    Cost Reduction..462
        Product or Service Costs..462
        Working Capital Costs 462
        Capital Equipment and Facilities...463
    Response Time Reduction 463
    Quality...465
    Customer Service Level...465
    Flexibility...466
    Agility..467
    Compatibility 467
    Integration...467
    Sustainability...468
    Risk Management...468
Industries Most Likely to Stress Continuous Improvement...................468
    Health Care: Hospitals and Wellness Centers................................468
    Pharmaceuticals...469
    Local Government...470
    Retail 470
    Education 470
Knowledge Management: Where Does It Fit?.....................................473
    From Data to Information 473
    From Information to Knowledge 475
    From Knowledge to Wisdom 476
Notes 477

Index 479

Foreword

We are now more than a decade past the turn of the millennium. Economic recession, a gridlocked American political system, unexpected weather extremes including Hurricanes Katrina and Sandy, and global economic instability precipitated by uncertainty over the future of the Euro and top leadership change in China have fueled an extraordinarily tough business environment. The modern economic world—shaped by the Industrial Revolution, followed by the information age, followed by a flattened global economy—is morphing again. Today, service enterprises that offer lower-paying jobs outnumber manufacturing enterprises that offer higher-paying jobs. There is high unemployment. Jobs that are available go unfilled because of the gaps between the focus of our education system and the skill sets now required for business success. There is a vanishing boundary between manufacturing and service, which has huge, unsettling implications, to giant multinational companies and tiny startups.

Those of us in manufacturing feel the uncertainty of these times. We are looking for new solutions, that new competitive edge, to ensure the survival of our enterprise and growth in the near term. Adoption of the latest trends, such as social media and cloud computing, scream through the headlines that our organizations must change … somehow. Manufacturing *is* slipping away, and value-added service *is* hard. But the pace of business is so intense that we do not have time to research and assimilate all the possibilities. In fact, it may not be clear where to focus our collective energies, or even how to get started.

Dick Crandall and his son, Rick Crandall, with all their considerable wisdom present a carefully researched volume that brings back the fundamental principle that any business, whether manufacturing or service, must create value to thrive. This book uses input–transformation–output modeling to extend the boundaries of traditional supply chain networks beyond logistics and distributive services to include the dimensions of producer services, social services, and personal services. It presents case studies highlighting the adaptation of manufacturing best practices into service and the adaptation of service best practices into manufacturing. This book focuses on the roles that technology, infrastructure, culture, and the formation of a knowledge corridor have to play to effect positive change.

Here in a single volume you will discover how to embrace every facet of the vanishing boundary between manufacturing and service to the future advantage of your organization.

William T. Walker, CFPIM, CIRM, CSCP
Resource Management Series Editor and Supply Chain Architect
Summit, New Jersey

Introduction

Business survival is not mandatory; you have a choice! However, when you make the obvious choice to survive, you must face the reality that you have chosen a most difficult challenge. However, it is not good enough just to survive; that will carry you only through the short term. To survive in the long term, you must prosper. And to prosper, you must do a lot of things right. One of the things you must do is to continue to improve. This is a book about the programs that help to achieve improvement.

Organization of the Book

Figure 0.1 shows how the book is organized. We begin with an introduction to the idea that the boundaries between manufacturing and services are vanishing. Consequently there is a need to enhance the knowledge transfer between the two areas. For the past century, manufacturing has led the way in developing ways to improve the management of a business, primarily in achieving lower costs and higher quality. However, as the economy and the number of employees continue to move quickly toward the service area, the service industries are becoming the source of a number of new management techniques, especially those that enhance customer service. In this book, we show how knowledge is flowing freely in both directions.

We arranged the chapters of this book in what we believe to be a logical sequence. We describe the basic concepts in Chapters 2, 3, and 4. In Chapter 2, we describe the role of critical success factors in improvement programs and strategic planning. In Chapter 3, we illustrate that an Input–Transformation–Output Model is the basic building block on which global supply chains depend. In Chapter 4, we describe management programs, especially those designed to provide improvements in organizations.

We then illustrate in detail how knowledge, in the form of management programs, flows from manufacturing to services in Chapter 5, and from services to manufacturing in Chapter 6. Not too long ago, it was clear that the information about improvement programs originated in manufacturing. Today, there is an

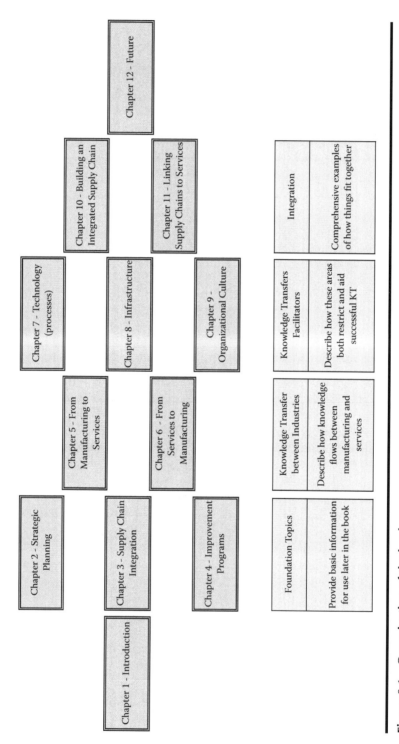

Figure 0.1 Organization of the book.

increasing flow from services to manufacturing. Tomorrow, there will be little distinction between manufacturing and services.

In Chapter 5, we have added two comprehensive case studies—General Electric and Hewlett-Packard—to show how manufacturing companies can add services to their product lines. The cases show that some service additions are successful, some less successful, and some are downright failures. While adding services can be good for a manufacturing company, they must be carefully selected to fit the strategic focus of the company.

In Chapter 6, we include case studies of two service companies—United Parcel Service (UPS) and Amazon—that have been successful in using improvement programs originating in manufacturing to improve their service operations. In some cases, they have added supply chain services that have moved them into manufacturing-type operations.

In Chapters 7, 8, and 9, we describe technology, infrastructure, and culture. Improvement programs depend heavily on new technology. However, technology is only an enabler; it is not sufficient by itself to carry improvement programs to successful conclusions. Although technology is often the driver of change, infrastructure and culture are initially barriers to improvement and have to be changed to become facilitators of improvement.

In Chapter 10, we describe the steps necessary to build an integrated supply chain. It is a journey that many companies have started but few, if any, have completed to their satisfaction. In Chapter 11, we show how external services are needed to enhance the operation of the entity we describe in Chapter 10. Although there are no simple answers, there are some stimulating challenges ahead.

Finally, in Chapter 12 we show that the past has shaped the present, but the present is just the prologue for what we all hope will be a bright future. However, the future will not be a linear extension of the past. The last chapter is filled with topics that are in your future. You either will use them to your advantage or be victimized by your competitors who use them.

How to Use This Book

We think you will benefit by reading the entire book. However, if you want to skip around, we have tried to make each chapter a stand-alone subject. We suggest that you read Chapter 1 to see what we have included in the book. This is the appetizer. If the topics are of interest to you, then move ahead through the succeeding chapters. In these chapters, we dig a little deeper into each topic and provide more substance and illustrations. If you think you do not need that much explanation, look at Chapter 12. See if you are ready for the future. If you are, great! If you decide you may not be, go back to some of the earlier chapters.

Who Can Use This Book

We believe this book will be of value to a large audience because it contains helpful information that we have assembled from a variety of sources, starting with our own experiences. We brought together a number of important concepts from disparate disciplines. We also tried to strip away the hype and facades of many "hot" topics to provide you with the basic ideas and concepts in a straightforward way.

If you are a manager in a large company, no doubt you are feeling the pressure to acquire more knowledge and apply it in a constructive way. If you are a manager or owner in a small company, you will find a wealth of practical information that you can adapt to your business. This book is about knowledge transfer and the need to look in new places for it.

In his book *The World Is Flat: A Brief History of the Twenty-First Century* (Farrar, Straus, and Giroux, 2005), Tom Friedman suggests that in the future, entrepreneurs will play an important role in creating jobs as the global market becomes more fragmented. Large companies may have difficulty in keeping pace with the changes taking place; small companies will have the flexibility to survive. Our book provides a picture of what it takes not only to survive but also to prosper. This is not a book about high-profile success stories in other companies; it is a book filled with ideas about how to make improvements in *your* business. We hope you will find it of value.

The Authors

Richard E. Crandall is a professor in the College of Business at Appalachian State University (ASU), in Boone, North Carolina. He holds CFPIM, CIRM, and CSCP certifications from the Association for Operations Management (APICS), and is a registered professional engineer and a certified public accountant. He has an undergraduate degree in industrial engineering from West Virginia University, a Master of Information Systems Management (MSIM) degree from Boston University, and a PhD in production/operations management from the University of South Carolina. Prior to joining ASU, Dick worked for over 25 years as an industrial engineer, and in management positions for manufacturing and service companies. He was a consultant with a major international firm, installing systems for both operations and financial applications. He is a co-author of *Principles of Supply Chain Management* (CRC Press, Boca Raton, Florida, 2010) along with William "Rick" Crandall and Charlie C. Chen. He currently writes the Relevant Research column for *APICS* magazine.

William "Rick" Crandall is Professor of
Management and Director of Accreditation
in the School of Business at the University
of North Carolina at Pembroke. He earned
his PhD in business administration with a
concentration in organizational behavior
and human resource management from
the University of Memphis. His primary
research interest is in the area of crisis management, helping organizations cope with
catastrophic events. He is lead author of
*Crisis Management: Leadership in the New
Strategy Landscape*, 2nd edition (with John
Parnell and John Spillan; Sage Publications,
Thousand Oaks, California, 2013). He is also
active in researching issues related to supply

chain management and educational effectiveness. Prior to entering higher education, Dr. Crandall worked in management for ARA Services (now ARAMARK), a
service management firm based in Philadelphia.

Acknowledgments

We have a number of people to thank for helping us to conceive of and then write this book. Most of the credit for the original idea goes to William T. Walker, or Bill to most of us who have had the good fortune to be acquainted with him. Bill suggested that there was a need for a book that drew on experience in both manufacturing and service operations. He felt that some of the "Relevant Research" columns that Dick had been doing for *APICS Magazine* had something to say in this area. That, combined with the practical experience that both authors have in several industries, both manufacturing and service, could provide a foundation for expanding the core idea into a worthwhile book. During our research and writing, Bill has been supportive of our efforts and a wonderful source of ideas about how to make the book more valuable to the reader. We probably wouldn't have tried, and I know we wouldn't have succeeded, without his guidance. Thanks, Bill, for all you've done.

We owe special appreciation to Jennifer Proctor, Editor-in-Chief, and Elizabeth Rennie, Managing Editor, of *APICS Magazine.* They encouraged Dick to write about topics that fit nicely in the book. We have drawn heavily from the following Relevant Research columns (portions of these articles are APICS copyright, and are used with permission):

1. Crandall, R.E., A fresh face for CRM: Looking beyond marketing, *APICS Magazine*, 16, 9, 20, 2006.
2. Crandall, R.E., Beating impossible deadlines, *APICS Magazine*, 16, 6, 20, 2006.
3. Crandall, R.E., Looking to service industries, new resources for learning a thing or two, *APICS Magazine*, 16, 10, 21, 2006.
4. Crandall, R.E., The epic life of ERP, where enterprise resources planning has been and where it's going, *APICS Magazine*, 16, 2, 17, 2006.
5. Crandall, R.E., Let's talk, better communications through IOS, *APICS Magazine*, 17, 6, 20, 2007.
6. Crandall, R.E., Dream or reality? Achieving lean and agile integrated supply chains, *APICS Magazine*, 15, 10, 20, 2005.

7. Crandall, R.E., Device or strategy? Exploring which role outsourcing should play, *APICS Magazine*, 15, 7, 21, 2005.
8. Crandall, R.E., Putting together a global sustainability movement, *APICS Magazine*, 20, 6, 26, 2010.

Thanks, Jennifer and Beth, for your help and encouragement.

We would like to thank all the companies we worked for during our time in industry prior to joining the academic ranks. Dick spent over twenty-five years in industry, where he worked for manufacturing companies, most notably Sylvania Electric and TRW in the electronic industries, for service companies in wholesale and retail, and for an international consulting company, where he worked with a variety of manufacturing and service companies. Rick spent over ten years in the food service industry, most of which was with ARAMARK in contract food services. We saw firsthand the differences and similarities between manufacturing and service businesses.

To supplement our own experience, we dug deeply into the extensive research that practitioners, consultants, and academics have done. We have reviewed hundreds of articles and dozens of books to uncover the concepts, principles, and techniques that are included in this book. We wish we could have shared more; however, in lieu of that, we have extensive references at the end of each chapter. We hope that you will find them of value and that the information in the book will be helpful to you in your work.

We would especially like to thank Lara Zoble, Acquiring Editor, Taylor & Francis Group, who has been the key person in this effort. She helped us focus our thinking through a careful evaluation of our proposal and has been a constant source of support, but has not tried to micromanage our efforts. Thanks, Lara; it has been a real pleasure to work with you.

Jessica Vakili is a project coordinator for the Taylor & Francis Group. More important to us, she is the project coordinator for our book. She has been helpful in providing guidance about all the things we needed to do and has answered any number of questions that we came up with. Jessica, we really did try to give you a "perfect storm" kind of book; however, despite our best efforts, I suspect that we didn't reach the Six Sigma level of performance. Thanks for covering for us, and for helping us through all the pre-production steps in preparing the book for publication.

When we submitted our original typewritten pages of the manuscript to the editor, we wondered how it could ever look as professional as most of the books we read in our daily work. Now we know it takes professionals like our project editor Amy Rodriguez to make that transformation possible. Thanks, but we think the least you could have done would have been to invite us to come to your office in Boca Raton, Florida, to review the proofs instead of working on them in the middle of winter in the mountains. We thank her for the unenviable job of trying to make our words sound enlightened, or at least less obtuse, as well as worrying about grammar, punctuation, pagination, location of exhibits, and all those other things that we were very willing to outsource to such an expert.

We are confident there are a number of other people who helped this book come alive. However, in this age of electronic communication, we do not always have a chance to meet everyone personally. Thank you; we hope you enjoyed it as much as we did.

Finally, we would like to give a special acknowledgment to Jean Crandall, wife of Dick and mother of Rick. She encouraged us, but even more important, she read every word (several times) and offered many helpful suggestions. Despite her best efforts, there may be errors. If you find them, blame us, not her or any of the people mentioned above.

Chapter 1

The Vanishing Manufacturing/ Services Boundary

The manufacturing/services boundary is vanishing. Increased global competition, extended supply chains, and increased customer awareness are forcing businesses to find new ways to compete effectively and efficiently. This quest to find ways to compete has created a somewhat ironic dilemma; manufacturers must become more customer-oriented and service companies must become more effective and efficient. Each side is learning from the other.

Differences between Manufacturing and Service

For a number of years, business scholars have been careful to point out the differences between manufacturing and services. As a result, the implication was that each segment has a different set of goals and means of reaching those objectives. On the contrary, both manufacturing and service companies have the same objective—to obtain customers and provide what those customers want. Their interests are mutually dependent. Over time, businesses have understood the need to view manufacturing and services as a complementary package to be provided to their customers.

This is not to suggest that manufacturing and services are the same. There is certainly evidence that service companies require different perspectives and management techniques from manufacturing firms. This does not mean that the two sectors are incompatible, but it does mean their management requirements are

1

different. However, this observation is true even looking within the manufacturing sector. Management of a job shop is different from the management of repetitive line flow. Further, line flow management is different from operating a process industry. It is also true for service operations. Managing a fast-food restaurant is different from managing a global management consulting firm.

Let's look at some major differences between manufacturing and service businesses. One major difference is that manufacturing companies produce a tangible product, while service companies provide an intangible service. Because of the intangibility of their output, service companies show differences in three areas—service organization assessment, service production strategy, and the service production process.[1]

In service organization assessment, the performance measures must include both objective measures of internal performance and subjective measures of customers' perceptions of the quality and value of the service, recognizing that one customer's perception may be slightly different from another customer's. Subjective measures of performance are difficult to develop but are helpful in providing direction for the business.

In service production strategy, the service product and the service process must be designed to allow customer participation in the design and execution of the service, such as in self-checkout terminals at a grocery store. The service production strategy also has to consider that the consumption of the service may be simultaneous with its production, such as in the attendance at a concert. Because service companies cannot produce inventory like manufacturing companies, they must maintain flexible capacity, usually in the form of the variable staffing of multi-skilled employees.

The service production process is also different. Although technology is widely used in service operations, there is no substitute for an appropriate setting in which knowledgeable and empowered employees can deal with the wide variety of customer encounters. Customers must be attracted, they must be pleased with the initial service experience, and they must be retained, hopefully over a long period of time. Sometimes, customers may experience unsatisfactory service occurrences. When this occurs, it is only through the intervention of knowledgeable employees that the customer will remain loyal.

Although there are differences between manufacturing and service objectives, it is becoming more important to combine product-producing activities with service-providing activities to sustain a successful business. Blending manufacturing and services requires the capability to combine different kinds of businesses. This blending raises the complexity of businesses and increases the need to continually improve.

This book looks at two major themes. The first is that the boundary between manufacturing and service businesses is vanishing and new composite businesses are emerging. The second theme is that these composite businesses continually have to improve their competitive advantage through the introduction, implementation, and assimilation of management improvement programs. The process of achieving this competitive advantage relies on the careful implementation of changes in technology, infrastructure, and organizational culture.

Forces That Are Eliminating the Boundary

What is driving business to recognize the need to eliminate the manufacturing/services boundary? Increased global competition is certainly a major factor. Businesses must find ways to be more competitive, and this requires exploring new avenues for developing a competitive advantage. This usually means adding services. In addition, various forces are making it possible to readily combine practices from the manufacturing and services arenas.

The Economy Is Going through a Natural Evolution

One general trend involves the evolution of the economy. Historically, the economy has moved from agriculture to manufacturing and, more recently, to services. The introduction of technology, often through automation, and modern management methods has made it possible to increase productivity dramatically in agriculture and manufacturing and thereby increase the standard of living for the general population. In advanced industrial countries, the increased emphasis on automation has eliminated many of the more routine jobs and has increased the need for more skilled workers, in both manufacturing and services. Emerging countries still rely more heavily on manual or less automated jobs to produce the goods and services needed. Job specialization, from the scientific management movement, is being replaced with job enlargement and enrichment, with greater responsibilities and variability of work for employees.

The increased prosperity has made it possible for consumers to have more disposable income and more leisure time, as working hours have also decreased. This trend is a natural progression and comes about because of the demand for newer services from consumers who can afford to pay for these services. Initially, the demand for these services was somewhat price insensitive; today, consumers are still demanding the service, but at a competitive price. Consequently, there is a need to focus improvement efforts in the areas of controlling service costs, which are usually the ones with the highest labor content. Services are segregated into industries, such as health care and financial services. Health care services can be further segmented into hospitals, extended care facilities, and research institutions. Financial services can include consumer services, mortgage lending, and investment advisory services, to name a few. Later in this book, we will discuss the concept of mass customization for both products and services. Businesses that provide improvement programs as a service will also begin to approach mass customization as they adapt improvement programs to meet specific objectives for individual businesses or organizations.

There Is a Need to Identify Critical Success Factors

Competition forces businesses to improve. As competition becomes tougher, businesses must seek to do that which is essential to their survival. Since no one company can do everything competently, each company must decide what tasks it "must" do in order to compete successfully. In Chapter 2, we describe the concept of critical success factors (CSFs). A CSF is a strategy, plan, or program that will provide the business with a competitive advantage, for a limited period. However, no matter how innovative and successful a CSF is, it has a limited life; therefore businesses must successfully identify and implement new CSFs on a regular basis.

Supply Chains Are Expanding to Integrate Manufacturing and Service

In this book, we extend the concept of a product–service bundle from a single entity to the entire supply chain. As businesses attempt to link their own organization with other entities, they will interact more formally with other types of businesses. Service businesses become partners with manufacturing companies as they move upstream in their supply chain, and manufacturing companies become dependent on service operations as they move downstream in their supply chain. This concept is developed fully in Chapter 3.

It Is Easier to Blend Tasks, Resources, and Techniques into Programs

As part of gaining a competitive advantage, companies are learning to view management concepts and techniques as modular units that can be combined in a variety of programs to meet specific needs. The primary steps include:

- Identification of basic management theories
  - Scientific (systematically analyze work to develop the best methods)
  - Bureaucratic (plan, organize, direct, and control)
  - Human relations management (develop employees as thinking participants in an organization's activities)
- Adding systems theory (combine the parts into a logical whole)
- Adding contingency theory (adapt available resources to existing conditions)
- Blending systems and contingency theories with management theories into improvement programs that fit the business need.

We describe this selection and adoption process for improvement programs in Chapter 4.

Service Businesses Are Looking for Ways to Operate More Efficiently

Service businesses are closer to the customer and have always been customer-centric. However, as services become a larger part of the economy and the workforce, they are being pressured to become more cost and quality conscious. Consequently, they are looking to the manufacturing area for tools and techniques that will enable them to become more effective (doing the right things) and efficient (doing things right). In Chapter 5, we describe a group of the improvement programs developed in manufacturing that are now being implemented successfully in service businesses. The following sections describe examples of ways in which service businesses are seeking efficiencies.

Health Care Providers Are Looking for Ways to Cut Costs and Improve Quality

Rising health care costs are of concern to leaders in business and government as well as the individual consumer. Everyone wants improved health care; however, not everyone is able to pay for it. Health care professionals realize that costs are high and must be reduced; the question is, how? One way of reducing hospital costs is to reduce the length of a patient's stay. This reduction requires not only an improved service process during the patient's stay at the hospital but also improved management of the pre-admission and post-recovery phases of the patient's experience. One area that is appealing for many health care entities is the use of lean production (a program to reduce waste) combined with Six Sigma (a program to improve quality) into a program known as Lean Sigma.[2]

Banks Are Facing Competition from Lower-Cost Banks

Some service businesses have offered increasing levels of services in the interest of attracting and retaining customers. As a result, these businesses are finding they have provided a greater variety of services than the consumer needs or wants. In an insightful look at this situation, Clayton Christensen, Scott Anthony, and Erik Roth call this "overshooting."[3] They point out that companies innovate faster than customers can change. As a result, products become too good, at least in their variety and complexity. A company that has overshot its market is subject to competition from a competitor that can offer reduced but still acceptable services at lower costs. Banking is an industry with this problem. Banks are also trying to reduce costs. They have increased the use of technology to reduce the labor component of their operations. Examples include ATMs, automated telephone responses, and online bill paying. They have also bought into the need to improve quality as some of the most ardent supporters of the Six Sigma movement include banks.

Investment Bankers Are Facing Competition from Commercial Banks and Discount Brokers

Individual investors have a great deal more information available to them today and often do not need or desire the extensive investment counseling that was once the norm. As a result, low-cost, minimum-level investing is increasing in popularity among investors. The reduced service requirements mean that a greater number and variety of businesses can enter the business of selling stocks. In this case, the business needs to introduce information technology (IT) to improve the flow of information while recognizing that this will reduce the level of personal service offered to the consumer.

The Retail Industry Is Moving in Two Strategic Directions—Low Cost and Focus

The retail industry is moving in two directions. One is toward low-cost, broad product lines, lean operations, and minimal customer service, often associated, for instance, with Walmart. A second direction is toward specialty stores, with a more limited product line and higher quality of service. This trend is luring customers away from intermediate stores that now face the problems of competing in this dichotomous atmosphere. The industry stalwarts—department stores—are under intense competitive pressures and are consolidating and redirecting their strategies, often by dropping product lines, such as electronics and appliances, which have been dominated by firms like Best Buy and Lowe's. However, the "big box" stores are also under competition from online retailers, such as Amazon. Amazon now sells hand tools, long the domain for such home improvement retailers as Home Depot and Lowe's. As a corollary, shopping malls are also under competitive pressures. Many consumers no longer have the time to "shop"; they want to "buy." Consequently, they prefer to go to stores with easy access, resulting in a new form of shopping area that is becoming popular. These take the form of outdoor strip malls with convenient parking and larger stores that have a deep product selection. These examples represent why service companies want to provide the "right level of service at a low cost."

Manufacturing Is Adding Services to Be More Customer Focused

It is no longer sufficient to manufacture a low-cost, high-quality product. A complementary service package is necessary. The product is rapidly becoming only the order qualifier; the service that accompanies the product represents the order winner.[4] As a result, manufacturers must focus more attention on the service sectors of

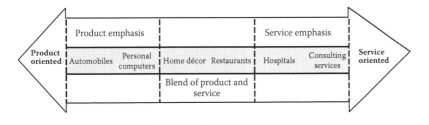

Figure 1.1 Product-service continuum.

their businesses. When they do, they find that some of the techniques they used in manufacturing work in the service sector, such as the organization of work flows or the need for capacity planning. However, they may also find that some functions are different, such as in measuring productivity or defining service quality in a tangible way.

With these realizations, manufacturing managers can acquire a greater appreciation for the services sector and the need to blend it into their total product/service bundle. Figure 1.1 illustrates the overlap between services and manufacturing.

The figure illustrates a continuum with businesses that have a heavy orientation on products shown on the left side of the diagram. Businesses that have a heavy emphasis on services are shown to the right, and businesses that sell a blend of product and service are depicted in the center of the figure. Even those businesses that have a heavy product orientation, such as automobiles and computers, are adding services such as those described in the following sections.

Post-Sales Services

Consumers are becoming buyers who are more knowledgeable and look beyond just the purchase price of the product. They want to consider the lifetime costs of products and, as a result, are concerned about the reliability and service requirements of products. In response, businesses are looking beyond the sale of the product and considering the customer's post-sale needs. For example, automobile manufacturers are working through their dealer networks to put more emphasis on their service shops and many have systems that enable them to contact customers, such as with a letter or an e-mail, to remind them their cars are due for service. These notifications include not only the major repairs or services needed but also routine procedures, such as oil changes. Many dealers have initiated a "fast lube" service to compete with competitors in their geographical area. Computer sellers are also providing such services as software upgrades and repairs. One of the more interesting examples of this is the repair facility operated by UPS for Toshiba in Louisville, Kentucky. This outsourcing program makes it possible for the manufacturer to provide faster responses to customers who need it.[5]

Financing of Purchases

Another form of added service is financing of the purchase by the seller. Automobile dealers provide the option of financing the customer's purchase, usually through the dealer's manufacturer. The rates are usually competitive with other financing sources such as banks. In some cases the rates are even lower, as the manufacturer uses the financing option as an incentive to the customer. This arrangement also provides a way to maintain contact with the customer during the lifetime of the purchase. Industrial equipment manufacturers have used this approach as a means of selling their equipment to small businesses. This type of transaction enables firms to buy the equipment if their creditworthiness is not adequate to obtain a loan from a bank.

Online Purchasing

Consumers can now buy virtually anything online. Companies often offer attractive return policies that reduce the risk of online purchasing. Books and clothing are two areas that have made great progress in attracting buyers. Although major purchases such as automobiles are still not widespread on the Internet, consumers can use the manufacturers' websites to gather information about potential automobile choices before going to a dealer to make the purchase.

Customer Input to Product Design

In the past, manufacturers often operated with the philosophy of producing whatever they could make well and then trying to find a party that would buy their product. Today, there is more awareness of the necessity to anticipate what the customer wants or needs and then make a product to satisfy that need. This added service may take the form of obtaining input from potential or existing customers through focus groups or surveys. This up-front service will enable the product to be more appealing to the customer. This change is described as moving from a "make and sell" attitude to a "sense and respond" approach.[7]

Enhanced Customer Relationship Management

Customer relationship management (CRM) is becoming more significant for manufacturing businesses. CRM is about retaining customers, and many manufacturers are finding that adding post-sales services is a good way to accomplish this goal. The use of CRM and the other examples identified above represent only a few of the ways that manufacturing companies are enhancing their competitiveness by adding services to their product package. We describe other applications more fully in Chapter 6.

Technology Development

New technology arrives at an almost overwhelming rate. It results in new materials that are lighter, stronger, and more environmentally friendly; equipment that is more reliable and accurate; production facilities that are flexible and employee-friendly; information systems that facilitate analysis and communication; and employees who benefit from improved training, availability of tools, and innovative management techniques. Some view technology as the ultimate means of achieving success in organizations. On the other hand, many view technology as a negative force and are quick to point out the adverse effects of the increased deployment of technology. Despite some resistance, the use of technology, especially in the form of information technology, is expanding rapidly and is forcing change in organizations. We describe these changes more fully in Chapter 7.

Although technology is often a driver of change, it is rarely sufficient to achieve lasting improvement unless the infrastructure and culture also change within an organization. Figure 1.2 is a hypothetical portrayal of the interaction among technology, infrastructure, and culture during the implementation of a continuous improvement program. Technology is a driver, but it may falter unless actively supported by adjustments to the organization's structure and acceptance within the culture of the organization. The infrastructure is important in facilitating communications and collaboration; however, it may be acceptance by the culture that eventually provides the extra support that the program needs to succeed. Eventually, all three change agents—technology, infrastructure, and culture—should come

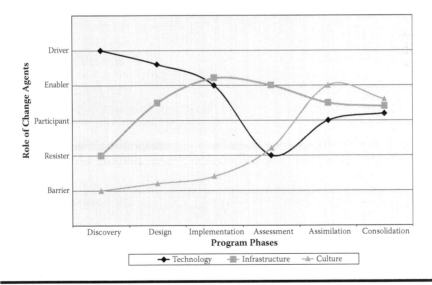

Figure 1.2 Relative participation of change agents.

together to work in synchronization to assimilate the management improvement program into the ongoing operations of the company.

A Movement toward a Process Perspective

Organizational concepts are changing and making it easier to blend different functional areas. The rigid, hierarchical line-and-staff structure that has been a mainstay of organizations since the Industrial Revolution is changing. Businesses are considering other ways to restructure to accomplish their objectives. This is especially true when businesses want to implement major strategic changes in their organization. They might use a matrix format to blend functional areas, or a project management approach to achieve a major objective successfully. They may use a virtual organization orientation to enable them to combine and disband groups as needed. Some of the underlying reasons for innovative organization structures include the following:

- Employee work assignments are moving from job specialization to self-directed work teams.
- Communications are moving from paper formats (snail mail) to electronic (e-mail, videoconferencing, and a variety of social media formats).
- Organizations are moving from hierarchical (vertical) structures to network (horizontal) formats.
- Businesses are changing their focus from improving transactions to improving processes.

In the past, management grouped similar functions together, such as cutting, sewing, and packing in an apparel shop; brake presses, drill presses, and lathes in a machine shop; and accounts receivable, accounts payable, and payroll in an accounting shop. Consequently, the emphasis was on the batch processing of transactions. This grouping worked well, but it was internally oriented and not designed with the customer in mind. As management learns more about what their customers want and need, they are redesigning processes to better serve the customer. Although the need to change is becoming more obvious, many organizations are not moving as rapidly in changing their infrastructure and culture as they are in implementing new technology. Technology without organizational change is a recipe for disaster. This discussion is covered more fully in Chapter 8.

Outsourcing Focuses More Attention on the Need for Good Service

Outsourcing requires adjustments in infrastructure and culture as well as technology. Outsourcing is a service function and embodies the essence of customer service. Both the manufacturing of goods and the providing of services are being outsourced. The drivers of outsourcing are to (1) reduce costs, usually with an off-shore partner, and (2) devote more attention to one's core competencies. While outsourcing is often associated with using foreign suppliers, there may be occasions when it is achieved with domestic suppliers. The electronic communication capability sometimes makes it easy to outsource services, such as software programming, call center activities, and routine accounting. However, outsourcing triggers the need for organization realignment, also a subject for Chapter 8.

Changing Personal and Organizational Relationships

Another factor that affects the manufacturing/services boundary is the change in the personal and organizational relationships that exist within and outside the organization, including relationships between employees and managers, the company and suppliers, the company and customers, and the company and its owners or stockholders. Also impacting the manufacturing/services boundary are the relationships between the company and its secondary stakeholders, such as regulatory agencies, service providers, financing institutions, insurers, health care agencies, and professional service firms. Relationships can exist between individuals and between organizations, although the latter is more difficult to establish and manage. Developing effective relationships usually requires changing internal attitudes and corporate cultures, both difficult but not impossible to accomplish.

Workers today possess increased technical knowledge. The result is a more effective person–machine relationship. As their education level increases, employees are able to handle a greater variety and complexity of duties. Many employees want more variety and empowerment, but it is up to the company to create the environment in which this increased knowledge can be used most effectively. These factors are both facilitators and barriers in situations where companies attempt to introduce change, such as in the following:

- Technology, in the form of equipment advances, evolved from hand tools to assist workers, to power tools to help workers accomplish the task faster, to automation to replace workers, to interactive equipment to optimize the human–machine interface.
- The workforce changed from immigrant workers with little formal education and lacking language skills at the beginning of the twentieth century to college graduates with a greater awareness of global events.
- Product lines evolved from companies with a single product line and only domestic locations to global companies with multiple products and services and locations.

The role of organizational culture and its potential to hinder or help improvement programs is covered more fully in Chapter 9.

The Need to Integrate Companies into Supply Chains

Building supply chains requires constructing relationships through communication and cooperation. These are attributes found in service businesses; therefore, building supply chains involves the development of increased service capabilities. In Chapter 10, we show how this goal requires a careful and comprehensive process that links entities together into an effective and efficient network.

Services Become More Relevant to All Types of Customers

Services are important not only to individual consumers but to manufacturing and service businesses as well. There are producer services in which one service company provides services to another business. There are consumer services in which service companies provide services to individual consumers. There are also services in which consumers provide services to themselves, or self-service. We describe these various services and how they fit with supply chains in Chapter 11.

Continuous Improvements Continue to Blur the Boundary between Manufacturing and Services

In Chapter 12 we describe the changes that we expect in improvement programs. Some of the changes will be linear and rather obvious extensions of the present. Other changes, however, are apt to be more revolutionary. We provide three examples in the following sections.

There Is a Movement toward an Industry Focus and away from a Functional Focus

With the emphasis today on customers or market segments, it follows that there will be a greater orientation of management improvement programs toward industry applications. Because industries include both manufacturing and services, there will be more of an interest in looking at the similarities along industry lines. This realization will lead to recognizing and exploiting these similarities and designing improvement programs to fit these areas of focus.

The Mass Customization Concept Requires a Blending of Manufacturing and Service

The mass customization movement is just getting started. Although the concept has been around for over a decade, it is still in the early stages, and companies are still developing ways to best achieve it. David Bowen and William Youngdahl believe that "as the defining characteristics of manufacturing and service technologies

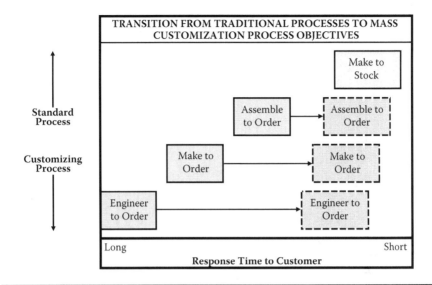

Figure 1.3   Transition to mass customization.

continue to blur, we see a 'common industrial paradigm' emerging in which a simi-lar logic is finding appeal in both manufacturing and service sectors."[6] They believe mass customization provides a way to blend these basic manufacturing and ser-vice principles.

Figure 1.3 shows the migration from classic manufacturing strategies to a strat-egy of mass customization. Regardless of a company's present make-to-order strat-egy, they will be looking for ways to incorporate more customer customization into their processes. It is unlikely that mass customization will become the universal way of doing things, at least not in the immediate future. Rather, it will be a selec-tive strategy for an increasing number of companies.

Now for a nonlinear example. In recent years, researchers have used chaos theory as an alternate way of thinking about decision areas. Traditionally, managers were trained to think in rational, sequential patterns. This thinking led to the viewpoint that future forecasts could be based on past data; therefore businesses could plan by statistically extending the past results. For more than three decades, scientists in fields other than business, such as meteorology, biology, and mathematics, have been developing a theory that addresses nonlinear progressions. They have labeled this field "chaos theory." Managers and forecasters are just beginning to recognize that chaos theory may have applications in the tumultuous environment of the business world. Chapter 12 offers additional insights on how chaos theory is being used to solve business problems.

In the previous sections, we have described some of the forces that are at work to reduce the differences between manufacturing and services. In the next section, we show how these eroding differences will eventually disappear.

The Vanishing Manufacturing/Services Boundary

In this section, we describe how manufacturing and services are moving from separate and distinct business entities to a scenario in which their mutual interests have caused them to become loosely integrated entities, on their way to becoming a composite entity. We use the Venn diagram in Figure 1.4 as an illustration of this transition.

Phase 1: Separated Disciplines

At one time, manufacturing and services were separate entities, as shown in the top portion of Figure 1.4. They had contact of course, because manufacturing produced goods that were sold to service organizations in distribution and retail. The manufacturing businesses had two major components—production (of goods) and support services.

Most of the improvement programs in manufacturing organizations focused on the production of goods. Gradually, however, the manufacturing firms began to recognize the need for improvement programs in their support service areas, such as order processing, product development, quality assurance, and distribution. Consequently, these manufacturers developed programs that focused on cost reduction and quality improvement. As an afterthought, they found that some of the programs could be applied to their support service areas as well.

Meanwhile, service businesses were primarily concerned with selling the goods they received from manufacturers. They became proficient at selling but were not as effective at managing the operations side of their business processes. For a while they survived and even prospered, as the market demands were growing faster than the services side could keep up with. After a time, however, supply caught up with demand, and these service companies began to recognize their need to become more efficient.

Phase 2: Internal Improvements in Costs and Quality

During the 1970s through the late 1980s, two major shifts in strategy occurred involving manufacturing and service businesses. Manufacturing companies realized they could gain additional cost reductions and quality improvements by adapting some of the same techniques they used in the production areas and applying them to their service areas. They also recognized they should add additional services to supplement their product package. As a result, they began to adapt their continuous improvement programs to service areas.

On the services side, businesses began to feel the pressure of competition, especially from "big box" retailers such as Walmart and Home Depot. Much of the pressure was to reduce costs, and many service companies recognized that manufacturing companies were proficient in implementing cost reduction programs.

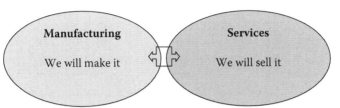

The boundary is clear
Traditional Interface of Manufacturing and Services

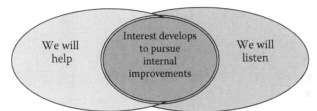

The boundary begins to blur
Increased overlap from internal improvements

The boundary further erodes
Increased overlap from interest in customer service

The boundary vanishes
Manufacturing and services with congruent interests

Figure 1.4   The vanishing boundaries.

Consequently, these service companies, especially wholesalers and retailers, focused more attention on inventory management, employee scheduling, automatic equipment, and the smooth flow of goods to reduce costs.

Service organizations also became interested in applying the quality improvements techniques used in manufacturing. Total quality management (TQM) became a popular topic of conversation among service industries, especially financial institutions and hospitals. Although TQM achieved only moderate success, this relative lack of success was probably more the result of indifferent implementation than a failure of the quality improvement techniques. TQM's successor, Six Sigma, takes a more disciplined approach to the process of quality improvement, and where properly applied appears to be making a positive contribution in a variety of service industries.

Phase 3: Customer Service Improvements

By the 1990s, businesses of all types recognized the need to consider more closely the role of the customer in their planning. Service companies had always maintained direct contact with customers but were now faced with the challenge of providing increased facets of customer service while at the same time reducing the cost and improving the quality of those services. Using the cost reduction and quality improvement techniques developed in manufacturing were no longer a novelty; they were a necessity.

Manufacturing companies also recognized they could no longer use a make-and-sell strategy—first deciding what products to produce and then deciding how to sell those products to their customers. Put differently, they had to learn what products their customers wanted and then how best to produce those products. They also found that customers wanted their products complemented with services, so manufacturing companies had to consider how to bundle services with products. In some cases, the complementary services became as important as the product itself. Thus manufacturers had to become more customer-oriented and learn that service organizations knew a good bit about how to manage the customer.

Many businesses experience this phase today. As service companies continue to learn about cost reduction and quality improvement, manufacturing companies are learning more about the care required for their customers. Stephen Haeckel presents a strong argument that no matter how adaptive an organization is today, it will be inadequate in the future. He advocates that businesses must develop a "sense and respond" approach to be able to cope with the increasing complexity of business and its increasing needs to develop social systems.[7]

Phase 4: Integrated Product and Service Functions

In the future, companies will no longer be classified as manufacturing or service. For example, is IBM a manufacturing company or a service company? What about

General Electric? Nike? Dell? Or a host of other companies that have component parts of both manufacturing and service? And what if they do not include needed products or services within their organization? They can simply outsource the creation of those goods or services or create a virtual organization to satisfy the need.

However, no matter how cleverly they arrange their supply-side constituents, businesses will always face the need to improve continuously. Today's standard of excellence will be tomorrow's minimum acceptable standard. To borrow from Terry Hill, today's order winner will be tomorrow's order qualifier.[8]

That is what we describe in this book—the rationale behind this blending of manufacturing and services. We do not have a simple, step-by-easy-step process to recommend, because effective transitions are not easy. We highlight four companies—two that are thought of as manufacturing companies and have added services, and two that are thought of as service companies—that are using techniques developed in manufacturing to improve their competitiveness. Each of these companies have encountered successes and failures in their transitions. However, we provide an array of information that will help managers find a way to cope with the transition to a composite company. In short, this book can help managers address the vanishing manufacturing/service boundary.

Summary

As you read further, if you are in manufacturing, look for ways you can extend manufacturing techniques into your service operations and, even more important, how you can learn from the service industries. If you are in a service operation, look for manufacturing techniques that you can use in your service operations. If you are in a nonprofit organization, we use the term *business* in most cases; however, much of what we describe has application in nonbusiness entities, such as government agencies and other nonprofit organizations.

The intent of this book is to suggest ways that any organization can become more competitive in uncertain times.

References

1. Bowen, J. and Ford, R.C., Managing service organizations: Does having a "thing" make a difference? *Journal of Management*, 28, 3, 447, 2002.
2. Snee, R.D. and Hoerl, R.W., *Six Sigma beyond the Factory Floor*, Pearson Prentice Hall, Upper Saddle River, NJ, 2005.
3. Christensen, C.M., Anthony, S.D., and Roth, E.A., *Seeing What's Next: Using the Theories of Innovation to Predict Industry Change*, Harvard Business School Press, Boston, 2004.
4. Hill, T., *Manufacturing Strategy* (2nd ed.), Irwin, Burr Ridge, IL, 1994.

5. http://www.pressroom.ups.com/pressreleases/archives/archive/0,1363,4421,00.html, Toshiba and UPS join to set new standard for laptop repair, *UPS Newsroom*, April 27, 2004.
6. Bowen, D.E and Youngdahl, W.E., "Lean" service: In defense of a production-line approach, *International Journal of Service Industry Management*, 9, 3, 207, 1998.
7. Haeckel, S.H., *Adaptive Enterprise Creating and Leading Sense-and-Respond Organizations*, Harvard Business School Press, Boston, 1999.
8. Hill, T., *Manufacturing Strategy* (2nd ed.), Irwin, Burr Ridge, IL, 1994.

Chapter 2

Critical Success Factors and Strategic Planning

Managerial Comment 2.1
So What Is Corporate Life?

Corporate life in today's business world is a mix of scarce resources, unrealistic deadlines, instantaneous communication, demanding customers, and fierce competition.[1]

Chapter 1 proposed that the manufacturing/services boundary is vanishing. Why is this happening? Perhaps the best answer is that businesses need to compete more effectively. So how can they more effectively compete? By identifying activities they must accomplish well to be successful. Those activities are called critical success factors (CSFs). This chapter describes how businesses can identify CSFs and then incorporate them into their strategic and operational planning and control processes.

What Are Critical Success Factors?

In this section, we show that CSFs are the result of focusing on the important things to accomplish in a business. CSFs help us see that although an organization can accomplish a vast agenda of activities, not all of those accomplishments are necessary for the success of the business. However, some are critical, and these become the CSFs.

Much of the literature on CSFs takes an information technology perspective. John Rockart is credited with being the first to use the term *critical success factor* in his article on how to improve reporting key information to top executives. He defined CSFs as "the few key areas where 'things must go right' for the business to flourish .... As a result, the critical success factors are areas of activity that should receive constant and careful attention from management."[2]

Herbert Simon also emphasized the need to separate the important information from the vast amount of information that is generated in a business. See Managerial Comment 2.2.

Managerial Comment 2.2   The Main Thing

The main requirement in the design of organizational communication systems is not to reduce scarcity of information, but to combat the glut of information, so that we may find time to attend to that information which is most relevant to our tasks— something that is possible only if we can find our way expeditiously through the morass of irrelevancies that our information systems contain.[3]

Christine Bullen[4] also associated CSFs with information reporting systems and described the spread of the concept to many industries worldwide. She pointed out that CSFs helped usher in the age of executive information systems by demonstrating that when managers have relevant computer-based information, they can use it to improve their decision making.

A broader description of CSFs states, "The Critical Success Factor (CSF) concept is a formal process of establishing and maintaining corporate priorities. CSFs are internal or external events or possible events that can affect the firm either positively or negatively and thus require special attention. CSFs provide an early warning system for management and a way to avoid surprises or missed opportunities."[5]

While a business needs to perform many tasks to be successful, the CSFs are the things it "must" do well. Each business must identify for itself those CSFs that are critical to its success. Management cannot decide on its CSFs in a vacuum. The environment in which they operate must be considered. In Chapter 3, we describe the environment more fully when we review the open systems concept. This concept describes how management must consider forces external to their business, such as economic conditions, competitors' actions, technology developments, social customs, environmental concerns, and government actions. Managers must also recognize that increased industrialization and globalization of a country carries with it an increase in the CSFs that must be met.

The Evolution of CSFs in the United States

Here we describe how global CSFs have evolved in the United States. The evolution described is representative and a way to define further the nature of CSFs. CSFs vary over time, among industries, among companies in the same industry, and among managers within the same company. Ultimately, CSFs must be specific to the need at hand.

To compete effectively, manufacturing companies should include those CSFs that relate to improved customer service. Likewise, service companies should include CSFs that relate to improvements in cost efficiencies and quality.

We will describe the effect of the competitive situation in the United States on CSFs; however, similar trends are occurring in other countries, particularly those engaged in global markets—the European Union, Russia, and the Far East (China, India, Japan, Korea), as well as relative newcomers, such as Indonesia and Vietnam. It is especially relevant to see CSFs in light of the economic impact of political revolutions, such as the transition of Communist countries to market-driven economies. In *The Commanding Heights,* Daniel Yergin and Joseph Stanislaw provide a fascinating description of how individual countries are making this transition.[6] Ian Bremmer describes a similar theme about the movement of nations from state-controlled to market-driven economies in *The J Curve.*[7]

The United States is an example of a country that has witnessed changes in CSFs throughout its history. Table 2.1 summarizes these changes. In the table headings, we use the terms *order qualifiers* and *order winners,* as introduced by Terry Hill in his book *Manufacturing Strategy.*[8] Order qualifiers are "those competitive characteristics that a firm must exhibit to be a viable competitor in the marketplace" and order winners are "those competitive characteristics that cause a firm's customers to choose that firm's goods and services over those of its competitors."[9]

- ■ *Beginning (from First Settlements through 1800).* In the first stages of the country's industrial development, there was little internal competition among businesses. Most of the competition for finished goods and services came from other countries; however, without some industry, there was little money available for residents to buy finished goods. Agriculture was the principal industry. Residents were largely self-sufficient and any goods they produced, such as furniture or fresh vegetables, were for their own use. If they produced more than they needed, they sold the excess or traded it to their neighbors. The capability to satisfy order qualifier requirements was the minimum

Table 2.1 A Look at Critical Success Factors in the United States

Time	Competitive Situation	Order Qualifiers	Order Winners
1840s–1890s: Industrial Revolution	Little: U.S. was primarily an importer of finished goods; an exporter of raw materials; developing the "American system" of manufacturing	Function	Function, availability
1890s–1930s: Growth and Recovery	Limited competition; U.S. was beginning to become an industrialized nation, next to Europe; little competition from Far Eastern countries	Function, availability	Price (Ford), variety (General Motors)
1940s–1950s: Mass Production	Little competition after WWII (Germany and other European countries rebuilding; Japan not a major factor; U.S. could sell all it could make)	Function, availability	Price, quality, delivery
1960s–1970s: Financial Management	Stirrings from Germany and Japan, in automobiles and basic industries (steel); U.S. living off successes of the 50s and acquisitions	Availability, price	Quality, delivery, service
1980s: Awakening	Inroads from Japan and other Far East countries, in consumer electronics and other basic industries; U.S. not prepared to compete	Availability, price, delivery (JIT)	Quality, service, product variety
1990s: Revival	U.S. firms countered in some industries; Japan and other Far East countries faltered; Europe solid; rapid movement to global market	Availability, price, delivery, quality (TQM)	Product variety, service, flexibility, integrated systems

Table 2.1 (continued) A Look at Critical Success Factors in the United States

Time	Competitive Situation	Order Qualifiers	Order Winners
2000+: Mass Customization	Global market; offshore outsourcing; strategic partners; financial integration; mass customization; integrated systems; lean manufacturing	Availability, price, quality, delivery, flexibility; variety, integrated systems	Mass customization, paperless transactions, integrated supply chains, virtual corporations

necessary to be considered "in the business"; order winners were additional capabilities beyond the level of order qualifiers. In the United States, settlers cleared the land and exported timber while farmers grew tobacco and cotton for export. In return, they imported tools, clothing, and equipment for use in developing increased industrial capability.

■ *Industrial Revolution (1840s–1890s).* During most of the nineteenth century, the United States developed its industrial capability to record levels and made the transition from a primarily agricultural economy to one that included a significant industrial capability. The Industrial Revolution saw the beginnings of major industries, such as the telegraph, telephones, railroads, textiles, oil, and steel. The country needed all of these basic industries to feed, clothe, and transport the rapidly expanding population. Although there was still competition from other countries, primarily Europe, American companies could be successful if they were able to produce the goods (e.g., steel) or provide a service (e.g., railroads). The market was expanding so rapidly that it was difficult for an organization's available capacity to meet the demand for its products.

■ *Growth and Recovery (1890s–1930s).* During this period, the United States made the transition to an industrialized country and became a major player in the global marketplace. The automobile industry played a major role in this transition. Henry Ford set the stage by emphasizing his company's ability to mass-produce a standard automobile and sell it at a low price. General Motors used a different strategy—that of offering an array of choices. The competition between the two companies created the need to meet new CSFs—that of lower prices and greater variety. It also provided for the transition of the United States from an agricultural- to an industrial-based economy. This growth period also set the stage for the origins of the service economy.

■ *Mass Production (1940s–1950s).* During this period, the United States emerged from the Great Depression and became involved in World War II.

As a result, the country increased in terms of industrial capability, especially in the areas of automobile and aircraft manufacturing as well as atomic energy. Immediately following the war, the United States was able to sell almost anything it could produce in excess of domestic demand. This feat was possible because other industrialized countries in Europe and Japan had their industrial capability destroyed during the war. Competition was almost nonexistent and U.S. companies did not have much to worry about except making sure they had the capacity to meet demand. As a result, companies did not strive to make improvements in their technology or management methods. Indeed, life was good!

■ *Arisings (1960s–1970s)*. As Europe and Japan recovered from World War II, they rebuilt their industrial capacity with the latest technology and looked for new markets to develop, once they satisfied their own domestic markets. Japan transformed itself from a maker of low-quality goods to a producer of high-quality products. Two areas of particular improvement were in the automobile and consumer electronics industries.

■ *Awakening (1980s)*. The need for quality products reached a new level during this period and became a CSF that was recognized in other countries. However, because of the lack of competition, producing quality products was not a CSF that many companies in the United States acknowledged as important. For example, the steel industry resisted modernization because it did not need additional capacity and could not justify new technology. Consequently, this was a period of financial wizardry as conglomerates abounded and synergy (the whole is greater than the sum of the parts) was the magic word. Manufacturing lost its luster and automobile companies tried to compete on styling and reduced prices. The need for new CSFs such as quality and response times became a reality.

■ *Globalization I (1990s)*. Although some degree of globalization had existed before this period, a movement to expand more aggressively was emerging. This movement progressed, despite the fact that some remnants of "Buy American" campaigns were still active. Walmart became the leader in low-price retailing, but at the expense of buying many products overseas. Many other companies watched and waited before finally deciding that buying overseas was not a bad thing, especially because Americans wanted low prices.

■ *Globalization II and Mass Customization (2000 and after)*. The United States is currently in a transition mode. Traditional manufacturing businesses are diminishing as basic production operations are being outsourced to offshore companies. The major U.S. automobile manufacturers—General Motors and Ford—are losing market share. The apparel industry is also disappearing as companies purchase more clothing from foreign countries that offer dramatically lower labor costs and generally acceptable quality. Understandably, different constituencies have mixed reactions to these trends. Some are pessimistic and want tighter controls on imports; others are optimistic and want industry leaders to help make the transition from traditional industries to futuristic industries. To accomplish this transition requires a change not only in the type of products and services offered but also in the attitudes and knowledge of the workers.

New CSFs are imperative. In addition to low prices, high quality, and reduced response times, businesses must be agile, flexible, and virtual. The ante is going up; successful businesses have to offer a larger variety of products and services. However, offering a wider variety of products or services is not always the ultimate answer. The ideal situation, at least in the minds of some, is to treat each customer as a market of one, and design and deliver products and services unique to each person. Although this scenario may appear overly idealistic, many see it as the direction of the future. What will this transition involve for the United States? It is a transition from one life cycle to a new one. In fact, that transition is already taking place when a country abandons the former life cycle built around traditional manufacturing, with service support, to a new life cycle—one that is built around a service/manufacturing package of goods and services that is systems-oriented and holistic with respect to customers' needs.

Other Changes during a Country's Economic Life Cycle

In addition to an increase in competition, the United States and other developed countries face the following changes during their national economic life cycles:

- *From agriculture to manufacturing to services.* There is a natural progression from an agricultural society to an industrialized one and then on to a service-oriented society. Although agriculture has always been a fruitful component of the U.S. economy, the number of workers in this sector has declined steadily. During the Industrial Revolution, factories attracted workers from the farms, which led to production facilities with large numbers of employees, many of them represented by labor unions. The advances in production machinery replaced many of these workers; their fate resulted in retraining into other fields of work, often service oriented. Today, offshoring and offshore outsourcing has further exacerbated this shift so that the U.S. economy is heavily service-based at present.
- *From simple to complex.* Large company-operated farms replaced the small family farm. Small businesses have grown from local units to large, global entities. Ironically, many formerly large companies have downsized as massive layoffs have resulted from international competition. Nonetheless, regardless of the size of the business, relationships within and between entities have increased in complexity. Technology and global business expansion are the key reasons why.

- *From tactical to strategic.* As complexity increases, the time horizon for decisions also increases. It is no longer possible to take a short-term perspective and focus on tactical decisions that look only weeks or months into the future. Management must address strategic issues that involve looking several years down the road. This long-term perspective can have huge impacts on the future survival of businesses. Unfortunately, many managers are still struggling with how to best plan for the long term.

- *From raw materials to integrated systems.* Many developing countries begin their globalization efforts by exporting their raw materials or specialty products. In exchange, they hope to achieve the means for more advanced production in the form of equipment and technology knowledge. As they advance in their level of industrialization, they become increasingly adept at making products with more advanced features and capabilities. These advances enable them to develop a greater range of services. Finally, when they can produce advanced products and provide extensive services, they can begin to assemble the components into integrated systems.

- *From labor-intensive to equipment-intensive to a blend of the two.* Industrialization requires equipment to assist the worker. In some areas, this led to attempts to use automation to replace the worker. However, as complexity increases and customization becomes a requirement, it is becoming apparent that an appropriate blend of workers and equipment provides the best combination. Workers in modern factories do less physical work (machines and robots handle this) and more monitoring tasks, such as the overseeing of those machines and robots.

- *From job specialization to self-directed teams.* Job specialization originated from Frederick Taylor's Scientific Management philosophy during the early 1900s. The goal was to increase worker productivity by requiring employees to perform a specific task in the production sequence. Job specialization became the human backbone of production up through much of the twenty-first century. However, as the demand for faster response times and increased customer service grew, businesses had to allow employees to make more decisions in the production process. The resulting movement led to employee-led production teams having the authority to make major decisions on the shop floor. This evolution requires not only a change in the workers' job functions but also a change in the structure and job content of managers.

- *From "bigger is better" to "small is beautiful."* Although it is true in the United States that manufacturing evolved from small business units to large industrial structures, the opposite has been true in the services realm. The move to a service-oriented economy requires managing smaller business units. Service businesses must be near their market and this requires more outlets. Manufacturers that incorporate more services in their offerings will also move closer to their markets to reduce lead times.

- *From local optimizing to global satisficing.* Decisions are getting too complex to make with optimizing technology alone; there has to be a combination

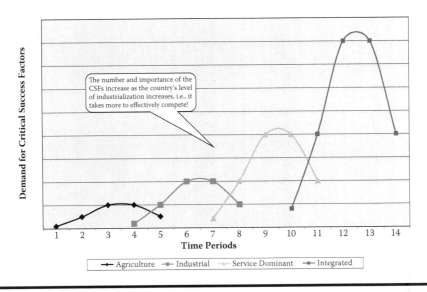

The number and importance of the CSFs increase as the country's level of industrialization increases, i.e., it takes more to effectively compete!

Figure 2.1 Phases of economic growth.

of information, intuition, and initiative. Herbert Simon (1997) coined the term *satisficing* to describe the need to select a course of action that is "good enough." Satisficing is necessary because there is not the time or capability to make an optimal decision. Even with this less demanding criteria, managers will require holistic and systems integrating skills.[10]

■ *From an obsession with "bottom-line results" to a growing awareness of the triple bottom line.* This is a concern not only with economic well-being but also with an orientation toward social responsibilities and environmental preservation.[11] Although not yet a widespread movement, the growing number of initiatives in the sustainability area is providing a clear signal to businesses that they must build this transition into their future plans.

Some view the evolution of a country's economic growth as progressing through several phases: agriculture, industrial, service, and integration. Each phase includes some of the past and requires higher levels of achievement. Figure 2.1 illustrates this concept graphically. As the economy moves from one phase to the next, the demand for higher performance increases. This increased demand means that the number of strategic CSFs must increase also.

The Need to Be Effective

CSFs help management to focus on the top priority decisions and actions necessary to keep the business operating as desired. The quote from Herbert Simon in

Managerial Comment 2.2 emphasized the need to concentrate most of the effort on sifting through the available information to find the right things to use. Three other renowned management thinkers add to this admonition to identify the important areas for management's attention—Peter Drucker, Peter Senge, and Joseph Juran.

Drucker points out that the manager's job is to optimize the yield from the available resources of employees, equipment, facilities, and capital. He makes a careful distinction between effectiveness and efficiency. One of his most famous quotes states, "Effectiveness is the foundation of success—efficiency is a minimum condition for survival after success has been achieved. Efficiency is concerned with doing things right. Effectiveness is doing the right things."[12] Thus, efficiency is what we commonly call a process characteristic, and effectiveness is the manifestation of efficient practices. Effectiveness is often labeled an outcome variable by management scholars.

Senge describes how systems thinking can organize the complexity of today's business environment. The goal is for management to be able to identify the causes of organizational problems and to formulate remedies that will be successful and enduring. One of the problems that managers face in their organizations is the overabundance of information, much of it generated internally. The challenge is to use the information that is most important and set aside the rest. As Senge states, "What we most need are ways to know what is important and what is not important, what variables to focus on and which to pay less attention to—and we need ways to do this which can help groups or teams develop shared understanding."[13]

Juran describes control as maintaining the status quo. In comparing business strategies to process control charts, he points out that businesses should not focus on control alone but should also be concerned with how to continue to improve. He emphasizes that although control may be desirable, it can be a cruel hoax, a built-in procedure for avoiding progress. Companies can become so preoccupied with meeting targets that they fail to challenge the target itself. He offers an alternative "breakthrough," which he describes as a dynamic, decisive movement to new, higher levels of performance. He gives a compelling description of the "vital few" versus the "trivial many."[14]

The term *efficiency* was developed first for commercial businesses and was associated with measurable inputs and outputs. It is more difficult to measure efficiency in noncommercial organizations where the outputs are less tangible. Simon suggests that this is best done by comparing actual progress against the goals or program objectives established by the organization.[15]

How do managers cope with these challenges? They must determine the critical success factors for their businesses. Once they have identified their CSFs, they can prepare plans at all levels of the business. CSFs help to identify the right things to do, the "vital few," as opposed to attempting to do everything well. With the availability of so much information today and the capability to create even more information quickly, it is easy to see that managers immerse themselves in too many details and have difficulty sorting out what is important from what is unimportant.

CSFs represent a way to identify the things that a business must do well. A good starting point is with the strategic planning process. If a business can select CSFs at the strategic level, the business can then develop a hierarchy of CSFs to integrate throughout the operational and program planning levels.

A Hierarchy of the Planning Process

CSFs are an important part of the planning process in a company. To show how CSFs fit into the company's planning, we will briefly describe a typical planning process for a company. Figure 2.2 shows a planning hierarchy in a business. The top portion of the diagram illustrates a structured planning process to develop strategic and business plans. However, the planning process in most companies is not likely to proceed in such an orderly fashion. More likely, management will acknowledge that while plans tend to be static, the operations of a business are dynamic. As a result, there is always the problem of how best to match actual results with planned. The bottom portion of the figure depicts the need for interaction among the various plans to keep them current and meaningful.

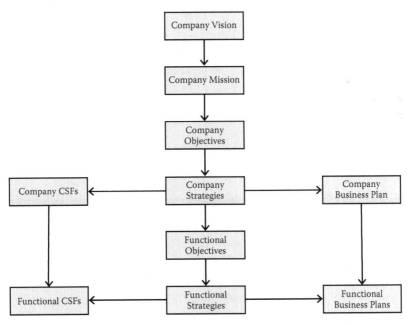

Figure 2.2   Strategic and business planning hierarchy.

What Is Strategic Planning?

Strategic planning focuses on managing for the future. Typically, companies are looking at a time line down the road that may range from two to five years, although some planning may account for longer periods. The basic questions that strategic planning addresses include:

- Where do we want our company to be in the future?
- What should it look like in three to five years?
- How will the company look the same? How will it be different?
- What new markets should the company seek?
- What current business ventures should be abandoned?

Management writer John Parnell sums up the essence of strategic planning in Managerial Comment 2.3, The Essence of Strategic Planning.

Managerial Comment 2.3
The Essence of Strategic Planning

Strategic management (planning) refers to the continuous process of determining the mission and goals of the organization within the context of its external environment and its internal strengths and weaknesses, formulating and implementing strategies, and exerting strategic control to ensure that the organization's strategies are successful in attaining its goals.[16]

Strategic planning begins by taking a hard look at the environment outside of the company. Specifically, top management needs to ascertain the opportunities and threats that face the business. Opportunities represent new markets to pursue. For example, a company may seek to expand its existing products to new markets outside of the country. On the other hand, it may seek to develop new products it can sell to existing domestic markets. Threats, on the other hand, are forces that can take away market share or impact profitability in a negative manner. Obviously, competitors are one example of a threat. Rising fuel costs, bad weather, the raising of taxes, or the possibility of war also represent possible threats. Determining the opportunities and threats is the catalyst that begins the strategic planning process. Reliable, long-range plans depend on this type of analysis.

Knowledge of the external environment must also be supplemented by an internal assessment, determining the strengths and weakness of the company. Strengths are things the company does well and are often referred to as distinctive

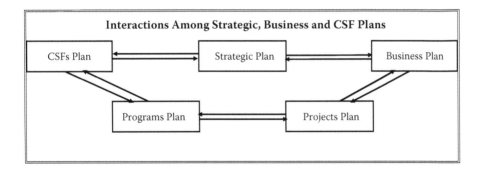

Figure 2.3   Interactions among plans.

competencies. These strengths are often unique and help set the company apart from its competition. Some of these strengths may originate from the personality of the founder. For example, Southwest Airlines is well known for its happy employees and corporate sense of humor. Having fun on the job was one of the values that the airline's founder Herb Kelleher stressed at the onset.

Knowledge of strengths is necessary in formulating strategy because these are qualities the company most likely wants to perpetuate. Weaknesses, on the other hand, may or may not be surmountable. Some weaknesses of a company should be overcome, such as poor training of employees or negative relations with certain customer groups. Including strategic plans to overcome these weaknesses would be advisable. However, companies may have to live with some of their weaknesses. For example, a company may not have the best location for attracting top graduates. Moving the company may not be feasible either, so the weakness may just have to remain a weakness. A company can develop a strategy to deal with that weakness, recognizing that it will be a problem. Consequently, strategic planning would focus not on relocating the company but on finding innovative ways to sell the appeal of the company and its locale to various types of college graduates.

Once the various plans have been prepared, management should develop links among the various plans. The individual plans should receive inputs from each other and provide results and changes to each other. Although it is impractical to change plans every time new inputs are received, the inputs should be available when the next plan is developed. Figure 2.3 shows these interactive relationships.

The Strategic Planning Process

In theory, most businesses develop a vision of why they exist, usually expressed in a few general statements. They expand their vision into a mission statement that adds some specifics to the general direction they want to pursue. Although these are important in providing a reference for subsequent planning, vision and mission statements evolve over time and do not necessarily result from a single meeting.

Developing company objectives is also necessary. Although the objectives should relate to the company mission, in practice most of the company objectives are probably developed by comparing past strategic plans with the actual results. The objectives should include all the goals needed for the strategic planning period; however, they should be somewhat general to allow for further definition later.

In developing company strategies, management must begin to get more specific. Strategies are the actions necessary to achieve the company's objectives or goals. They are necessary to provide direction and performance expectations for the functional areas of the company. Working with the guidance from the company strategies, each functional area of the business—such as sales and marketing, finance and accounting, engineering, operations, human resources, information technology, and others—prepares its own objectives and strategies.

With inputs from all areas of the company, top management puts together a strategic plan. It is important that all managers responsible for achieving the plan's objectives be fully involved in preparing the key elements of the plan. Top management defines the planning process and facilitates the assembling and preparation of the plan. Although it sounds straightforward, it is often difficult and requires a great deal of interaction, reconciliation, collaboration, and adjusting among those levels of management that are involved in this part of the process.

A plan should consider a platform for growth and profits as well as take into consideration the following CSFs:

- Financial factors: Positive cash flow; debt ratios
- Marketing factors: New customers; new products and services to offer
- Production factors: Capacity availability; responsiveness to demand; trained workforce
- Logistics factors: New supplier networks; transportation issues—getting the product to the customer and getting supplies and raw materials to the facility
- Quality factors: Satisfied customers; standards of excellence for products and services
- Human resource factors: Increasing intellectual capital; attracting and retaining excellent employees.

The above description is logical; however, it is likely more theoretical than practical. In recent years there has been a growing awareness that strategic planning is less a hierarchical, top-down process and more of a horizontal, interactive process that reflects an increased realistic perspective. Figure 2.4 indicates that strategic plans must reflect the effect of past actions and the resultant beginning position at the time the strategic planning process begins.

Figure 2.4 requires that companies recognize their present position is a result of how well they performed in meeting their previous plans within the open system influences in which they operate.

Progressive Strategic Planning

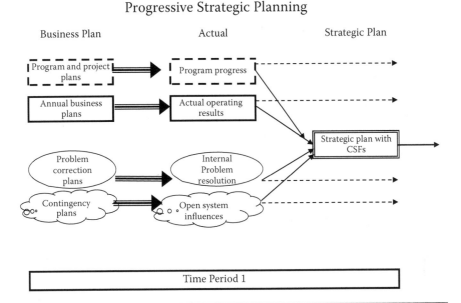

Figure 2.4   A contemporary view of strategic planning.

The Business Planning Process

In general, strategic plans will have more words than numbers. Because most strategic plans are multiyear plans, it is unrealistic to plan at the micro level. Most business plans, sometimes called annual plans, begin with financial expectations and provide the detail, especially in the financial area, in the form of sales forecasts and expense budgets on a monthly time scale. As a result, business plans often have more numbers than words. This difference is both a strength and a weakness of business plans. It is a strength in that it provides specific targets against which to measure progress. It is a weakness because the assumptions on which the plans are built often change throughout the year, and revising the business plan can be a major interruption. As the fiscal year progresses, the business plan becomes increasingly misaligned from the actual activities of the business. As a result, it is important that management have other ways of measuring performance and a seamless strategy to adjust the business plans if necessary. No doubt our readers have experienced situations where the business plan was rendered inaccurate because of unforeseen circumstances. In one of our own experiences, a weeklong weather crisis closed down a facility unexpectedly, causing a major shortage in revenue. Needless to say, the business plan was inaccurate for the remainder of the fiscal year.

Project and Program Plans

In addition to the strategic and business plans, it is important to identify the specific programs and projects implemented during the planning period. The terms *programs* and *projects* are sometimes used interchangeably. Although it may not be essential to differentiate between them, we will consider *programs* as broader than *projects*. For example, a company may decide to implement a Six Sigma program.

Within the program, there may be a number of separate projects, such as training employees in the use of statistical process control (SPC) techniques or holding a series of kaizen blitz sessions. Some companies may not prepare separate program and project plans; instead, they embed these activities within the strategic or business plan. The value of having separate plans is that it focuses more attention on their completion and separates the physical improvement activities from the day-to-day operations, which are often wrapped within the bundle of financial budgets. It is possible, and even desirable, to prepare project schedules that can be used to execute the project plans and measure performance against the project schedule.

Another justification for formulating a separate project schedule lies in the reality of day-to-day operations. Project implementation often stalls when an operational problem arises. When this occurs, the tendency on the part of many managers is to put the project on the back burner "for today" and run the business because, after all, that is the priority. For example, there may be a need for additional staff to teach employees new statistical process control techniques. However, if a key employee is out sick, a decision may be made to delay the training "to another day." As the days become weeks, momentum for the project declines and before long, the project stalls. Project scheduling should thus include contingency plans for these types of situations.

CSFs Plan

The plans described above should be comprehensive because they need to include all of the components a company hopes to accomplish during the year. Out of that all-inclusiveness, they need to identify those things that they must do well—their CSFs. In the following sections, we will describe how CSFs fit into each level of the planning process.

A Hierarchy of Critical Success Factors

CSFs should be planned at all levels of a company:

- Strategic: At the corporate or strategic business unit (SBU) level
- Functional: By function within the company
- Project: By major project or program.

Table 2.2   Examples of Interrelated CSFs

Level of Plan	Critical Success Factors
Strategic Planning	Increase the overall quality of products
	Strategic Business Unit (SBU)—Increase the quality level of the Chicago plant
Operations Planning	Department—Increase the quality level of molding
	Individual—Increase the quality level of each operator
Program Planning	Introduce a Six Sigma program
Project Planning	Provide statistical process control (SPC) training to all operators
	Schedule six kaizen blitzes during the year

To be effective, CSFs should be linked together to provide effective coordination among the different plans. Developing this hierarchy makes it possible to link strategic, functional, and project planning at all levels of an organization. This framework makes it feasible to set priorities and to utilize resources in the most effective way.

Table 2.2 shows an example of this kind of hierarchical planning. The strategic plan provides an overall objective to improve quality. At each level below the strategic plan—operation planning, program planning, and project planning—activities that are more specific are designated to achieve an increased quality level.

This example offers a straightforward approach to developing a hierarchy of CSFs. However, in practice it is a challenging task because it requires a parallel planning process. Not everything can be a CSF; to attempt this would be equivalent to making every order a rush order or viewing all customers as equally important. Indeed, the essence of a CSF is that it must be done successfully. There must be a way of setting priorities in the planning process, and identifying CSFs is one way of doing it.

Background of CSFs in Strategic Planning

Determining the CSFs for a company is part of the strategic planning process. CSFs fit within strategic planning at the macro level; they can also be used at any level of functional planning, such as within systems (IS), and at the program planning level, such as in the implementation of a continuous improvement program.

CSFs become a subset of the strategic plan's goals and strategies. They compose the activities of the company to be performed successfully. There might be other goals and strategies that enhance the welfare of the company, but these are not necessarily CSFs. For example, a goal of the company may be to reduce absenteeism among its employees, but this goal is not a CSF in most cases. Once CSFs have been identified at the strategic level for the entire company, it is then possible to use them as guidelines so that the functional areas of the company can develop CSFs.

The term *critical success factors* originated in connection with executive information systems. Rockart[17] identified procedures for providing executive management information needs:

- The by-product technique: Regardless of managerial needs, all reports are essentially by-products of a particular system, such as sales or payroll, designed primarily to perform routine paperwork processing. This is probably the predominant method and often provides management with more information than what they need.

- The null approach: Advocates of this approach conclude that all computer-based reports—no matter how they are developed—are useless, and that management must dynamically acquire information through personal contact and informal means.

- The key indicator system: This approach involves (1) selecting key indicators to measure, (2) exception reporting of results, and (3) expanding availability of better, cheaper, and more flexible visual display techniques (for example, providing graphs in a boardroom).

- The total study process: This approach looks at all the existing information systems, compares the results with the information needs, and designs changes to fill in the gaps. The amount of data and opinions is staggering and the approach is expensive.

Rockart described the CSF approach as occurring in two steps: (1) interviewing key executives to identify the goals and the underlying CSFs, and (2) deciding how to measure progress. With this approach, he believed the information would be of greater value to top management. He credits Daniel[18] and Anthony, Dearden, and Vancill[19] with providing background to the CSF approach. Rockart offered examples of CSFs for different industries. For the automobile industry, he suggested excellent automobile styling, quality dealer systems, effective cost controls, and meeting energy standards. In a service industry, such as a government hospital, he proposed successful integration with other hospitals in the area, efficient use of scarce medical resources, and improved cost accounting procedures.[20]

Although Rockart's work was early, the concepts have found widespread use over the decades. In another example, Berry et al.[21] showed how the use of CSFs could aid in the linking of marketing and service operations strategies. In a study of retail banks, they explored industry CSFs along two dimensions: market-oriented and competitor-oriented. The choice of operations strategy is dependent on the marketing strategy.

Characteristics of Strategic CSFs

How do companies determine their strategic CSFs? Some top managers begin by looking at their outside threats and opportunities—a competitor has introduced a new product or the federal government has passed a law that requires compliance with financial reporting for public companies. Some CSFs result from an examination of a business's internal strengths or weaknesses—the development of closer supply chain relationships or the need to extend the employee empowerment program. Some strategic CSFs remain the same over time; they become part of the vision and mission of the company. Other strategic CSFs change over time; they become part of the dynamic plans of the business.

CSFs should be developed through the interaction of key employees at all levels—individual, department, division, and corporate. Once top-level managers in an organization have identified their CSFs, these can be compared to other managers' CSFs in the company to determine where there are overlapping concerns, related issues, and obvious conflicts. Put another way, individual managers can think through their personal CSFs, groups of managers can think through their organization's CSFs, and groups of organizations can think through their industry CSFs. Eventually, it should be possible to compile a hierarchy of CSFs for all levels of a company. Sometimes the CSF development process can help in aligning the thinking of all levels of management, as described in Managerial Comment 2.4.

Managerial Comment 2.4
Churning Out the CSFs

Christine Bullen describes an example of how the CSF process can help to resolve conflicts. In one high-technology engineering company, the CEO called a meeting of his top managers to discuss their progress in developing their CSFs. In the interest of moving the program along, the CEO offered his list of CSFs. Although initially upset at being presented with what they interpreted as a mandate from the CEO, the top managers soon saw that the CEO's critical success factors were very similar to their own, and operational in nature. During the subsequent discussion, it became obvious that the CEO, who was also the entrepreneur who founded the firm, was reluctant to give up hands-on control of the day-to-day activities. The CSF process, by focusing on what each person viewed as critical, revealed the conflict and made it possible for this management team and their CEO to move ahead in a productive manner.[22]

The initial success of CSFs led to their application in other ways. The CSF method helps individual managers think through the vast number of activities and issues they must perform, reduce that number to a manageable number, and set priorities on the activities so they can focus on the most critical ones.

To be most useful, CSFs must be:

■ Important to achieving overall corporate goals and objectives
■ Measurable and controllable by the organization
■ Relatively few in number—remember, not everything can be critical
■ Expressed as things that must be done well—not the endpoint of the process
■ Applicable to all companies in the industry with similar objectives and strategies
■ Hierarchical in nature—some CSFs will pertain to the overall company, and others narrowly focus in one functional area.[23]

CSFs play an important role in identifying the areas of strategic importance to businesses.

How do you distinguish CSFs from strategies? A CSF identifies an area of emphasis; a strategy defines an action that relates to the CSF emphasis. CSFs are not the same as strategies—strategies are ways to achieve CSF goals.

Critical success factor analysis is most effective when conducted from the top down:

■ Analyze the corporate mission, objectives, and strategies to pinpoint the success factors of the overall business.
■ Determine the CSFs for each business unit's functional areas.
■ Develop strategies to leverage competitive strengths and overcome weaknesses in each area.
■ Develop measurement tools that will enable managers to monitor performance against the plans.
■ Finally, establish processes and procedures to report performance information in a timely fashion.[24]

CSFs are the characteristics, conditions, or variables that can have a significant impact on a company's success. Their analysis can aid in the strategy development process. Possible sources of CSFs include environmental analysis, analysis of industry structure, industry/business experts, analysis of competition, analysis of the dominant firm in the industry, company assessment, temporal/intuitive factors, and profit impact.[25]

The Temporal Nature of CSFs

CSFs change over time. Among other factors, they change over a product's life cycle. Figure 2.5 shows a typical product life cycle for both a new product and a new service. In general, the emphasis shifts during each phase of the life cycle and

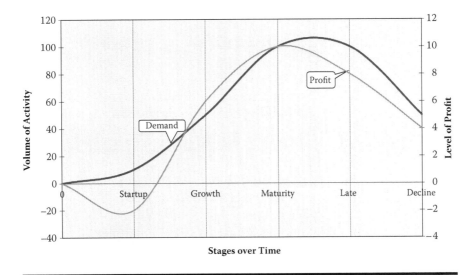

Figure 2.5 Product life cycle.

the functional areas of the organization need to adapt their own CSFs to keep pace with the changes.

Startup

In the early, or startup, stages of a product's life cycle, the primary CSFs will be concerned with product function and its appropriate presentation to the prospective buyer. A company may be test marketing the product in a small market; it is essential that the product function as designed and be acceptable in appearance. It is also important that the product be presented in an attractive, or marketable, form.

Growth

If the product gains acceptance and begins to move into the growth part of the product life cycle, the CSFs shift to availability and quality. The manufacturer must have the product available in sufficient quantities to meet the demand. This is often a treacherous time because it is difficult to forecast demand for a new product, yet if there is unsatisfied demand, the product may never reach its full potential. In addition to having the capacity to meet demand, the manufacturer must be diligent in providing a high-quality product over longer production runs. Just as lack of availability will kill a new product, so will a record of defective products, especially if they are hazardous to the user, such as toys for children or medications for adults. In addition, the manufacturer must have the flexibility to adjust to the refinements needed in the product as customer reactions and feedback become available.

Maturity

At some point, the demand for the new product or service will begin to level off, often because other competitors enter the marketplace and gain a share of the demand. When this occurs, the CSFs shift again, this time to cost containment and reliability. Although there is a continuing need for availability, most of the companies have developed sufficient capacity to handle the demand. With increased competition, there is increasing pressure on prices and, as a result, on product costs. Because of global competition and the move to outsourcing, the need for competitive costs becomes a vital CSF. In addition, product quality enters a new phase toward enhanced reliability. The product must not only work as expected but also operate over extended periods without an undue number of failures.

Decline

Many products eventually reach the decline stage of their life cycle. At this point, the CSFs that are emphasized may vary from one company to another. Some may continue to focus on reducing the price of the product, such as in clothing styles or older appliance models. Others may maintain or even increase the price, but emphasize availability, such as with replacement parts for older-model automobiles. Other companies may hasten the decline of the existing product by promoting a replacement product that is "bigger and better" or, in this day of mobility, "smaller and better," such as for cell phones or electronic readers.

The product life cycle is an inevitable sequence for all products and services. Some life cycles are short; others are long. Even though some products may appear to have an indefinite life cycle, such as aspirin or chewing gum, a closer examination would reveal that these products experience a series of life cycles, where the existing product is replaced with a newer version of the old product and, as a result, enters a new life cycle.

While the product life cycle is changing, the process that is used to make the product or provide the service must also change. We describe this transition later in the chapter.

The Role of CSFs in Operational Planning

Once companies develop CSFs at the strategic level, they can extend them into operational plans for each functional area, such as marketing, accounting, finance, information technology, and operations. As shown in Figure 2.2, the CSFs become more specific as they move lower into the organizational structure.

We will look at some of the CSFs recommended for specific functional or industry areas, including purchasing, human resources, accounting, information systems, and hospitals.

- *Purchasing.* Most purchasing programs concentrate on reducing costs and use purchase price variance as a measure of how well this is being accomplished. However, it may be equally important to try to reduce the purchase price variability of items. This requires a cross-functional effort among purchasing, marketing, and finance to form an agreement on the specifications of the product or service. In addition to agreeing on the common requirements, each function can consider other factors. Purchasing can use such techniques as risk sharing between customer and supplier or smoothing formulas to average the fluctuations in prices. Marketing can set up an indexing scheme in which the price to the customer varies with the price of the critical supply item, such as in retail gasoline sales. Finance can use hedging techniques, such as future options, for commodity items like energy, metals, and agricultural items.[26]

- *Human resources (HR).* This area is evolving into something called organizational effectiveness (OE), in which HR professionals are full partners in their companies' decision-making processes. Many companies see the relevance of linking OE to business strategy, and they have begun to use a relatively simple model to reshape their organizations. The model includes such components as a corporate mission, a strategy, and the organization's CSFs. The overall performance of a company may depend on how well it fits its CSFs to its strategy and how well it blends the HR implications to the technology, structure, systems, policies, and skills of the company. As HR evolves into OE and becomes more strategic and less reactive, its practitioners must become more business-focused to help their organizations deal with the business side of employee issues. Successful OE professionals must possess the talents and mind-sets to be strategists, consultants, and facilitators. In their emerging role, OE professionals must upgrade their skills and capabilities to meet the demands of their new jobs.[27]

- *Accounting.* In the past, managers have often tried to use financial accounting information to manage their organizations. Financial accounting is more externally oriented to meet regulatory, financial analysts, and shareholder needs, and often is not satisfactory for internal, or management, purposes. Although financial accounting information is important and somewhat helpful in reporting on the financial status of a firm, it does not provide the right kind of information to help managers identify problems and to take corrective action. In the past, accounting-based information was the only form of management report available in organizations. CSFs changed that. This method encouraged line management, accounting, and information technology to settle on the information that would be most beneficial in the management process. The use of CSFs to determine what should be reported and how it should be measured revolutionized computer-based information reporting systems. As a result, the phrase has become widely used and the concept has been applied in many industries around the globe. To a certain extent, CSFs helped in the development of executive information systems by demonstrating that when managers are provided with relevant computer-based information, they can use it to improve their decision making.[28]

■ *Management information systems.* The CSF method works well at the policy, operational, and strategic planning levels of information resource planning. Although CSF methods applied to requirements analysis help build conceptual models of roles in organizations, some managers have trouble using conceptual thought processes. Case studies show that the CSF technique helps identify information infrastructures and gives insight into information services that impact a firm's competitive position.[29] Another study of planning participative information systems extends CSF methodology to facilitate participation by many people within and around the organization. The resulting new methodology, called critical success chains (CSCs), extends CSF explicitly to model the relationships between IS attributes, CSFs, and organizational goals. A practical procedure is defined for data gathering and analysis to uncover and model CSC in the firm.[30]

■ *Operations.* Production or service operations management is a functional area that is greatly influenced by the CSFs selected by other functional areas. In selecting its own CSFs, the operations function affects other functional areas.[31] As the product moves through its life-cycle stages, CSFs change and the operations function must adapt to keep pace with these changes. Accordingly, changes in the manufacturing processes may be necessary. They may change their positioning strategy, their manufacturing flows, or their manufacturing strategies. Figure 2.6 describes how a company chooses among alternative processing strategies.

Positioning Strategy

The positioning strategy reflects the position of the product on the life-cycle curve. In the early stages, the manufacturer uses a process-focused strategy that groups like functions together to enable the production of a wide variety of products. This configuration is necessary when trying to determine the acceptance of different product variations in the marketplace. As the product demand increases for some varieties, the company can set up additional capacity in what is known as a product focus, where all the equipment and support services needed to make a single product are grouped together. At this stage, they have a combination of processes, some geared to small volume and variety, the remainder making high-volume, low-variety output. If the progression continues, they may eventually move completely to a product process strategy, making high volumes of a small variety of products. The dilemma comes when the demand begins to decline. At this stage, there may also be a need to begin customizing products for selected customers. This process may require an adaptation of the product focus to make it more flexible in making a wider variety of customized products.

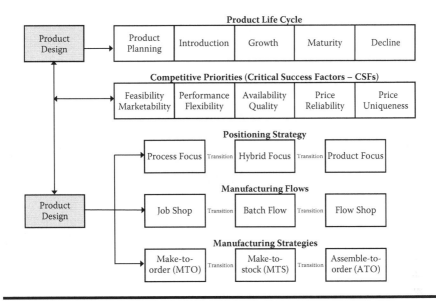

Figure 2.6 Selection of processing strategies.

Manufacturing Flows

A process focus is usually associated with a job-shop environment. Job shops make a wide variety of products at relatively low volumes. As the volume increases, there is a need to move to a batch flow process, where batches of like products move through the production process. The equipment can still be grouped by function and there is an intermittent flow of goods from one function, or department, to the next. The batch flow process has been successful and durable in most production environments. However, it is not as efficient as a flow shop, where the goods are processed along a production line or in some form of organized flow, an approach that is the lowest-cost form, but limited in flexibility and variety.

Manufacturing Strategies

The manufacturing strategies selected by companies may not be tightly related to the product life cycle; however, there should be some relationship. Most companies that make low-volume, high-variety products and use a process positioning strategy use the make-to-order (MTO) strategy. As they move along the life cycle toward high-volume, low-variety production, a make-to-stock (MTS) strategy becomes more practical. Today, the assemble-to-order (ATO) strategy is associated with a maturing of the MTS strategy, where the manufacturer can maintain its high-volume capability but begin to provide greater variety by using modular subassemblies and adding the customizing features after an order is received. This strategy will be described more fully in later chapters under the mass customization label.

Role of CSFs in Selecting Management Programs

As a result of identifying their CSFs, companies may decide to implement incremental or radical improvement programs. We discuss these programs more fully in Chapter 4. Each program requires that CSFs be developed to guide the program. For example, a Six Sigma program would have as a CSF the need to improve quality. It may also include a CSF to increase the empowerment of employees.

The following studies on CSFs illustrate their relationship with various management programs that have been implemented in recent years:

- Lean manufacturing. Achanga et al. recently looked at the CSFs needed in implementing lean manufacturing. They found six CSFs that were essential: leadership, management, finance, organizational culture, skills, and expertise, among other factors.[32]
- ERP. Stephen King and Thomas Burgess also developed CSFs for ERP implementations. They identify the top ten ERP CSFs:
  1. Top management support
  2. Project team competence
  3. Interdepartmental cooperation
  4. Clear goals and objectives
  5. Project management
  6. Interdepartmental communication
  7. Management of expectations
  8. A project champion
  9. Vendor support
  10. Careful package selection.[33]
- Web-based supply chain management systems (WSCMS). A survey revealed five major dimensions of CSFs for WSCMS implementation, namely, communication, top management commitment, data security, training and education, and hardware and software reliability.[34]
- Information center (IC). A study of an information center identified potential CSFs that ended up as composite CSFs: commitment to the IC concept, quality of IC support services, ease of end-user computing, role clarity, and coordination of end-user computing. Tests indicate that these composite CSFs tend to vary in importance among themselves, but they are comparatively constant individually across IC stages.[35]
- Inter-organizational information system (IOS). A case study examined the benefits of the IOS for the companies studied. The research revealed seven CSFs for the IOS: intensive stimulation, shared vision, cross-organizational implementation team, high integration with internal information systems, inter-organizational business process re-engineering, advanced legacy information systems, and infrastructure and shared industry standard.[36]

- Technology absorptive capacity. Lin, Tan, and Chang found a significant relationship between technology absorptive capacity and the following CSFs: technology diffusion channels, interaction mechanisms, and R&D resources. Different organizations will experience different technology transfer performances.[37]
- IBM is another company that employs the use of CSFs. Hardaker and Ward described the use of CSFs at IBM as follows: "Like the mission, CSFs are not the how-to of an enterprise, and they are not directly manageable. Often they are statements of hope or fear." They go on to stress that the CSFs would be necessary for the success of the company and sufficient for the needs of the company.[38]

Almost every project manager can list the main factors—the CSFs—that distinguish between project failure and project success. However, despite the fact that CSFs are well known, the rate of failed projects remains high. This may result when CSFs are too general and do not contain knowledge specific enough to better support project managers' decision making. A field study that involved 282 project managers found that the most critical planning processes that have the greatest impact on project success, are "definition of activities to be performed in the project," "schedule development," "organizational planning," "staff acquisition," "communications planning," and "developing a project plan." The study also found that project managers usually do not divide their time effectively among the different processes, diminishing their influence on project success.[39]

Organizations today are constantly searching for even the slightest competitive advantage and cost savings. The performance of an organization is important in achieving either outcome. Unfortunately, most organizations take a tunnel-vision approach to improving performance. Employees often resist the implementation of costly enterprise resource planning implementations. Despite training and management's efforts to explain the benefits of the program to the company, such programs are often viewed as "this year's" attempt to frustrate the employees. Accordingly, process improvements are rarely rigorous enough to provide sustained results. Yet, when improvement efforts for these key factors—culture, process, and technology—are managed in an integrated manner, success can be achieved and sustained over the long term. Only by understanding the interrelationship between these business-critical factors can lasting performance be sustainable.[40]

Performance Measurement and CSFs

When selecting CSFs, it is important to measure the results, or performance, against the CSF goals. This means that the CSF must have measurable goals. Quantitative goals are easier to measure; however, managers should also attempt to measure the progress against qualitative goals.

CSFs often take the form of a project or a program. Although there may be some financial goals that will tie into the company's financial reports, it is more likely that the goals will be expressed in terms of expected outcomes, or events, such as the completion of a program to adopt an activity-based costing system. In this case, the performance will be based on completion times or quality of the program outcomes, as opposed to financial results. Most accounting systems are not designed to measure the effectiveness of programs. Consequently, supplemental measurement programs are needed, especially in measuring less quantitative performance indicators, such as those represented by CSFs.

A way to measure this type of performance is to use key performance indicators (KPI). KPIs are not the same as CSFs; they are a way to measure the accomplishments of a CSF. KPIs are flexible and can be used in any kind of environment. Often they are the way that operating managers measure the performance of their own responsibilities. Although they cannot always be made to relate directly to the firm's financial results, they are a good way to monitor the progress of CSF projects.

Summary

In this chapter, we have described one of the basic elements of a business—strategic planning—and showed how this extends into operational planning and program planning. At each level of planning, management needs to identify those strategies, or actions, that are essential for the company's success. These essential goals or strategies are the CSFs and they are unique to each business. We build on this concept in Chapter 3 when we describe the evolution of supply chains, because building effective and efficient supply chains is a CSF for practically every business today. We further develop the idea of CSFs in Chapter 4 as we describe the evolution of continuous improvement programs, also a necessity for successful businesses.

References

1. Bullen, C.V., Productivity CSFs for knowledge workers, *Information Strategy,* 12, 1, 14, 1995, p. 14.
2. Rockart, J.F., Chief executives define their own data needs, *Harvard Business Review,* 57, 2, 81, 1979.
3. Simon, H.A., *Administrative Behavior: A Study of Decision Making Processes in Administrative Organizations* (4th ed.), The Free Press, New York, 1997, p. 23.
4. Bullen, C.V., Productivity CSFs for knowledge workers, *Information Strategy,* 12, 1, 14, 1995.
5. Dickinson, R., Ferguson, C., and Sircar, S., Setting priorities with CSFs, *Business,* 35, 2, 44, 1985, p. 44.

6. Yergin, D. and Stanislaw, J., *The Commanding Heights: The Battle for the World Economy*, Touchstone, New York, 2002.
7. Bremmer, I., *The J Curve: A New Way to Understand Why Nations Rise and Fall*, Simon & Schuster, New York, 2006.
8. Hill, T., *Manufacturing Strategy*, Irwin, Homewood, IL, 1989.
9. Blackstone, J.H. and Cox, J.F. III, *APICS Dictionary* (13th ed.), APICS—The Association for Operations Management, Falls Church, VA, 2010.
10. Simon, H.A., *Administrative Behavior: A Study of Decision Making Processes in Administrative Organizations* (4th ed.), The Free Press, New York, 1997.
11. Elkington, J., *Cannibals with Forks: The Triple Bottom Line of 21st Century Business*, Capstone, Oxford, United Kingdom, 1999.
12. Drucker, P.F., *Management: Tasks, Responsibilities, Practices*, Harper & Row, New York, 1974.
13. Senge, P.M., *The Fifth Discipline, The Art and Practice of the Learning Organization*, Doubleday, New York, 1990.
14. Juran, J.M., *Managerial Breakthrough, A New Concept of the Manager's Job*, McGraw-Hill, New York, 1964.
15. Simon, H.A., *Administrative Behavior: A Study of Decision Making Processes in Administrative Organizations* (4th ed.), The Free Press, New York, 1997.
16. Parnell, J., *Strategic Management: Theory and Practice* (2nd ed.), Atomic Dog, Cincinnati, 2006.
17. Rockart, J.F., Chief executives define their own data needs, *Harvard Business Review*, 57, 2, 81, 1979.
18. Daniel, D.R., Management information crisis, *Harvard Business Review*, 39, 9, 111, 1961.
19. Anthony, R.N., Dearden, J., and Vancil, R.F., *Management Control Systems*, Irwin, Homewood, IL, 1972.
20. Rockart, J.F., Chief executives define their own data needs, *Harvard Business Review*, 57, 2, 81, 1979.
21. Berry, W.L., Hill, T., Klompmaker, J.E., and McLaughlin, C.P., Linking strategy formulation in marketing and operations: Empirical research, *Journal of Operations Management*, 10, 3, 294, 1991.
22. Bullen, C.V., Productivity CSFs for knowledge workers, *Information Strategy*, 12, 1, 14, 1995.
23. Freund, Y.P., Planner's guide: Critical success factors, *Planning Review*, 16, 4, 20, 1988.
24. Freund, Y.P., Planner's guide: Critical success factors, *Planning Review*, 16, 4, 20, 1988.
25. Leidecker, J.K. and Bruno, A.V., Identifying and using critical success factors, *Long Range Planning*, 17, 1, 23, 1984.
26. Bruning, B. and Lustig, M., Weathering the storm, *Inside Supply Management*, 17, 7, 26, 2006.
27. Edwards, M.H. and Hyer, R.M., The organizational effectiveness evolution: Wizardry or humbug? *The Human Resource Professional*, 5, 1, 28, 1992.
28. Bullen, C.V., Productivity CSFs for knowledge workers, *Information Strategy*, 12, 1, 14, 1995.
29. Boynton, A.C. and Zmud, R.W., An assessment of critical success factors, *Sloan Management Review*, 25, 4, 17, 1984.
30. Peffers, K., Gengler, C.E., and Tuunanen, T., Extending critical success factors methodology to facilitate broadly participative information systems planning, *Journal of Management Information Systems*, 20, 1, 51, 2003.

31. Hanks, G.F., Rx for better management: Critical success factors, *Management Accounting*, 70, 4, 45, 1988.

32. Achanga, P., Shehab, E., Roy, R., and Nelder, G., Critical success factors for lean implementation within SMEs, *Journal of Manufacturing Technology Management*, 17, 4, 460, 2006.

33. King, S.F. and Burgess, T.F., Beyond critical success factors: A dynamic model of enterprise system innovation, *International Journal of Information Management*, 26, 1, 59, 2006.

34. Ngai, E.W.T., Cheng, T.C.E., and Ho, S.S.M., Critical success factors of Web-based supply-chain management systems: An exploratory study, *Production Planning & Control*, 15, 6, 622, 2004.

35. Magal, S.R., Carr, H.H., and Watson, H.J., Critical success factors for information center managers, *MIS Quarterly*, 12, 3, 413, 1988.

36. Lu, X.H., Huang, L.H., and Heng, M.S.H., Critical success factors of inter-organizational information systems—A case study of Cisco and Xiao Tong in China, *Information & Management*, 43, 3, 395, 2006.

37. Lin, C., Tan, B., and Chang, S., The critical factors for technology absorptive capacity, *Industrial Management + Data Systems*, 102, 5/6, 300, 2002.

38. Hardaker, M. and Ward, B.K., How to make a team work, *Harvard Business Review*, 65, 6, 112, 1987.

39. Zwikael, O. and Globerson, S., From critical success factors to critical success processes, *International Journal of Production Research*, 44, 17, 3433, 2006.

40. Marquardt, M., Smith, K., and Brooks, J.L., Integrated performance improvement: Managing change across process, technology, and culture, *Performance Improvement*, 43, 10, 23, 2004.

Chapter 3

The ITO Model

The ITO Model

The Input–Transformation–Output (ITO) Model is the DNA of the supply chain. The supply chain is composed of a series of linked entities, including manufacturing and service businesses. This chapter describes the ITO Model and how these elements can be linked together to become a functioning supply chain. The following topics are discussed here:

- Types of models
- Uses and benefits of these models
- The basic ITO Model
- Extensions to the basic model
- The concepts of open and closed systems
- Supply chain formation
- The ITO Model and its application to management functions

A single ITO module is equivalent to a transaction, such as the receipt of an order from a customer. A linkage of several ITO modules is a process, such as the complete handling of an order from receipt to delivery. A supply chain involves the linking of a process from one company to an input or output process in another company. To help us understand the ITO Model, it would be beneficial to examine why models are used in the first place. Managerial Comment 3.1 by Martin Starr begins our discussion on why models are useful.

Managerial Comment 3.1   The Nature of Models

Operations management is a many-problem field. In other words, the manager concerned with operations encounters one problem after another. These many problems arise because many operations must be accomplished to assure that the result of the operations is of agreed-upon quantities, on schedule, with specific qualities and costs. Clearly, these operations create many problems, and it is not obvious how to treat such great variety. Seldom is the manager faced with one big problem for which one major strategy decision will suffice.

By simplifying reality in a systematic and organized fashion, it is possible to study systems that are too complex to be understood by intuition. Then, for such systems, we can diagnose errors to bring about useful remedial changes. The idea of a model is crucial to this reasoning, because a model is a simplified representation of reality. It is constructed in such a way as to explain the behavior of some but not all aspects of that reality. The reason that a model is employed is that it is always *less complex* than the actual situation in the real world. It must, however, be a good representation of those factors or dimensions that are strongly related to the systems objectives; otherwise, it will not be a useful model and, therefore, it should not be used.

Models have always existed, but the recognition of model properties is just yesterday's achievement. This recognition of *properties* has made an enormous difference to operations management as a field because the properties of models were found to be very general, so general, that all products and any services could be studied.[1]

Introduction to Models

This chapter shows how we can use ITO models to describe the basic elements of operations management in any type of business. It also shows how managers use a variety of models to help them in their decision making and management of their companies.

Types of Models

Three types of models exist—physical, mathematical, and conceptual. Physical models are smaller objects that represent larger objects. Wind tunnels test prototype airplanes to predict how the full-size airplane will react in various flight conditions.

A model building and its surrounding parking area can help planners decide the size and layout of the full-scale facility. A model of a sales counter for a retail store can be set up for use as a training facility for new sales personnel. Physical models are widely used in almost every kind of business to help us better understand the composition and relationship of the elements that compose a given situation.

Mathematical models represent relationships among variables that we can quantify. Some of these relationships include the following:

> total costs equals fixed costs plus variable costs, or TC = FC + VC;

or

> the percentage of process defects equals the total number of defects divided by the total number of units tested, or % p = d/n;

also

> the number of inventory turns equals the annual cost of goods sold divided by the average inventory, or Turns = COGS/AI.

As with physical models, we use mathematical models to help us understand and improve our business performance.

Conceptual models represent ideas. They are usually more difficult to design and understand than physical and mathematical models. However, conceptual models are often simple when compared with the real world they represent, and are therefore subject to more liberal interpretation. Do you believe that increased knowledge leads to improved performance? If you do, how would you show this relationship in a model? You could simply state the idea in equation form as:

> Increased Knowledge → Improved Performance.

You might also display the idea as a graph, with Knowledge as the independent variable represented on the horizontal axis with increases to the right. The dependent variable, Performance, is shown on the vertical axis with increases moving from bottom to top. When the two variables are graphed together, an upward-sloping line would result, starting in the lower left part of the graph and moving toward the upper right. How rapidly would it increase? Would it be linear or curvilinear? Would it continue to increase indefinitely or would it reach a peak and start down? You can see that the curve could vary under different conditions. Yet, all conditions support the basic idea that improved performance follows increased knowledge.

In this chapter we introduce a conceptual model that describes the basic operation of a business. We use this model throughout the book; the remaining chapters explain the model in more detail and show how to use it in managing a business.

The model is also useful in explaining the similarities and differences among different kinds of businesses.

Why Are Models Used?

Operations managers face many problems. They must address some problems rather quickly; others, involving strategic decisions and larger investments, require more critical analysis before making a decision. The use of models can help in this latter kind of decision. An effective model can help answer several questions:

- **What are the important parts of the problem?** A model identifies important components of a problem. Recognizing these allows decision makers to focus on the components that are most critical from those that are less critical. For example, the manager who is experiencing high employee turnover would want to identify the causes of the turnover. These causes are the important components of the model. One such cause may be job dissatisfaction.
- **How do these parts relate to each other and the problem at hand?** A model also helps to specify relationships among components of a problem, especially the relationship between causes and effects. Models often designate independent variables that represent causes of particular symptoms. Dependent variables represent events influenced by the independent variables. For example, job dissatisfaction, an independent variable, can influence employee turnover, a dependent variable, in an unfavorable way. In other words, high job dissatisfaction can lead to high employee turnover.
- **Why are these parts important?** In the example of employee turnover, many models have been offered by management researchers that indicate which independent variables are the most important influencers of turnover. The manager who wishes to decrease employee turnover will seek to identify the variables to change in order to reduce turnover.
- **Under what conditions will this problem exist?** Models specify conditions such as when events occur, where they occur, and who makes things happen. For example, there are employee turnover models that vary according to the industry under study as well as the particular type of job.

Answers to these questions make it possible for us to use models to study other business problems such as identifying the key variables in the application and training of inexperienced employees. We can also examine the model itself for opportunities to make improvements in its structure and operation.

Benefits of Using Models

Managers find several benefits from using models. Models can help them control and manipulate the large number of details that surround business problems.

A model clarifies the distinction between critical elements that require close attention and extraneous factors that often distort or confuse real relationships between inputs and outputs. Models allow managers to experiment with changes to see what effects they would have in real management situations. In other words, managers can benefit from asking "what if" questions. Without models, managers would have to make changes in real systems, a time-consuming and expensive process with a high cost of failure. Models can help problem solvers to evaluate the effectiveness and efficiency of various alternatives, separating poor solutions from better ones. Models can improve intra-organizational communications, as managers often discuss operating systems using models. Models help to focus on information-gathering activities. By identifying the key factors of a real situation, a model directs managers' attention to information they need to make decisions.[2] As with any management tool, the value of the model depends on the competence of the user.

The Basic ITO Model—Inputs, Transformation, and Outputs

Figure 3.1 depicts the basic ITO Model. It shows the transformation of things, such as wood or steel, into goods, such as furniture or automobiles. It also shows the transformation in service industries, where the inputs are often people.

The transformation process can be as simple as building a table out of pieces of wood or cashing a check at a bank. It can be as complex as building an airplane or changing a customer's anger over poor service to delight when an employee resolves the complaint with courtesy and promptness. The models described in this chapter examine how the transformation process applies to a wide variety of business situations.

Table 3.1 shows a number of examples of the ITO Model in service applications. Inputs can be both things and people. However, measuring the outputs is more difficult in services. For example, the output of a hospital can be counted as the number of patients processed, but there is greater variation in the patients than the output of a manufacturer making wood desks. This variation makes it

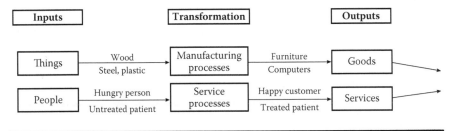

Figure 3.1   Basic Input–Transformation–Output Model.

Table 3.1   Examples of ITO Modules

Type of Business	Inputs	Transformation Process	Outputs	Customers
Retail supermarket	Unserved customers	Provide facility, inventory, system, and employees to make food and other items available	Served customers	Individual consumers
Electric utility	High voltage, undistributed power	Transport power from generating plant to distribution points; reduce voltage for use	Low voltage, distributed power	Individuals and industrial customers
Hotel	Arriving customers	Provide food, accommodations, and related items to a variety of customers	Departing and satisfied customers	Individuals and groups
Hospital	Sick or injured patients	Provide diagnosis, tests, and treatment for a variety of illnesses	Treated patients	Individuals
Public accounting firm	Unaudited companies	Provide a series of professional services, as needed	Audited company	Public companies

more difficult to measure both the effectiveness and efficiency of service transformation processes.

The General Systems Model

We can extend the simple linking relationships shown above as a building block for a more general integrated system. Outputs are the goods or services produced by a business. If the output is a tangible product, the business usually fits into one of the extractive or transformative industries. In any of the other industries, outputs are less tangible; consequently, the business usually fits into a services category.

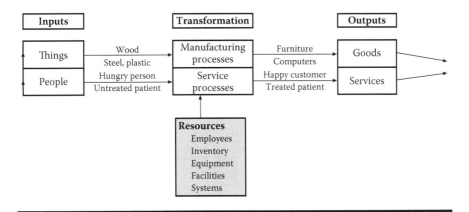

Figure 3.2   Addition of resources to the ITO Model.

Outputs, as shown in the general systems model, often become inputs to another ITO module. A general system consists of several ITO modules linked together.

Because inputs are often outputs from another part of the system, inputs are also both tangible and intangible. Iron ore, for example, is both an output from a mining operation and an input for the refining process that produces iron as its output. The iron then becomes input for the next process that adds other materials to produce outputs of steel. Steel becomes an input for other processes and the linking continues until we have a final product for the consumer.

On the services side, high school graduates (output) become an input to a university, where faculty and staff transform them into first-year students (output). Following a series of intermediate transformations, they become college graduates (output). Each graduate becomes an input to a company where additional transformations take place in an organizational setting. These individuals will also experience transformations in the social and political environments of which they are a part.

The transformation process requires the use of resources. The resources include the employees who perform the work, the equipment they use, the facilities in which they work, plus the systems and information they use. Figure 3.2 shows this addition to the ITO Model.

To expand the model a bit, we can add customers as the recipients of the goods and suppliers as the providers of the input materials. These additions show the beginning of a supply chain, or the linking of different entities into an integrated unit (Figure 3.3).

Traditionally, manufacturing and services have been viewed as unrelated processes. However, in practice, they are almost inseparable. Most manufacturers find they must provide some degree of services to accompany their goods. For example, an automobile manufacturer sells a product but also provides services to the consumer through the dealer network, or a computer company provides an online help center for customers having problems with their computers.

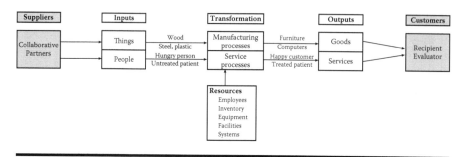

Figure 3.3 Addition of customers and suppliers.

Let us look at another example in a service business. Consider a distribution company that receives an order from a retail store. The transformation process begins with the order as an input. However, the customer is also an input, in the sense of having an unmet need and expecting to become a satisfied customer (output). What does the distribution company do to transform the customer? The distribution company may send an order acknowledgment confirming the order when the customer wants it and then ensure that it is shipped on time. Suppose the distribution company dissatisfies the customer because the order shipped late with two of the ten items ordered missing. The company may send a message to the customer to explain that the back-ordered items would be shipped on a certain date at no extra charge to the customer and then follow up to ensure that the items were shipped to and received by the customer. This example illustrates how parallel processes can work on different inputs and produce different, but related, outputs at the same time. Consequently, most businesses are transforming people at the same time they are transforming things.

The Basic Model Extended

In Figure 3.4, we extend the model to show that the output from one transformation becomes an input to another ITO module. Suppose the first input is an order

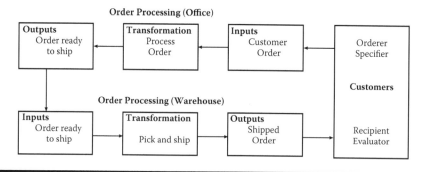

Figure 3.4 Order processing links.

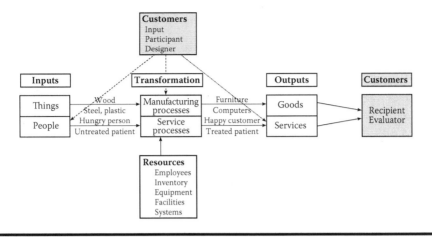

Figure 3.5   Additional customer roles in ITO Model.

from the customer of a distribution company. The distribution company transforms the customer's order into an approved order ready to be shipped (output), accomplishing this task through a series of actions that include approving the customer's credit, verifying that the ordered items are in stock and available for shipping, and scheduling the order's shipping date. The approved order then becomes the input for the warehouse, where the items are retrieved from stocking locations, loaded onto the delivery truck, and delivered to the customer. The output from this transformation process is the completed order. This linking of ITO modules illustrates a common series of activities, or a process, in any kind of business.

Linking two ITO modules is just a beginning. If we can link two ITO modules, we can link a third module, a fourth, a fifth, and so on, until we complete the desired configuration that we need.

Now, let us add another dimension to the model. We have shown the customer as the recipient and evaluator of the goods and services provided. Customers can also be inputs when they are waiting to be treated at a hospital or waiting for a haircut at the hair salon. However, in recent years, customers have also become participants in the transformation process, such as when they select their own groceries at a supermarket and use the self-checkout terminal. Customers can also help design their educational program. They can select the format in which they will receive instruction—by traditional attendance in a classroom or online in a distance education environment. Figure 3.5 shows this extension of the model.

Components of the Model

The models described above show how we can link simple ITO modules together to make a more complex model. This integration process is an important idea—one that we will discuss more fully later in the chapter. At this point, however, we will disaggregate the basic ITO Model into its components and look at each one separately.

Customers

The objects of all service activities are satisfied customers. In a sense, services require that a business transform its customers, or potential customers, into satisfied customers. The transformation process may affect customers minimally, such as in the delivery of electric power, or substantially, as when a surgeon performs heart bypass surgery. Because customers vary, most companies need to develop processes for the best handling of each type of customer.

An important consideration is whether the customer is adaptable and willing to participate in the service process. Customer participation can make the process more productive, but more importantly, the customer helps determine the quality and customization of the final service. Examples of customer participation include the use of salad bars at restaurants or planning a two-week vacation online by selecting various travel and entertainment options.

Customers will possess different levels of knowledge about the services they receive. Some customers will have little knowledge, like the new member who joins a fitness club and needs assistance in learning to operate the various exercise machines. Other customers will possess a great deal of knowledge about the services but will still need some minimal amounts of assistance. An example is the fitness club member who knows how to operate all of the machines but needs advice on how to set up a specific exercise program. This example illustrates how the service plan must satisfy both extremes, even in the same business.

Customers may receive services as individuals or as part of a group. Usually, the level of customer contact and customization of service is greater for individuals than for groups. The effect of the service will vary for each person within the group. This variation makes it more difficult to satisfy all of the members of the group. In a tour of the Gettysburg battleground, tour group members who are not familiar with the battle may be very interested in the tour because the information being presented is new. Other members may be bored by the presentation if it is information with which they are familiar.

Outputs

Outputs can take the form of goods, or things, which we can specify and measure to see if they conform to specifications. Thin plastic paper is an example of an output that is used to wrap after-dinner mints. It must be attractive and meet other physical specifications; simultaneously, it provides part of the service package at a restaurant. The other form of output is the customer, each of whom is unique. Measuring the output for a customer may be a complex undertaking. The measure of a completed output may be as indefinite as the perception of the customer. For example, attending a training course on the use of a word processing software package transforms a person with little or no knowledge to a person with more knowledge. However, the knowledge gained varies with each person. Clearly, defining the output for a pure service is difficult and the achievement of a satisfactory output requires variations in the process used. An understanding of the expectations of the customer before providing the service may preclude subsequent customer dissatisfaction or rework of the service.

The amount of customization required of a service varies among outputs, especially when the output is a person instead of a thing. Another variation is the rate of demand by the customer. Demand is rarely at a constant rate; therefore, the transformation process has to be able to accommodate both customizing the service for the customer and adjusting the demand rate.

The success of any business depends on how well the outputs reflect the critical success factors (CSFs) of the business. This requirement includes the cost of service, the timeliness of service, and the quality of services. Careful planning is required in providing the service package to maximize the benefits of CSFs.

Transformation Process

Much of the research in management has focused on the transformation process. Some variables in the transformation of services are tangible, such as determining the number of store sites in a retail business. Other variables are intangible, such as the level of judgment required by a display designer in a custom flower-arrangement shop.

The transformation process can be viewed as a technical function, capable of standardizing or customizing the service. Here, we attempt to adapt the transformation to any of the variations that may be encountered from the operating environment. For example, the service transformation can be rigid, such as in the use of ATMs, or flexible, such as with the providing of human bank tellers.

Automation is an example of the active use of technology in the providing of services. Automation usually increases over time as the service becomes more standardized, as in the case of travel insurance policies sold through vending machines. The amount of automation also increases as the technology capability increases; for instance, airline reservation systems are now available to every travel agent and online to the Internet user.

The transformation process must be adaptable to variations in tasks and rates of service. The service can be continuous (e.g., an electric utility) or discrete (e.g., a lawyer preparing a house purchase agreement). Companies can service customers at their place of business (e.g., a bakery), in the customer's house (e.g., an interior decorator), and over a communications system (e.g., a home shopping channel). The presence of many variables in the transformation process precludes adopting a "take two aspirin and call me in the morning" type of approach for most businesses. Each business has a uniqueness that results from a combination of variables that works for that particular service and owner.

Inputs

Inputs can be things or people. With things as inputs, businesses attempt to standardize their inputs to reduce undesirable variations in the transformation process. With people as inputs, businesses attempt to build adaptability into the transformation process, usually with trained employees, to accommodate the increased variations that often arise. Such variations may result when we introduce additional changes in tasks and work flows. Consequently, the ideal transformation process must have sufficient flexibility to adapt to input variations.

We can combine transformation variables in hundreds of ways, each of which may require a different approach in the design and management of the resulting service business. Although inputs, processes, outputs, and customers are presented as separate elements, they operate within the framework of an integrated system. As a result, a change in one element will impact other elements as well. For example, an adjustment in input, often the result of a changed customer need, requires a subsequent change in the transformation process. Furthermore, these changes will also result in changes in one or more of the output variables.

Although describing a separate case for each possible combination of variables is impossible, Table 3.2 shows three possible combinations. The top section of the table describes a simple service, the processing of customer checks, where things (the checks) are the primary input. Even in some operations where there is no direct customer encounter, some people are involved, mainly the employees who process the checks. Many manufacturing plants also fall in this category, particularly those that are isolated geographically from their customers. In terms of technology, the transformation process is standardized and geared toward economies of scale.

The middle section of the table describes a combination of a product and service output, such as the delivery and installation of a refrigerator. In this transformation, a large technical core (the factory) manufactures a small technical unit (the refrigerator). The service function is illustrated through the delivery, installation, and demonstration of the refrigerator. This combination of output, both things and persons (or alternately, product and service), is common in most business transactions; what varies is the amount of service to add to the sale of the product.

The bottom section of the table depicts an example of a pure service: the design and performance of an accounting audit at a small business. The technology core includes a vast amount of individual and group knowledge, and interaction with the customer is extensive.

The general approach that we can glean from Table 3.2 is that businesses need to (1) develop a total service concept, (2) design the service package, and (3) manage the transformation processes. Later chapters present each step in more depth.

Table 3.2 Examples of the ITO Model for Services

Inputs	*Transformation*	*Outputs*	*Customer*
Things (Checks)			
Consistent task content Consistent work flow	Processing mortgage payments in the back room of a commercial bank Large technical core with low customer contact and dedicated resources to do standard services	Product—notation in customer's file that payment has been made; no formal report to customer	Knowledgeable (at least for this task) individual who has adapted to the bank's system of payment
Things (Refrigerators)			
People (delivery crews who can also install)	Delivery and installation of a refrigerator Large technical core to make; small crew to install The product is transformed and the customer is also transformed, from an expectant buyer to a delighted new refrigerator owner	Delivered, installed, and demonstrated product Apprehensive prospective owner becomes a delighted new customer	Individual with moderate knowledge, at least as user of the product Discriminating buyer of service Some flexibility in timing of delivery
People (Accounting Audit at a Small Business)			
Uncertain arrival rate Uncertain work flow	A CPA must design and conduct an audit of a small business that has never been audited Small technical core with high customer contact by servers who must perform highly customized services	A completed audit with a variety of accompanying work papers and financial statements	Relatively lacking in knowledge of the audit process and with limited resources to participate in the audit process and the results

Extending the Basic ITO Model into Supply Chain Configurations

The preceding discussion shows how the basic ITO Model works. It also provides a glimpse of the ideal supply chain by showing the addition of customers on the receiving end and suppliers on the providing end. This section describes how basic ITO models can be linked together to form a supply chain. The clothing industry is a good example. It includes a combination of both manufacturing and service businesses. It also illustrates the complexity of many supply chains in that a number of independent businesses must work together to make the supply chain effective. At the end of this chapter, we discuss the difficulty in making a complex supply chain work smoothly.

A Look at the Supply Chain

Figure 3.6 illustrates how a number of ITO modules can be linked to each other in that one ITO output becomes the input for another, and so on. In this example, the process begins with the growing of cotton and ends with the recycling of worn clothing.

The Closed and Open Systems Model

Up to this point, we have been discussing the ITO Model and its linking with other ITO models in what is referred to as a closed-system model. That is, essentially all transformation activities are under the control of the parties designing the system. In the real world, such a scenario does not exist because businesses operate in an open-system environment. The open-system concept acknowledges that external factors, such as competitors, the economy, technology, society, the environment, and government, may have a significant effect on the operations of the supply chain.

In addition to the differences between closed and open systems, general systems theory includes several other attributes of systems, including holism (viewing the whole system, not just the parts), goal seeking (the need for tangible objectives), entropy (the tendency of a system to fall into disorder over time), regulation (the need for management), hierarchy (components within subsystems within systems), differentiation (the uniqueness of a system to fit its application), and equifinality (alternate ways to reach the same objectives). "Of all the distinctions noted, that of open and closed systems appears to be most fundamental and most useful. This classification rests upon the basis of resource availability."[3]

Closed Systems

The closed-system model is a concept that is useful in describing a system because it makes it possible to screen out a number of variables that are difficult to consider. It may be thought of as saying "all other things being equal" to confine the analysis

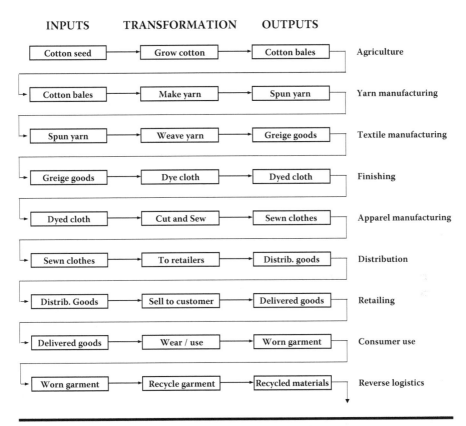

INPUTS **TRANSFORMATION** **OUTPUTS**

Cotton seed →	Grow cotton →	Cotton bales — Agriculture
Cotton bales →	Make yarn →	Spun yarn — Yarn manufacturing
Spun yarn →	Weave yarn →	Greige goods — Textile manufacturing
Greige goods →	Dye cloth →	Dyed cloth — Finishing
Dyed cloth →	Cut and Sew →	Sewn clothes — Apparel manufacturing
Sewn clothes →	To retailers →	Distrib. goods — Distribution
Distrib. Goods →	Sell to customer →	Delivered goods — Retailing
Delivered goods →	Wear / use →	Worn garment — Consumer use
Worn garment →	Recycle garment →	Recycled materials — Reverse logistics

Figure 3.6 Linking of ITO models to form a supply chain.

and planning to those activities within the control of management. One assumption is that all of the system's resources are present within the system under study and no additional resources will enter the system from the outside environment. The resources include the employees, equipment, facilities, technology, processes, and information used in impacting the transformation process. "Problem solving by business executives is often done in a quasi-closed system with the aim of simplifying the situation enough so that a rough estimate can be arrived at."[4]

Entropy is a useful concept that relates to closed systems. Entropy suggests that a system will eventually fall into disorder and become less effective over time unless it is reenergized with additional resources. This means that the resources must come from outside the system, so although the closed system concept is useful, it is not completely adequate for understanding today's business environment.

Open Systems

The general systems model shows that every system has a boundary that limits its sphere of application. However, the environment outside the system, no matter

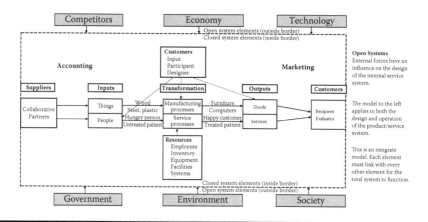

Figure 3.7 Open-system environment.

how closely constrained, affects the transformation processes of all systems. For example, a university may have considerable control over curricula, course content, classroom procedures, and acceptable student academic performance. However, external factors outside of the university, such as the student's family, economic conditions, funding sources, social pressures, books, and many other factors, also influence the university and its relationship with the student.

Figure 3.7 shows the relationship of the ITO Model to external factors that may influence the effectiveness of a system. Elements in the open system include competitors, the economy, technology, the government, environmental concerns, and societal forces. It also illustrates the idea that the organization has little control over these factors but does have some ability to anticipate their effects in the planning process. As a result, these factors should be considered in the design and operation of a business.

Here are some ways these external forces could affect a business. Competitors may introduce new products, raise or reduce prices, offer sales promotions, or introduce a new kind of store. The economy may reflect increased interest rates or a slowdown in housing starts. Technology advances could provide new software for inter-organizational communication, a new material that is stronger and lighter than steel, or a breakthrough in the use of radiofrequency identification (RFID) tags. The government could raise the minimum wage, enact legislation that would limit sales of critical items, enact new regulations on reserve requirements for banks, or reduce spending on weapons systems. Environmental considerations could enhance sustainable development efforts, reduce the use of toxic materials, or encourage the use of recycling or remanufacturing processes. Society could lose its interest in high-fat foods, increase its acceptance of new technologies such as a preference for online buying, or begin to save money instead of spending it. Any business, regardless of size, faces at least some of the challenges presented by an open-system environment.

Closed-System Strategy

James Thompson[5] studied the effect of closed and open systems on the organization. The system within the organizational boundary represents the closed system and the opening of that system to the environmental factors represents the open system. The rational model results from a closed-system strategy, and the natural-system model, discussed below, results from an open-system strategy.

The closed system is comparable to a determinate system in which we know all relevant variables; therefore, we can accurately predict the future performance of the system. *Scientific management*, *administrative management*, and *bureaucracy* are examples of rational models. We discuss these in more detail in Chapter 4. They suggest that the ingredients of the organization are deliberately chosen for their necessary contribution to a goal and the internal structures established are those deliberately intended to attain the highest efficiency.

Managerial Comment 3.2   Isolation of the Technical Core: Classical Management Thinkers

James Thompson developed a series of propositions to explain how the technical core of the organization must be isolated as much as possible from the environment to ensure maximum efficiency.

Proposition 1. Organizations seek to seal their core technologies from environmental influences.

Proposition 2. Under norms of rationality, organizations seek to buffer environmental influences by surrounding their technical cores with input and output components. If an organization can buffer its technical core from the environment, then it presumably can enjoy the conditions for maximum efficiencies. However, if buffering the technical core is not possible, then the organization has to seek other devices for protecting its technical core.

Proposition 3. Organizations seek to smooth out input and output transactions. This means that if the company cannot use inventory to buffer its technical core, it must attempt to smooth the demand in a way that reduces the fluctuations in the environment. This would involve changing the demand patterns that exist in a company's business, and because this is not always possible, the technical core must settle for a lower level of efficiency.

Proposition 4. Organizations seek to anticipate and adapt to environment changes that they cannot buffer or level. This means that organizations must forecast the demand and attempt to set their technical capability (capacities) to gain

much of the efficiencies of a closed system, though with fluctuations in the environment.

Proposition 5. When buffering, leveling, and forecasting do
not protect their technical cores from environmental fluctuations, organizations resort to rationing. Rationing represents the
attempt by the organization to schedule demand or to allocate
available output to the demand. Rationing is an unhappy solution, for its use signifies that the technology is not operating at
its maximum, yet some system of priorities for the allocation of
capacity under adverse conditions is essential if a technology
is to be instrumentally effective.[6]

Open-System Strategy

The rational model, or a closed-system strategy, works when we can control the
variables and relationships at hand. This is seldom the case in practice; therefore,
most businesses must resort to a natural-system model and employ an open-system
strategy. This assumption occurs when we expect uncertainty, when the number
of variables exceeds comprehension, or when we cannot control or predict some
variables because of outside influences.

Rational models produced important insights but did not provide an adequate
understanding of complex organizations. Herbert Simon proposed that in dealing
with complexity, the organization must develop processes for searching, learning,
and deciding. He called this process making decisions in bounded rationality. This
requirement involved replacing the maximum efficiency criterion with one of satisfactory accomplishment. Put another way, the perfect outcome did not have to be
attained; instead, a decision that was sufficient would be satisfactory. Thus, decision
making now involved "satisficing" rather than maximizing.[7]

How do we use these ideas in business operations?

1. The notion of closed systems and open systems suggests that most businesses
   conform to the open-system model because their interactions involve external customers.
2. Although the total business requires an open-system model, isolating the technical core may be possible and viewed as operating under a closed-system model.
   For a fuller explanation of the technical core, see Managerial Comment 3.2.
3. Service organizations can employ strategies to isolate the technical core when
   the environment is uncertain or fluctuating. These strategies can include
   buffering, smoothing, anticipating, and rationing.[8]
4. The activities of the organization can be viewed as gathering inputs, transforming the inputs by some technology or process, then distributing the outputs of the transformation process in an effective manner.

Table 3.3 The Effect of Variation on Process

Condition	Inputs	Transformation	Outputs
Closed system with an isolated technical core	Small variation, standardized	Controlled, routine	Standard, high volume
Open system with an external focus	Large variation, nonstandardized	Flexible, less controlled	Custom, low volume

Both Perrow[9] and Thompson[10] incorporated the ITO Model in their work. They pointed out that minimizing variations in inputs allows greater control in the transformation process and results in standardized outputs. Allowing large variations in inputs requires greater flexibility in the process and results in more customization of the output. Table 3.3 shows a summary of these relationships.

Both Perrow and Thompson placed heavy emphasis on the inputs used and the transformation process; they did not focus as directly on outputs and the direct contribution of the customers. They portrayed the customer as part of the outside environment, not as a direct participant in the company's organizational structure.

Managerial Comment 3.3   Core Competency

What is a core competency? The answer I hear from many executives is, "It is what we do well." My response is, "What if you are really good at running the company cafeteria?" This may be well done, but it does not help the company make money or achieve advantage over the competition. A more accurate answer to this question is, core competencies are the thing we have to do well to achieve competitive advantage. This may include superior R&D, superior production, superior marketing or, you guessed it, superior supply chain management. In the last case, this is a process of identifying what your company has to do well and, at the same time, identifying what your supply chain partners have to do well to see the overall supply chain is successful."[11]

Feedback

In the interest of keeping our diagrams as simple as possible, we have not mentioned an essential element of any system—feedback. Feedback completes the loop in the closed-loop system. All transformation processes produce an output, and managers of the transformation process are interested in the quality of that output. The first feedback should come from the managers within the company as they evaluate their own performance, based on the results of their performance measurement systems. Additional feedback should be solicited from the customers in both formal and informal communications. Another source of useful feedback is from employees, who can often spot problems or provide useful suggestions even before members of management. Although feedback can provide useful information, it is often difficult to gather and analyze.

The Fine Art of Building Relationships

We have shown how individual ITO modules can be linked together to form complex supply chains. There is a challenge in the actual process of linking these ITO modules together. What transformation process is required to accomplish this monumental task? It is not manufacturing or services in the traditional sense. Figure 3.8 illustrates this new kind of transformation, a process we call "relationship building."

We have transformation processes for manufacturing and for providing services. But do we have transformation processes for building relationships? Probably not at the present time; however, we will address this topic later in this book. But first, we will describe a hypothetical example of why it is both necessary and difficult to link individual ITOs into a complex supply chain.

Our example includes three separate entities—a manufacturer with one supplier and one customer. Rather simple, but if we can create it for this configuration, we can extend it to multiple configurations. Within each of the entities, we have three separate processes, and between the entities, we have two interfaces. For illustration purposes, we depict them as shown in Figure 3.9. The arrows represent the process steps and generally point to the right. However, they are not exactly

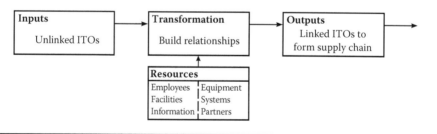

Figure 3.8 Relationship building.

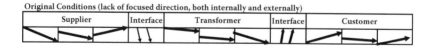

Figure 3.9 Original conditions in relationship building.

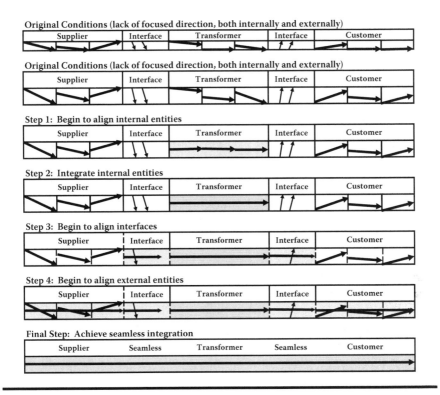

Figure 3.10 Progression to seamless flow.

aligned, which illustrates that a smooth flow of goods or services from supplier to producer to customer is not always possible.

Figure 3.10 shows a progression of alignment steps. First, there is a need to align the internal functions within a single business entity. Lest this appear too obvious and simple, one of the authors recalls a discussion with a VP of Operations for a tool company. When asked what his major problem was when he first joined the company, he replied, "Getting the production department to make what marketing was selling." He went on to explain this seeming incongruity. It seems that the production department employees, including the supervisors, were paid on an incentive system based on output. Consequently, they wanted to produce large volumes, regardless of demand, to maximize their piece rate incentives and production bonuses. Usually the production department favored manufacturing the older products, while marketing was pushing newer products to keep the company competitive. This example illustrates how conflicts of interest caused by compensation or performance measurement systems are commonplace; they must be resolved to smooth the flow of materials along the internal supply chain.

Once a reasonably good alignment is achieved within a company, it can begin to work on aligning the interfaces between itself and its customers and suppliers.

Often, this realignment includes not only the flow of information, such as orders, but also the flow of physical goods. To attain an adequate flow of orders, there must be compatibility within the information technology functions and this process often does not occur easily. Ideally, all businesses are moving toward electronic information flows to achieve speed, accuracy, and lower costs; however, not all businesses are moving at the same pace or with the same kind of hardware and software. In addition to the information flow alignment, there is a need to decide on how best to transport the goods. These decisions require agreements on timing, cost, and selection of intermediaries.

In this realignment process, a customer may ask for changes within a supplier's operation. Factors to be resolved include the quality level of the product, the amount of capacity to be allocated, the response time for orders, the performance level of the product, the type of packaging to be used, and the level of readiness for use. If these issues are not reconciled, inventory levels will build up and service expectations will not improve.

Another model describes this progression as moving through the following steps: Ad Hoc to Defined to Linked to Integrated to Extended.[11a]

We have shown the difficulties in smoothing the flow of goods and services through a relatively simple supply chain. Think of the difficulties in extending this example through all the possible supply chain configurations that are likely even in small companies. (An analogy can be drawn to human DNA, whereby each person is unique because of the way his or her DNA strands are linked together.) If that were not enough, we have only been talking about the forward flow in one supply chain. Most companies will have multiple supply chains to smooth. Developing well-defined supply chains for all products or services is probably impractical, if not impossible. Consequently, a company must choose which supply chains will have priority. In a sense, they will decide which supply chains will be critical to their success.

Up to this point, we have described only supply chains moving toward the customer. In the next section, we will describe the added complexity of the reverse logistics concept.

The Concept of Reverse Logistics

A reverse logistics supply chain is defined in the *APICS Dictionary* as "a complete supply chain dedicated to the reverse flow of products and materials for the purpose of returns, repair, remanufacture, and/or recycling."[12] Like forward supply chains, few companies are very far along in the implementation of their reverse logistics programs. If building the reverse supply chain is as difficult as the forward supply chain, and most writers suggest this is true, it represents another challenge for most businesses.

A distinction should be made between "reverse logistics" and "green supply chains." Reverse logistics refers to the process for handling returns and how best to reintroduce the returns, as whole units or component parts, back into the forward supply chain, through repackaging, remanufacturing, or reprocessing. The intent is to avoid costs or, in more forward-looking companies, generate added revenue. A green supply chain implies a concern with the eventual disposal of the materials in the finished products and the effect on the environment. Although both elements may be part of the same system, they have different drivers.

Why the Interest in Reverse Logistics?

One of the major trends in warehousing in the twenty-first century will be reverse logistics. Brockmann identified key drivers of this movement including increasing regulatory pressure to recycle and reduce landfill waste, the growing focus on customer relations, and the potential for lowering costs by reusing packaging and reclaiming unsold or damaged product.[13] Another trend is the increase in returns, especially at the retail level. More liberal returns policies and the rapid rise of e-commerce contribute to these increases. In light of these increases, retail businesses will need to develop more formal processes to handle them. In addition, they, in turn, will seek to return some of their product to their suppliers, the manufacturers. What results is that the product moves backward along the supply chain, in other words, in reverse. The interest in sustainable development practices is a strong trend that encourages reverse logistics.

In his fact-filled book, Blumberg[14] lists scenarios where products will need to be recovered and pushed back into the supply chain:

- Products in the field that have failed and need to be repaired or properly disposed of
- Parts and subassemblies of products that can be reused, either because they are perfectly good (no trouble found) or because they can be repaired or reworked
- Products that are perfectly good, but nevertheless have been returned by the purchaser as well as products sitting on retailers' shelves that have not been sold
- Products and materials that have been recalled or are obsolete, but still have a useful life
- Products, materials, and goods that have been thrown away, but can be recycled and reused
- Products at the end of a lease, but not at the end of life.

Blumberg points out that reverse logistics is essentially the responsibility of the original manufacturer of the product. In this situation, a closed-loop supply chain (CLSC) exists because the manufacturer both makes and recovers the produced product.

In addition to the economic implications, public awareness of environmental concerns is motivating governments to enact legislation mandating companies to meet stringent requirements in the handling of hazardous materials presently used in products, such as automobiles and electronic products. The European Union (EU) has enacted two major pieces of legislation—the Waste Electrical and Electronic Equipment (WEEE) Directive and the Restriction on Hazardous Substances (RoHS) Directive.

The WEEE Directive applies to a huge spectrum of electronic and electrical products. It sets criteria for the collection, treatment, recycling, and recovery of waste equipment and makes producers responsible for financing most of these activities.

The RoHS Directive bans placing on the EU market new electrical and electronic equipment containing more than agreed-upon levels of lead, cadmium, mercury, hexavalent chromium, polybrominated biphenyl (PBB), and polybrominated diphenyl ether (PBDE) flame retardants, effective July 1, 2006. It is the responsibility of the manufacturers to ensure that their products and their components comply.

One of the significant requirements of both of these laws is that the supplier of the product (usually the manufacturer) has responsibility for the eventual disposition of the product. Although U.S. manufacturers may avoid this requirement over the short run, it appears likely they will have to meet this requirement at some point in the future. Efforts launched collaboratively by the Environmental Protection Agency, by the U.S. Commerce Department, and at the industry level seek to lessen the environmental impact of small and midsize manufacturing suppliers while boosting efficiency, productivity, and profitability. Called the Green Supply Network, this program offers assessment reviews and advice in implementing "Lean & Clean" methodologies at all levels of the manufacturing supply chain.[15]

Benefits of Reverse Logistics

The most direct benefit from reverse logistics is cost avoidance. Ironically, upper management is often the main obstruction to achieving that benefit because they do not understand the significant costs that returns represent to the company. Even the recycling of buildings is considered in the reverse logistics process. The Pillowtex complex of buildings in Kannapolis, North Carolina, was demolished in 2006 to make way for a research campus. The wrecking company estimated it would recycle 10 million bricks (enough for 625 brick houses), 10,000 tons of steel (enough for 6,667 sedans), and 5,000 pine beams (enough to floor 500 living rooms) during the recovery phase of this project.[16]

There are also environmental benefits to reverse logistics. The private sector will move to recycling, both for profit and for public relations purposes. Dave Rosenfield of Roman Industries, Inc., comments, "I have no doubt that every industrial product will be subject to recycling and remanufacturing. At some level, state or federal, guidelines will be established to influence people to do it. The environmental political capital achieved by pressing on recycling issues can't be ignored. Politically it's hot to support recycling."[17]

Barriers to Reverse Logistics Implementation

As with all programs, there are barriers to their successful implementation. The Reverse Logistics Executive Council found the following major barriers:

- Perceptions that reverse logistics is relatively unimportant
- Lack of company policies that support reverse logistics
- Lack of systems that can implement reverse logistics
- Competitive issues that make reverse logistics cost prohibitive
- Management inattention
- Lack of personnel resources to carry out reverse logistics
- Lack of financial resources
- Legal issues.[18]

In a more recent study, Andel[19] found similar barriers:

- No executive overseer or champion who takes responsibility for reverse logistics and drives it
- The attitude that reverse logistics is "just a cost"
- The potential of missing the customer satisfaction and service differentiation opportunities
- An inability to quantify cost of returns
- Lack of systems/information visibility
- A silo versus cross-functional mentality
- Poor manual processes.

Perhaps the greatest barrier is thinking that returns can be handled on an ad hoc basis instead of using a systems approach.

System Design and Implementation

Figure 3.11 depicts a typical system of both forward and reverse supply chains. The top row shows the forward stream of raw materials to the final product for the consumer. The second row illustrates typical steps involved in the backward stream from consumer to final disposal.

To get the most out of a reverse logistics channel, excellent information management resources are necessary. The emerging importance of reverse logistics requires the execution and management of new inter- and intra-organizational processes. Given the complexity and uncertainty of these processes, information and communications technology (ICT) support is necessary, perhaps even more so than in traditional (forward) supply chain management. Moreover, the handling of reusable materials requires the involvement of new chain partners, such as material recovery organizations and e-marketplace intermediaries.

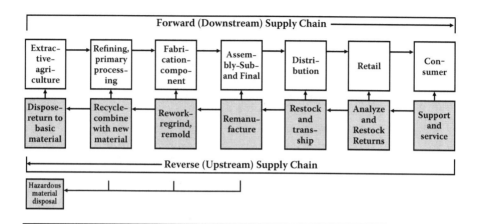

Figure 3.11   Forward and reverse supply chains.

ICT vendors have not yet given high priority to incorporating reverse logistics in their systems. The importance of managing reverse flows in the supply chain, coupled with the availability of Internet-based technologies have created new opportunities for progressive companies. Besides the development of new design, planning, and control components in existing ERP and APS systems, electronic marketplaces can be developed to manage return flows in the supply chain effectively.[20]

A recent study looked at the timing of reverse logistics programs (early or late adoption) and the commitment of resources. The research team found that early entrants have an advantage over late entrants because it is difficult for companies to catch up, although late entrants can somewhat offset their delay by committing additional resources to the effort. They recommend that:

- Organizations that have not yet started a formal reverse logistics program should do so.
- Organizations that have implemented reverse logistics need to be sure resource commitments are adequate to gain the potential benefits.
- Organizations that are unable to commit sufficient resources to reverse logistics should consider outsourcing.[21]

Michigan State University professors Diane Mollenkopf and David Closs provide their insights on the importance of reverse logistics in Managerial Comment 3.4.

Managerial Comment 3.4
Reverse Logistics Update

Reverse logistics has often been viewed as the unwanted stepchild of supply chain management. It has been seen as a necessary cost of business, a regulatory compliance issue, or a "green" initiative. But more companies are now seeing reverse logistics as a strategic activity—one that can enhance supply chain competitiveness over the long term. To understand how reverse logistics can create value, it is necessary to understand both the marketing and logistics components of this process. Companies can use the returns process to enhance marketing efforts by analyzing reasons for returns and conducting ongoing defect analysis. In many cases, the returns management process had been ad hoc, meaning that the strategies, operations, and guidelines were not well defined or thought out. However, all the organizations surveyed have quickly seen compelling benefits, to the point where they have all opted to build reverse logistics capabilities into their overall logistics and supply chain strategies and processes.[22]

Summary

This chapter introduced the basic Input–Transformation–Output (ITO) Model. We showed how to expand the basic ITO Model into more complex models through linking individual ITO models, both in a series and in parallel. We described open and closed systems, recognizing that most businesses operate in the open-system mode. Finally, we showed how the basic ITO module could be linked together into an integrated supply chain, although with considerable difficulty. We also showed that supply chains are like automobiles; companies need both forward and reverse capabilities.

In Chapter 4, we will show how a business performs a variety of transformation processes. In addition to the normal transformation processes of running the day-to-day business, they have to solve problems and make continuous improvements. All of these processes require using the ITO process.

References

1. Starr, M.K., *Systems Management of Operations,* Prentice-Hall, Englewood Cliffs, NJ, 1971.
2. Melnyk, S.A. and Denzler, D.R., *Operations Management, A Value-Driven Approach,* Irwin, Chicago, 1996.
3. Schoderbek, P.P., Schoderbek, C.G., and Kefalas, A.G., *Management Systems Conceptual Considerations* (4th ed.), BPI Irwin, Homewood, IL, 1990, p. 47.
4. Schoderbek, P.P., Schoderbek, C.G., and Kefalas, A.G., *Management Systems Conceptual Considerations* (4th ed.), BPI Irwin, Homewood, IL, 1990.
5. Thompson, J.D., *Organizations in Action,* McGraw-Hill, New York, 1967.

6. Thompson, J.D., *Organizations in Action,* McGraw-Hill, New York, 1967, pp. 19–23.
7. Simon, H.A., *Administrative Behavior, A Study of Decision-Making Processes in Administrative Organization* (5th ed.), The Free Press, New York, 1997.
8. Thompson, J.D., *Organizations in Action,* McGraw-Hill, New York, 1967.
9. Perrow, C., A framework for the comparative analysis of organizations, *American Sociological Review,* 32, 194, 1967.
10. Thompson, J.D., *Organizations in Action,* McGraw-Hill, New York, 1967.
11. Mentzer, J.T., Achieving competitive advantage through supply chain management, http://www.industryweek.com/PrintArticle.aspx?ArticleID=13355, January 10, 2007.
11a. McCormack, K.P., Johnson, W.C., and Walker, W.T. *Supply Chain Networks and Business Process Orientation,* St. Lucie Press, Delray Beach, FL, 2003.
12. Blackstone, J.F. III, *APICS Dictionary* (13th ed.), 2010, APICS—The Educational Society for Resource Management.
13. Brockmann, T., *21 Warehousing Trends in the 21st Century,* IIE Solutions, 31, 7, 36, 1999.
14. Blumberg, D.F., *Introduction to Management of Reverse Logistics and Closed Loop Supply Chain Processes,* CRC Press, Boca Raton, FL, 2005.
15. Anonymous, Lean and clean auto industry pilot expands to other verticals, *Manufacturing Business Technology,* 23, 2, 11, 2005.
16. Bell, A., Rubble full of riches, *The Charlotte Observer,* Sunday, February 12, 2006, p. B1.
17. Andel, T., Rethinking recycling, *Material Handling Management,* 59, 5, 35, 2004.
18. Caldwell, B., Reverse logistics, *Information Week,* Apr 12, 1999, issue 729, pp. 48–52.
19. Andel, T., Rethinking recycling, *Material Handling Management,* 59, 5, 35, 2004.
20. Von Hillegersberg, J., Zuidwijk, R., and van Nunen, J., Supporting return flows in the supply chain, *Communications of the ACM,* 44, 6, 74, 2001.
21. Richey, R.G., Chen, H., Genchev, S.E. and Daugherty, P.J., Developing effective reverse logistics programs, *Industrial Marketing Management,* 34, 8, 830, 2005.
22. Mollenkopf, D.A. and Closs, D.J., The hidden value in reverse logistics, *Supply Chain Management Review,* 9, 5, 34, 2005.

Chapter 4

The Role of Management Programs in Continuous Improvement

In Chapter 2 we described how critical success factors (CSFs) help a business in its strategic planning process, and as a result enable the company to become more effective by doing the right things. In this chapter we show how the selection of continuous improvement programs will help a business become more efficient by doing things right.

We examine several questions:

- What are management programs? How do they differ from normal, every-day operations?
- What is the life cycle of a management program? What happens to management programs at the end of the life cycle?
- Why are management programs important? What are their benefits? What are the obstacles to their implementation?
- Where do management programs come from? Who invents them? For what purpose? Are they original or adaptations from some other management concepts?
- Why are some programs successful and some not? Although there have been a number of well-publicized successes, many companies have achieved limited success.
- What does the future hold? Will management programs continue to be an active and worthwhile source of ideas to stimulate improvement in businesses?

By addressing these questions, we expect to provide managers with a better understanding of how to use management programs to meet their companies' needs.

What Are Management Programs?

In most entities, management is concerned with several different kinds of activities. On one hand, there is the normal, everyday operation of the business. In this book, we do not try to tell you how to run the normal activities of your business. Instead, we describe how the use of management programs—those activities outside the course of your normal business—can play a vital role in keeping your business prosperous.

Management programs are those coordinated activities in a business that are usually assigned a name to distinguish them from the normal operations of the business. They are often known by an acronym, such as Just-in-Time (JIT) or Total Quality Management (TQM) programs. In most cases, they are concentrated efforts to improve some part of the business operation, such as to reduce costs, improve quality, shorten response time, or provide greater flexibility and agility in meeting product or service customer requirements. They may involve a part or all of an organization. Usually, they are of a project nature, with a beginning, a life cycle, and an end. Table 4.1 lists examples of management programs grouped by their primary focus area.

Management programs usually originate as an attempt to introduce improvement into a business. They may be original for a particular company or they may be an adaptation of an existing program that has become popular, or at least reasonably successful, in another company. For example, Toyota started a program to reduce inventory and improve cash flow by revamping their production system. This program was first known as the Toyota Production System (TPS) and then by a variety of names, such as stockless production or zero inventories. Eventually, this program achieved widespread acceptance and became known as the JIT system, which today has morphed into the most popular program, known as lean production or lean manufacturing.

Most programs begin small to address a specific need. If successful, they usually expand into a broader program to the point that it becomes a management philosophy. For example, JIT started as a kanban scheduling system to reduce in-process inventory. Later, it broadened to a waste reduction program. From that, it expanded to a continuous improvement system, and in recent years has blossomed into a management philosophy whose primary mission is waste elimination. Even with all this growth, lean production has replaced JIT as the lean concept has grown in scope and magnitude since its inception in the early 1990s. Lean production has the basic objective of creating a smooth and coordinated flow of value-adding goods and services.

Table 4.1   Examples of Management Programs

Process Improvement	Quality Improvement
Just-in-Time (JIT)	Statistical process control (SPC)
Lean production or manufacturing	Total quality control (TQC)
Business process re-engineering (BPR)	Total quality management (TQM)
Agile manufacturing	Quality function deployment (QFD)
Computer-integrated manufacturing (CIM)	Six Sigma
Theory of Constraints (TOC)	**Supply Chain Management**
Mass customization	Quick response system (QRS)
Production Planning	Efficient consumer response (ECR)
	Vendor managed inventory (VMI)
Materials requirements planning (MRP)	Collaborative planning, forecasting and replenishment (CPFR)
Manufacturing resources planning (MRPII)	Supply chain management (SCM)
Enterprise Resources Planning (ERP)	**Performance Measurement**
Warehouse management system (WMS)	
Manufacturing execution system (MES)	Activity based costing (ABC)
Advanced planning and scheduling (APS)	Activity based management (ABM)
	Balanced scorecard (BSC)
Sales and operations planning (S&OP)	

Note: Representative list of the management programs that have been developed over the past two decades, grouped by their primary focus, or objective.

Source: Adapted from Crandall, R.E. and Crandall, W.R., An analysis of management programs: Origins, life cycles, and disappearances, *Proceedings of the 2007 Southeast Decision Sciences Institute Conference*, Savannah, GA.

Where do management programs fit in with running a business? Let us look first at the "other" kinds of activities in a business. Figure 4.1 shows three major types of transformation activities for a business—normal (or sustaining) day-to-day operations, problem-solving activities, and improvement programs. The middle section of the diagram illustrates the normal, or sustaining, operations in which the business transforms inputs into goods or services for sale. This is the concept introduced in Chapter 3 as the ITO Model. The bottom portion shows that a business must correct problems that occur, or, in the ITO vernacular, transform problems into solved problems. The top portion illustrates the concept of transforming ideas into improvement actions.

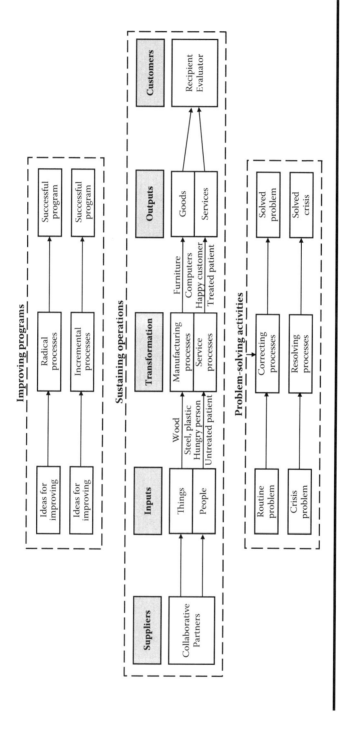

Figure 4.1 Types of business activities.

Normal or Sustaining Day-to-Day Operations

This is not a book about the sustaining operations (the middle part of Figure 4.1) of a business. While sustaining or normal operations are an essential part of a business, each industry is different, and each company within an industry is different. If you are a member of a successful business, you already know a great deal about how to run the normal operations. Therefore the bulk of this book describes the other two types of business activities—solving problems and making improvements. They are equally important to the normal operations but require a somewhat different approach (see Figure 4.1). William T. Walker provides a detailed explanation of how to build normal or sustaining operations in his book *Supply Chain Architecture.*[1a]

Problem-Solving Activities

Consider the problem-solving part of the diagram first. There are two problem-solving activities. Ideally, all of a company's problems could fall into a category of everyday or routine types of problems—machine breakdowns, late orders, unsatisfied customers, and product defects. Although some of these can be serious, their solutions are within the capability of the existing employees, facilities, and infrastructure of the organization. Companies typically use their own resources, such as employees and facilities, to solve these types of problems. Management programs, such as those listed in Table 4.1, are not normally used to address this category of normal operating problems.

However, there is another type of problem that can destroy a company. A crisis is a situation that develops unexpectedly; most companies do not have the experience or in-house capability to handle these types of problems with ease. Crises do not occur often, but they are high-impact events that can seriously hamper the smooth running of the organization. It can be an oil spill from an oceangoing tanker, food poisoning of multiple customers, tampered medications, tsunamis or earthquakes, or accounting scandals. When crises are handled promptly and properly, the company may actually prosper, such as in the case of Johnson & Johnson's Tylenol cyanide incident in the 1980s. In this example, Johnson & Johnson received positive publicity for the effective management of the crisis. Consequently, sales of the product eventually rebounded. If crises are neglected or denied, the company may be destroyed, such as in the case of Arthur Andersen and Enron. This is a special type of management activity and is not considered a program in the context of this chapter. There are a number of practitioner books on this subject, such as Steven Fink's *Crisis Management: Planning for the Inevitable,*[1] *Crisis Management in the New Strategy Landscape* by Crandall, Parnell, and Spillan,[1b] and Timothy Coomb's *Code Red in the Boardroom: Crisis Management as Organizational DNA.*[2]

Improvement Programs

This book is primarily about management programs that lie outside the realm of normal operations and crisis-related activities. We focus on those management programs that address improving some aspect of the business. Such programs can be classified as incremental or radical. Although both types of programs are in popular use, most of the programs probably would fit best in the incremental or continuous improvement category. The use of Six Sigma is an example of an incremental improvement program. These programs enable a company to improve by implementing small-step interventions alongside their normal operations.

Some programs are major, or radical, and are capable of transforming a company in substantial ways. They are often disruptive and usually require changes in the normal way of doing business. Such examples include Business Process Re-engineering (BPR) or lean production. When they are effective, they can save a failing company or make a winner out of a mediocre company. These types of programs also fall within our classification of improvement programs. Childe, Maull, and Bennett make the following distinction between continuous and radical change programs: "TQM-led programmes are usually associated with continuous improvement efforts and not with the radical change brought about by re-engineering. TQM may thus represent a necessary but insufficient condition for a successful BPR programme."[3] Another comparison of incremental versus radical change is described in the following excerpt from Richard Luekue.

Managerial Comment 4.1   Discontinuous versus Continuous Incremental Change

The company, having done its thing for many years, suddenly throws the cards in the air, everyone gets involved in reform and improvement, and then it's over. We call this "discontinuous change"—a single, abrupt shift from the past. The momentum of the organization is shifted, hopefully to a higher level of performance or in a more promising direction.

But the benefits of a successful single fix don't last forever. Change initiatives that accomplish stated goals often lead to complacency in senior management. Units that developed market-beating products and services during the change gradually shift their attention from innovation to defending their turf. Employees settle into routines and become more inward-looking. On top of all this, with each passing day the environment of competition and technology is altered. This combination of complacency, defensive behavior, routines, inward focus, and ever-evolving competition is the enemy of progress; ultimately it creates a situation in which major reform is needed once again.

If change programs are eventually followed by periods of organizational complacency and stasis—as described above—the alternative situation is one in which the organization and its people continually sense and respond to the external environment. Their radar is attuned to signals of change from customers, markets, competitors, and technologists. And they respond in appropriate ways. Simultaneously, they monitor internal activities to assure continuous improvement in key processes. Open communication assures that new ideas have a forum in which they are heard and objectively evaluated. Change is ongoing and takes place through many small steps—that is, through continuous incremental change.

On the surface, the advantages of continuous incremental change are many: small changes are easier to manage, and small changes enjoy greater probability of success than big ones.

Disruption is short-term and confined to small units at any given time the organization and its people are kept in a constant state of competitiveness and change-readiness.[4]

We can also show how to restructure the normal transformation process of a business through problem solving and the improving process. Figure 4.2 illustrates how a solved problem or an incremental improvement can lead to a restored or improved transformation process. A further extension to a restructured transformation process could result from resolving a crisis or implementing radical improvements. These cycles could repeat as a company continues to adjust to its changing environment.

In Figure 4.2, we start with the normal transformation process, converting inputs into outputs, shown as Condition I. The company begins an incremental improvement program, shown as Activity B. The program is successful, and when completed, the company assimilates the improvement into the normal operation of the business, shown as Activity D. On the other hand, suppose a problem develops, as depicted by Activity A. It could be that a customer's order is late being shipped or the food is served at the wrong temperature in a restaurant. The problem-solving process is used to correct the problem. As a result of solving the problem, the company discovers a way to improve the process. This link is shown as Activity E. In the case of the late order, management determines it can implement an early warning system to prevent late orders. By working the idea through the incremental improvement process, the problem can actually lead to an improvement. Even without Activity E, if the problem-solving process works as expected, the situation returns to a restored normal operation, shown as Condition II. If the incremental

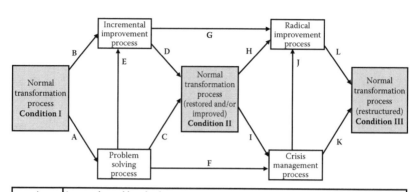

A	An everyday problem develops
B	A program of incremental improvement is implemented
C	The problem is solved and normal operations are restored
D	The improvement program improves the normal transformation process
E	The problem solving process feeds ideas to the incremental improvement process
F	The known problem is not solved and a crisis develops
G	The incremental process triggers a radical improvement program
H	A program of radical improvement is implemented
I	A crisis develops from an unexpected source
J	The crisis stimulates the need for radical improvement program
K	The crisis is resolved and normal operations are restored
L	The radical improvement program restructures the normal transformation process

Figure 4.2   A continuous transformation process.

improvement process works as designed, the improvement is implemented into the improved transformation process through Activity D.

The restored normal transformation process—Condition II—may find that competitive conditions require change. As a result, management may decide on the need for a radical improvement, such as the decision to completely restructure their production processes from a batch flow to a line flow. Activity H provides this link. The idea for a radical improvement could also have originated during an incremental improvement process, as shown by Activity G. As a result of the radical improvement process, the normal transformation process is restored, through Activity L, into a restructured process, shown as Condition III.

However, suppose that a number of customers at a restaurant eat improperly prepared food and incur food poisoning. The small problem becomes a crisis, as shown by Activity F. Now the crisis management process takes over to restore the normal transformation process through Activity K to Condition III. The crisis could have arisen from the normal transformation process, shown as Activity I, necessitating the activation of the crisis management process. Crisis management seeks to address problems that disrupt the normal operational flow of the business. Bringing the business back to normal and seeking to prevent future occurrences of the event may result in a transformation of some aspect of operations.

Figure 4.2 should not suggest that there is a rigid sequential process for correcting problems or introducing change. Most businesses do not operate in such a controlled environment. The diagram should reinforce the idea proposed in Figure 4.1 that there are three streams of transformation activities in most businesses: (1) normal transformation processes, (2) problem-solving processes, and (3) improving processes. Our emphasis is to describe how improving processes are a necessary part of business operations.

Management Program Life Cycles

Perhaps the most common point that researchers agree on is that management programs follow a life cycle. Scholars use bibliometric data—number of articles written about a particular program—to study the shape of these program life cycles. Most research indicates a bell shape as the most common cycle form.[5,6] However, Ponzi and Koenig have indicated that an S-shape is also a possible pattern.[7] Carson and associates acknowledge that shapes will vary in terms of slope rates because of the existence of other management fashions on the market that may impact the particular fashion under study.[8] However, a shape of some kind is plausible, most likely one that resembles a bell curve.

Life-Cycle Stages

In addition to the shape, the stages of the life cycle are also of interest. Abrahamson wrote extensively about management fads and fashions.[9] The widely cited model by Ettorre shows fashions progressing through a five-stage life cycle:

1. Discovery: "A buzzword is born."
2. Wild acceptance: "The idea catches fire."
3. Digestion: "The concept is subject to criticism."
4. Disillusionment: "The idea does not solve all problems."
5. Hard core: "Only true believers remain."[10]

The Gartner Research Group developed a "hype cycle" with the following phases: Technology Trigger (beginning), Peak of Inflated Expectations (growth), Trough of Disillusionment (decline), Slope of Enlightenment (revival), and Plateau of Productivity (sustained level).[11]

In Table 4.2, we compare the stages of program life cycles for publication and for implementation in businesses. The length of the life cycle varies. In the management research literature, programs with short life cycles are deemed fads and the more durable programs are considered fashions.

What, or who, are the drivers of a program's life cycle? Many of the popular management programs originated as a focused effort within a company, such as JIT at Toyota or Six Sigma at Motorola. The program may be designed internally

Table 4.2 Publication and Management Program Life Cycles

Publication Life Cycle	Management Program Life-Cycle Phase
Startup: The initial articles published on a program will reflect the birth or startup of that program. Usually, trade publications will dominate the articles first published.	Recognize a need for the program. Select and develop the program. Gain acceptance of the program with employees and other key stakeholders. Implement pilot projects.
Growth: The number of articles on that management program increases. On a graph, this would be represented by a positively sloping curve. While trade publications may still dominate, scholarly journal articles are beginning to appear.	Implement the program on a larger scale. Analyze and adapt the program to other industries. For example, JIT moves into the service industries.
Maturity: The management program is written about in a number of publications. Although there is no longer a growth phase of articles, practitioner and scholarly interest remains high.	Capitalize on the program and bring in subsequent versions or technologies related to that program.
Decline: The number of articles related to that program decreases substantially. Often, there is a more significant decrease in trade publication articles than in scholarly journal articles.	Rationalize or refine the program. The program may be abandoned and replaced by a different program. Or, the program may be so entrenched in the management philosophy and operations that it is no longer necessary to keep reading new articles on that program.
Reorientation: The program spins off related programs and applications.	Extend the scope of the management program into new applications.

or with the aid of a consultant. Often, consultants package the program as an addition to their product line and promote the program to other potential clients. In the early stages of a program, consultants and trade publications are often the primary sources of information about the program. A typical way for other practitioners to find out about the program is to attend conferences or workshops offered by consultants or trade associations. In the early stages, the reports about the program are usually positive and describe the benefits of implementing such a program. As time goes on, academic researchers begin to study the program and view it with greater objectivity. They begin to analyze the program elements and identify the major causes of success or failure. They often compile survey information that summarizes the actual results achieved, often reflecting a range of results, from high success to low success or even failures of the program. At this stage, many companies that are considering the program may become reluctant to even try the program, even though it still has many enthusiastic supporters. The keys to success may be in selecting the correct program and then implementing it correctly. We discuss both of these decision areas later in this chapter.

At the End of the Life Cycle

What happens to programs at the end of their life cycle? Those considered fads fade away quickly, although some remnants of the program may remain as traces of something worthwhile. Other programs may also appear to fade away because the number of articles written about them diminishes. This may mean that the program was not of value and companies discontinued its use. However, it is also possible that the program did have value but had lost its premier status because of the appearance of new programs.

Many programs morph into a new program, such as JIT being succeeded by lean or TQM by Six Sigma. This morphing shows up in the changes in the number of publications about the programs. Figure 4.3 shows this transition for JIT and lean, and Figure 4.4 illustrates the same transition for TQM and Six Sigma.[12]

Some companies assimilate management programs into their normal practices. Although they may not have a definite identity as originating in a specific

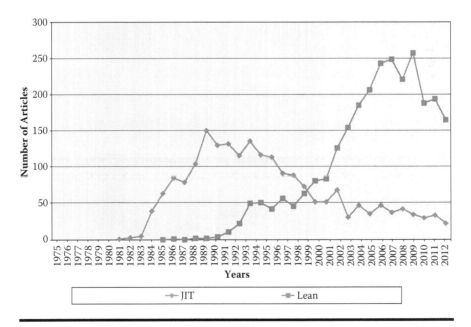

Figure 4.3   Number of JIT and lean production articles.

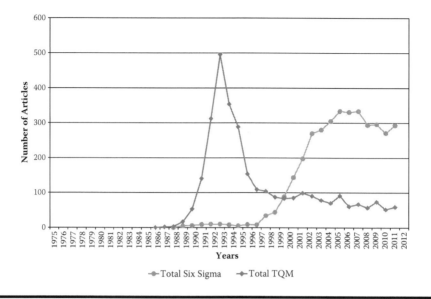

Figure 4.4 Number of TQM and Six Sigma articles.

program, basic elements of the program remain as standard practice. For example, some companies may introduce self-directed work teams as part of either a TQM or a JIT program and continue the use of those teams even after discontinuing the formal program.

Implications of Program Life Cycles for Management

What are the implications for managers? Why should they care about program life cycles? Several reasons seem most evident:

■ It is beneficial to know where a program is in its life cycle. Is it just getting started? If so, many of the articles will address the positive attributes of a program with little about the potential problems or hidden surprises. At this stage, companies should view the program optimistically, but with caution. As the program develops, more insightful articles will appear and help make an evaluation of the program more rational. As the end of the life cycle nears, management may replace the program with an even more appropriate program.

■ As programs develop along their life cycle and mature, successful implementation and further research make them more beneficial. A company can successfully implement a program early in its life cycle; however, it should continue to look for ways to improve that program. Often, improvement ideas are addressed in the more thoughtful articles that emerge later in the publication life cycle of a program.

■ New programs will replace old programs. Just as lean production has succeeded JIT, or Six Sigma supplanted TQM, most programs spawn new programs that are designed to correct problems or capture opportunities not addressed in the original program design. Although it is important to be progressive, it is also important to recognize that true differences between the old program and the new one will exist. What are the incremental differences? What are the incremental benefits? What are the incremental costs or inherent problems?

Implementing a program is a major project. It requires a company to commit employees and other resources to the effort. There is a great deal of information available about the most popular management programs, from the very beginning of the program's life cycle. Companies need not enter into a program uninformed.

Companies need management programs to help them improve. It is difficult to operate in a normal fashion and make improvements at the same time; therefore, many companies find it useful to mount special efforts—programs—to focus attention on their improvement efforts.

Why Are Management Programs Important?

Why do companies like to use programs to implement improvement efforts? Is it because they are "driven" to use them by developers and promoters, such as consultants or major customers, or do managers "pull" programs because of their needs?

As described earlier, most businesses deal with three types of activities: (1) conducting normal day-to-day operations, (2) correcting problems (and even crises) that arise, and (3) making improvements in their operations. Joseph Juran, a noted author and scholar in the quality improvement movement, presented an interesting insight into the functions of a business. He used the terms *control* and *breakthrough*. Control represents the normal operation of a business when things are running smoothly, or in control. If the process goes out of control, it must be fixed, or a problem solved. However, Juran warns that control implies maintaining the status quo, and that is never enough to satisfy the competitive pressures on a business. Instead of being satisfied with control, a business should be looking for breakthroughs, or improvements, in the way it does business. His book *Managerial Breakthrough* contains ideas about how to achieve breakthroughs in a business.[13]

The day-to-day operations represent the control part of the business in which managers perform the required functions of the business in a relatively stable environment using accepted practices without major variances in either the demand or supply side of the business. An example of this could be a retail store that serves customers with only normal variation among the customers and the services they require.

If there is a problem such that normal practices are insufficient to handle it, and the process goes out of control, there is a need to exercise the problem-solving elements in the business. Using a retail store again as an example, this scenario could be illustrated by running out of a heavily advertised sales item, a leak in the roof that is damaging merchandise, or a customer who slips and falls while shopping. Resolution of these problems is essential to the continued prosperity of the store. The actions required to solve a problem—restore the process to a controlled situation—are somewhat or even dramatically different from the normal operation itself.

However, there is another type of management problem that does not lend itself to normal, everyday problem-solving methods. This type of problem occurs when a company encounters a crisis. The crisis may be operational in nature, such as when an industrial accident occurs and an employee is severely injured or killed. It could also be strategic, as when a key executive dies unexpectedly. The crisis could be public relations oriented, such as when a product malfunctions and a subsequent recall is issued. In some crises, it may be necessary to call in a consultant with crisis management expertise.

Finally, there is the type of problem where a business needs to improve. Most businesses are fighting a constant battle to remain competitive, and to accomplish that, they have to keep improving. This is where management programs enter the picture. These programs can help a business to improve.

Where Do Management Programs Come From?

Are management programs original or do they have earlier origins in management practice? McMahon and Carr offer a growing concern that students today are reading less from the early writers and more about what today's scholars are attributing to the early writers. They suggest this may lead to a distortion of the original intentions, especially if the current writings have been assigned a different label. They recommend a return to reading the original writings as a part of the business educational process.[14]

In line with the concerns expressed by McMahon and Carr, we will show that some of today's most popular management programs had their origin in earlier management theories. A few basic management theories have evolved over the past century as separate streams of research and application. Businesses have learned to apply these theories to management programs. This section describes the process of packaging management theories into effective improvement programs. Let us look first at some of the foundational management theories.

Overview of Employee Management Theories

Prior to the end of the nineteenth century, job design and the management of workers was informal and left to the individual employee and the "boss" to figure out. Little attention was paid to how the work was accomplished; the only concern was that it was done. At the end of the nineteenth century and beginning of the twentieth century, a body of management knowledge began to emerge that sought ways to improve productivity. Henry Towne, of Yale and Towne, is usually credited with moving the study of management to the forefront in his presentation "The Engineer and the Economist," given to the American Society of Mechanical Engineers (ASME) in 1886.[15]

Scientific Management

Frederick W. Taylor is often referred to as the father of scientific management because of his research in work methods studies. His approach was based on the theory that any job can be improved by breaking it down to its basic elements and finding ways to improve work methods. However, Taylor was not alone in his quest for productivity improvements. Henry Gantt developed management principles and procedures that began to take a humanistic approach to management. Frank and Lillian Gilbreth studied methods and motion techniques. Harrington Emerson championed efficient operations as well as advocating a bonus pay plan for employees.

Taylor considered his work an attempt to develop basic principles as a systematic approach to management. This approach later evolved into a philosophy of work productivity known as "scientific management." Eventually, these principles became the foundation for the field of industrial engineering.

Taylor's philosophy led to job specialization: a job with a narrow range of tasks that have high repeatability, resulting in greater efficiency. Jobs designed in this manner are simple, require little training, and are easy to measure and manage. The scientific management approach made possible the high-speed, low-cost production that plays a great part in the standard of living we enjoy today. However, job specialization carried to the extreme can have significant adverse effects on employees, such as absenteeism, lack of motivation, and employee turnover. These may in turn cause serious problems in other parts of the production system.

Deciding on the degree of specialization is one of management's toughest decisions. Even though high specialization has resulted in tremendous productivity gains for many companies, there are still widespread opinions regarding the optimum degree of specialization. In service operations, work content seems to be less specialized using broader job designs. Today there appears to be a definite movement toward the expansion and enrichment of jobs to provide greater flexibility in responding to product and service variety, as well as providing more meaningful work for employees.

Human Relations Management

Because of the adverse effects of job specialization, some behaviorists and managers began to look for alternative ways to design and manage work. In the 1950s Eric

Trist's studies in Great Britain's coal fields led to his theory that work has both technical elements and behavioral elements. This dichotomy implies that management should seek a balance between the technical and behavioral aspects when designing jobs. Analysts should study the entire work system, not just individual tasks. Most writers believe that the Hawthorne studies at Western Electric provided the framework for the human relations movement. Although the results of the Hawthorne studies have been debated for decades, one area that seems to represent a consensus is that job design is an important element of a business operation.

Over time, more humanistic job design techniques evolved—job enlargement, job rotation, and job enrichment. Job enlargement gives an employee more tasks to perform, thereby increasing the need for more skills and presumably reducing boredom and increasing job satisfaction. For example, instead of just assembling a product, the employee may also perform preventive maintenance on the work equipment.

Job rotation allows employees to exchange jobs with fellow employees from time to time, usually on a predetermined schedule. This practice provides some diversity for the employee, enlarges job skills, and makes the employee more valuable to the company. The benefit for management is a more highly skilled, flexible workforce. An example of job rotation involves rotating bank employees between working as tellers and as loan processors. Restaurants have long recognized the need for job rotation as a training practice for aspiring managers. Typically, managers in training will learn to cook, wait tables, wash dishes, and perform other functions relevant to the restaurant.

Whereas job enlargement is the horizontal expansion of a job, job enrichment expands an employee's tasks vertically into aspects of managerial functions. Job enrichment not only expands tasks upward but also expands responsibility. It is the most comprehensive of the humanistic approaches to job design and embodies the three factors that Frederick Herzberg's research indicates enhances job satisfaction: achievement, recognition, and responsibility.[16]

In modern management, aspects of the human relations approach are well entrenched in the manufacturing of products and the providing of services. Regardless of the job design approach used, operations managers retain their responsibility to increase their company's competitive advantages. Business publications and research literature suggest that top-performing companies organize to meet the needs of their people; consequently, they attract better applicants, and their employees are self-motivated to do an excellent job. Chapter 9 describes how improvement programs depend on the successful involvement of employees and consideration for the organization's culture.

Administrative Management

Scientific management started with a shop-floor emphasis and gradually expanded over the years into a more general management theory. The administrative management stream of research was more of a top-down approach to management. One of the early thinkers in this area was Henri Fayol, a French engineer who progressed through the management ranks in the coal and iron industry during the latter part of the nineteenth century and the early part of the twentieth century. He considered the following groups of activities applicable to all general managers:

- Technical (production/manufacturing, adaptation)
- Commercial (buying, selling, and exchange)
- Financial (finding and using capital)
- Security (protection of property and persons)
- Accounting (stocktaking, balance sheet, costing, statistics)
- Managerial (planning, organizing, commanding—later modified to directing—coordinating, controlling).

He believed that the managerial functions needed further study, and expanded his view of this area by defining fourteen principles of management that seem so obvious to us today (see Managerial Comment 4.2).

Managerial Comment 4.2
Fayol's List of Management Principles

1. Division of Labor. Work should be divided to permit specialization.
2. Authority. Authority and responsibility should be equal.
3. Discipline. Discipline is necessary to develop obedience, diligence, energy, and respect.
4. Unity of Command. No subordinate should report to more than one supervisor.
5. Unity of Direction. All operations with the same objective should have one manager and one plan.
6. Subordination of Individual Interest to General Interest. The interest of one individual or group should not take precedence over the interest of an enterprise as a whole.
7. Remuneration. Rewards for work should be fair.
8. Centralization. The proper degree of centralization/decentralization for each undertaking is a matter of proportion.
9. Scalar Chain. A clear line of authority should extend from the highest to the lowest level of an organization.
10. Order. A place for everything and everything in its place.
11. Equity. Employees should be treated with kindness and justice.

12. Stability of Tenure of Personnel. Turnover should be minimized to assure successful goal accomplishment.
13. Initiative. Subordinates should be allowed the freedom to conceive and execute plans to develop their capacity to the fullest.
14. Esprit de Corps. Harmony and union build enterprise strength.

The familiar ring of Fayol's ideas suggests how thoroughly they have penetrated current managerial thinking. While many of them may seem relatively self-evident today, they were revolutionary when first advanced. They remain important not only because of his enormous influence on succeeding generations of managers, but also because of the continuing validity of his work. His concepts continue to have a significant impact on managerial thinking. For this reason, Fayol is known as the Father of Modern Management.[17]

Another pioneer in building the administrative management theory was Max Weber, a German sociologist who published his work at the end of the nineteenth century but was largely unknown in English-speaking circles until the 1920s. He outlined the characteristics of what he called "bureaucratic management." He used the term *bureaucratic* as an ideal modern and efficient method of organizing, not in the modern sense of a tangled, inefficient governmental organization. He described the following characteristics for an ideal bureaucracy:

- *Division of Labor.* Divide labor to define authority and responsibility clearly.
- *Authority Hierarchy.* Offices or positions are organized in a hierarchy of authority.
- *Formal Selection.* All employees are selected on the basis of technical qualifications demonstrated by formal examination, education, or training.
- *Career Orientation.* Managers are professionals rather than owners of the units they administer. They work for fixed salaries and pursue "careers" within their respective fields.
- *Formal Rules and Controls.* All employees are subject to formal rules and controls regarding the performance of their duties.
- *Impersonality.* Rules and controls are impersonal and uniformly applied in all cases.[18]

Although Weber considered his model to be efficient, recent experiences may weaken its effectiveness because of the rigidity of its structure. However, as with other pioneering work, some of its basics are still valid. Chapter 8 describes more fully how improvement programs require the appropriate selection and use of organizational structure concepts.

Systems Theory

Systems theory provided a way to blend elements of the major management theories into packages, or programs. Systems theory was formalized in 1954 when the Society for General Systems Theory, later renamed the Society for General Systems Research, was founded under the leadership of biologist Ludwig von Bertalanffy, economist Kenneth Boulding, biomathematician Anatol Rapoport, and physiologist Ralph Gerard.[19] Prior to that time, most researchers and practitioners used a reductionist approach in which they broke a large problem into small parts and attempted to solve the small problems and then reassemble the components into a more workable process. Systems theory encouraged viewing not just the components but also the relationships among those components. It has had wide application in medicine, as in the development of vaccines, gene splitting, DNA analysis, and organ transplants. Science and technology provided applications in space travel, weather forecasting, and digital data transmission. The computer has facilitated the design and implementation of systems not only in the sciences but also in business applications. Systems theory has evolved over the latter part of the twentieth century into an ever broader and more complex topic.

We consider some of the more recent developments in systems theory in Chapter 12, such as catastrophe, chaos, and complexity theory. These ideas are still in the formative stages but will no doubt become important components of business management in the coming years.

In the early part of the twentieth century, scientific management, administrative management, and human relations management were viewed as complete in themselves and independent of each other. Proponents tended to subscribe to one of these philosophies as a primary managerial approach to running their businesses. Applying systems thinking made it easier to select applicable elements from the different management theories to form a complete systems approach. It made it possible to select the best ideas from each of the basic management theories.

Contingency Theory

Contingency theory makes it possible to apply a concept, technique, or program in a modified format to a particular company to fit their specific needs. Contingency theory originated in the information systems area of management and has been widely extended to other management areas. For example, it supports the position that no single organizational structure—bureaucratic, decentralized, free form—is best. Instead, the structure should be adapted to the situation. We will show that

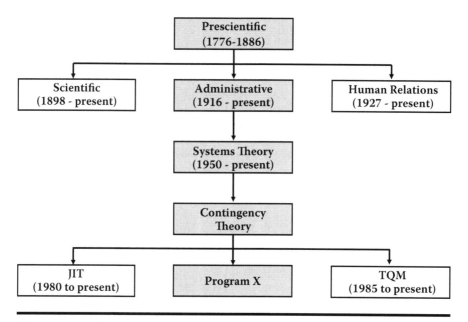

Figure 4.5   Evolution of management thought.

the most effective applications of management programs are to design and implement them to fit the specific needs of a business.

Figure 4.5 outlines the evolution of management thought described above. It illustrates how the popular management programs of today have received much of their content from earlier management theories. This approach suggests that a business could package an assortment of basic ideas into a program to fit its own specific needs. Western Electric developed statistical process control in the 1920s to improve quality processes. Toyota developed what became JIT in the 1960s to reduce waste. Motorola developed its Six Sigma program in the 1980s to set new targets in quality. These are just a few examples of businesses that introduced programs that would achieve widespread acceptance and application.

In Tables 4.3 and 4.4 we show a number of examples of the specific concepts or techniques that originate from each of the main management theories and how they fit together in a program such as JIT and TQM. Table 4.3 lists the evolution and goals for each program. Although they evolved as separate programs, they have a number of similar objectives. Table 4.4 shows how program elements of JIT and TQM could be derived from the basic management theories. These examples are illustrative and not meant to be all-inclusive.

What can we conclude? Many of the basic components of "new" management programs are elements from aspects of earlier management theories. Does this make these new programs less valuable? No, because they have been adapted (an application of contingency theory) to represent the best answer to an organization's specific needs.

Table 4.3 Comparison of JIT and TQM: Evolution and Objectives

	JIT/Lean Production	*TQM/Six Sigma*
Evolution of the program over time	Scientific management: 1890 Ford's River Rouge plant: 1905 Industrial engineering: 1940 Toyota production system: 1970 Just-in-Time (system): 1980 Lean Production: 1990s	Statistical process control: 1925 TQC (statistical): 1930 TQC (meet specs): 1950 TQC (satisfy customer): 1980 TQM (integrated system): 1990 Six Sigma: 1990s
Major objectives of these programs	Reduce inventories Reduce lead times Eliminate waste Pursue continuous improvement Recognize customer needs	Reduce cost of defects Offer competitive advantage Eliminate waste Pursue continuous improvement Recognize customer needs

Table 4.4 Program Concepts Derived from Management Theories

Management Theory Source	*JIT/Lean Production*	*TQM/Six Sigma*
Scientific management	Pull method of material flow Standardized work methods Uniform workstation loads	Continuous improvement Cost-of-quality Problem-solving process
Administrative management	Product focus Close supplier ties Group technology	Quality as a competitive weapon Benchmarking Quality as customer's perception
Human relations management	Flexible workforce Horizontal organization Teams/employee empowerment	Self-managing teams Quality at the source Cultural change

Why Are Some Programs Successful and Some Not?

Some management programs enjoy great success as they are widely heralded in trade publications and scholarly journals as leading-edge evidence of how companies can become more competitive. The same program in another company may achieve only limited success or even be considered a failure. Why is there so much discrepancy in success among companies that allegedly implement the same program?

The movement of a management program from one company to another involves a great deal of knowledge management (KM). Knowledge management systems (KMS) require liberal amounts of technology; however, their eventual success depends on supplementing technology with systems and people skills. In Managerial Comment 4.3, Robin Wakefield articulates this point further.

Managerial Comment 4.3
Knowledge Management Strategies

The development of KM strategies for knowledge transfer is a dynamic and complex undertaking. A principal belief within organizations is that the ability to compete based on knowledge depends primarily on people, rather than processes or technology. Strategies that guide the sharing of internal knowledge represent great challenges.[20]

The originators of individual programs considered them a success; otherwise they would not have promoted them. Have followers, or later adopters, of these programs been as successful? Some have, but many organizations have had limited success or considered the programs as failures. Why is this so? Is it the program or the situation? Figure 4.6 suggests that the answer must be in the fit between the program, as a proposed solution, and the problem or need of the business. Lack of success must arise from the incorrect matching of the program to the need, or the failure to implement the program correctly. Let's look at each of these possibilities.

Failure to Match Program with Need

Figure 4.6 shows conceptually what happens when programs are applied. The originator (or first company) is successful; a close follower using that program is also successful, most likely because it has an operation that is very similar to the originator. Additional followers similar to the first company can implement the programs and also achieve success. As the program grows in popularity and the success stories abound, other companies implement the program. As the program is extended into businesses that are different from the originator, however, the level of success varies

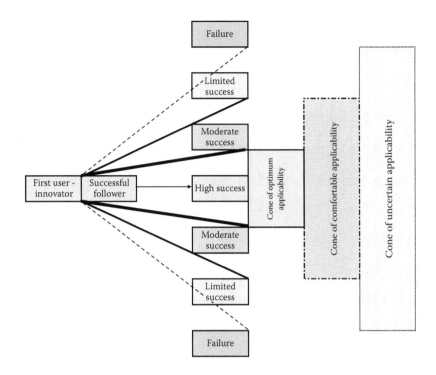

The first users or innovators of a management program do it successfully because they apply it to a specific set of conditions. As the program is applied in more remote conditions, the likelihood of success diminishes.

Figure 4.6 Program extensions and their chances for success.

and, in some cases, the program may actually be considered a failure. Conclusion: managers should match the program carefully with the needs of the organization.

As always, there is an alternative, and that is to adapt the basics of the program to the different conditions of the business. Figure 4.7 shows a conceptual model of this approach. As the conditions change, the adopter modifies the program elements or selects those elements that fit the new need. In general, the greater the difference in conditions in the new environment, the greater the need for adaptation.

Conditions vary for a number of reasons. We describe several that are among the most likely differences that could affect the eventual success of a management program:

■ *Strategic objectives.* Strategies vary among companies. One company may want to focus on cost reduction, and another company may focus on quality improvement. Each has different needs; therefore each company may utilize a different program.

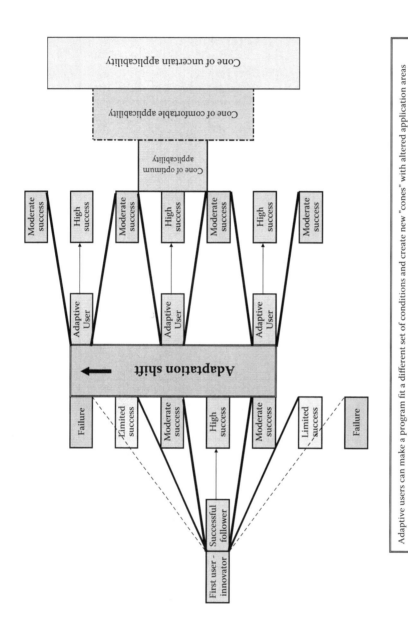

Figure 4.7   Adaptation of programs to new conditions.

Adaptive users can make a program fit a different set of conditions and create new "cones" with altered application areas

- *Types of products or services.* Product volumes and variety vary. A business that thrives on high-variety, low-volume products should not expect that a program developed for high-volume, low-variety products would fit their needs, at least not without some modification. The same is true for a line of service offerings. An inventory management program developed for widely fluctuating demand patterns, in a seasonal retail business, may have more variations than needed for a stable-demand business, such as the bread and milk departments of a grocery store.

- *Types of processes.* Manufacturing processes can be generally classified as job shop, batch, repetitive, and process. Service processes have been classified as a function of the degree of customer contact, such as high contact (hospitals) versus low contact (computerized banking). The wide range of requirements makes it difficult to apply the same program equally well to all types of processes. As with product variations, some modification to the program would be necessary to address different conditions.

- *Centralization versus decentralization.* Decision making is generally classified as centralized or decentralized. There are somewhat opposing trends evident today. The influence of the human relations movement is moving companies toward empowered employees, which suggests decentralized decision making. However, the widespread availability of information at all levels of detail is making centralized decision making feasible as well. It is important that management programs recognize the possible differences that can exist between these two approaches.

- *Cultures.* Cultures among businesses range widely. This topic is covered in more depth in Chapter 9. Internal factors may cause some of the differences, such as level of employee empowerment, level of job scope, types of wage payment systems, and management styles. External factors that influence the culture of a business include the demographics of the region in which the business is located and social trends. However, it is essential that the management program be adapted to the existing culture, or the existing culture modified to be compatible with the management program. It should be noted that modifying an organizational culture may be more challenging than modifying the management program.

- *Top management support.* Almost unanimously, authorities stress the need for top management support in the design and implementation of management programs. However, what top management support means can vary widely from business to business. This type of support is dependent on management styles, the relative importance of the program, the background and experience of top managers, and the relationship of top management to the external stakeholders in the program, such as consultants, customers, and suppliers.

- *Industry traditions.* Some industries have unique origins, practices, language, and peculiarities. Many improvement programs arise and are successfully implemented within a particular industry. For example, JIT and lean manufacturing are associated with the automobile industry, Quick Response systems with the retail industry, and MRP/ERP systems with repetitive manufacturing of all types. To implement a management program in one industry closely related to another may be a small step, such as from manufacturing washing machines to refrigerators. However, to move the same management program from automobile manufacturing to railroads or the health care industry may be a major transition because of the embedded practices within each of these industries. In some cases, a program will not gain any degree of acceptance until the terminology is changed to fit the industry where the program is being introduced.

Because of the above reasons, a management program needs to be adapted to its proposed application area if it is to have a chance of success. Even if it fitted correctly, it then must be correctly implemented, as we will see in the next section.

Figure 4.8 lists a number of management programs. The groupings are intended to show their primary objective, although most programs have multiple objectives. Figure 4.8 also shows the time period in which each program had a high level of interest.

Implementing the Program Correctly

Limited success or failure can also result from an inability to implement the program correctly. In the following sections, we outline some basic steps that are important in successfully implementing a management program.

Planning and Preparation

Implementing a program consists of two phases: (1) planning and preparation, and (2) execution and evaluation. Many businesses are eager to see results from a management program and, as a result, jump into the execution phase too quickly. Management programs are long-term projects that warrant a careful and systematic approach. The planning and preparation phase is essential, and each of the following steps is important for the program to be successful.

- *Identify the need for improvement.* The first step is to identify the need correctly. If the need is to improve quality, then a program designed to reduce cost through reduced inventory may not be appropriate. Certainly, a program should not be attempted just because another company has reported great success with it.
- *Select, or design, the right program.* We indicated earlier that many companies fail to achieve the desired results because they choose a program that was designed for different conditions. If that is the case, take the time to

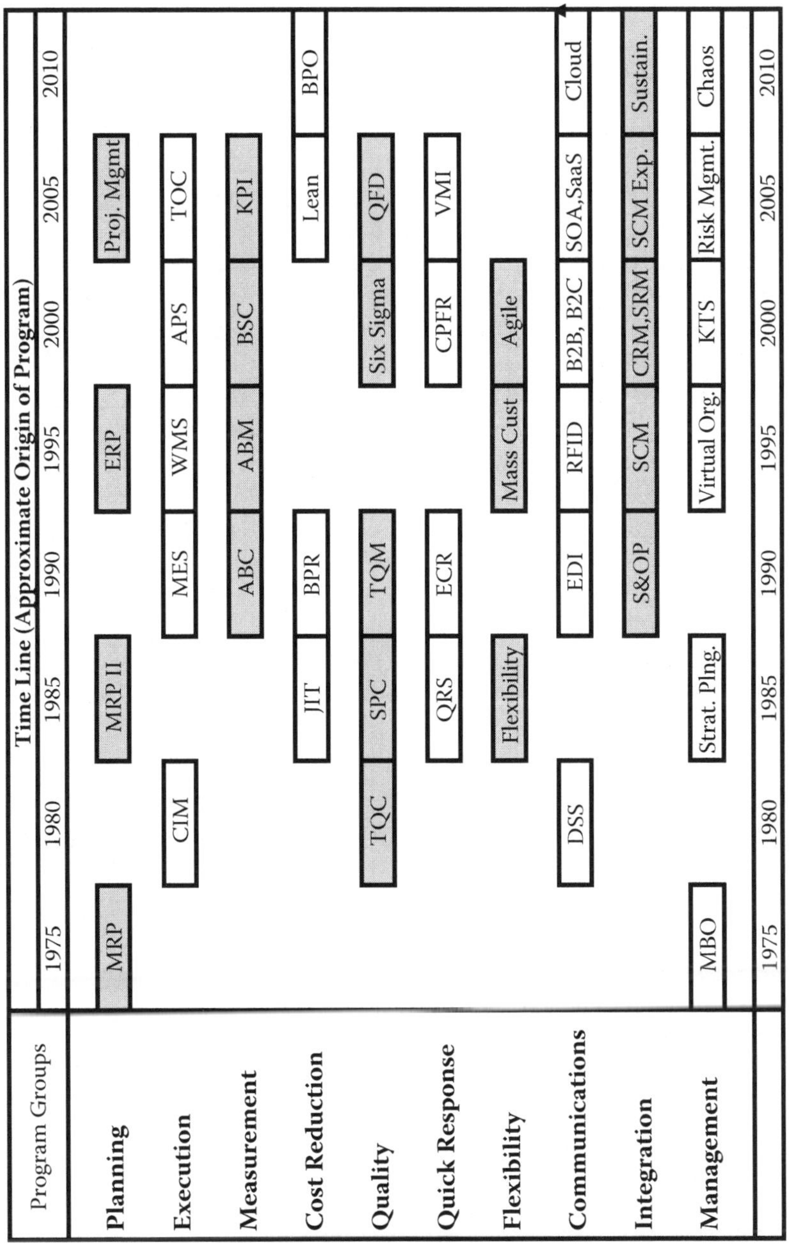

Program Groups	1975	1980	1985	1990	1995	2000	2005	2010
Planning	MRP		MRP II		ERP		Proj. Mgmt	
Execution		CIM		MES	WMS	APS	TOC	
Measurement				ABC	ABM	BSC	KPI	
Cost Reduction			JIT	BPR			Lean	BPO
Quality		TQC	SPC	TQM		Six Sigma	QFD	
Quick Response			QRS	ECR		CPFR	VMI	
Flexibility			Flexibility		Mass Cust	Agile		
Communications		DSS		EDI	RFID	B2B, B2C	SOA,SaaS	Cloud
Integration				S&OP	SCM	CRM,SRM	SCM Exp.	Sustain.
Management	MBO		Strat. Plng.		Virtual Org.	KTS	Risk Mgmt.	Chaos

Time Line (Approximate Origin of Program)

Figure 4.8 Evolution of management programs over time.

select those elements of the program that will fit your company's needs, even if you have to omit some of the more attractive parts of the program. For example, the concept of problem-solving teams has widespread application. On the other hand, a kanban scheduling system designed for repetitive production may not be appropriate in a process industry. If you can see a way to adapt other program elements or develop new approaches, do not hesitate to do that. Just avoid trying to fit a square peg in a round hole.

- *Create a receptive environment.* The first two steps of planning and preparation do not necessarily involve the part of the organization that will be responsible for the actual implementation. Now is the time to get that part of the company involved. Most people resist change, either actively or passively, unless they have been involved in its design.[21] Many well-designed programs have failed because the employees affected by it resisted its implementation. It is better to prevent a potential resistance problem early than to try to solve it later when the program is supposed to be up and running.
- *Organize for successful implementation.* You may have the right program and have conducted a good job of involving the employees; however, do not forget to organize the implementation effort to make it work. This phase could involve the organization of teams, deciding whether project team members will be full or part time, deciding whether to use consultants, determining the level of training required, or selecting the area in which to test and pilot the program.

Execution and Evaluation

If the planning and preparation steps have been managed well, the execution phase has a better chance of succeeding. This is not to say that it will not be time consuming, costly, and fraught with its share of problems and frustrations. However, these problems are normal in the execution of new programs and are surmountable. The following guidelines should be considered in the executive and evaluation phase:

- *Outline the steps to be accomplished in the implementation process.* This phase involves proper scheduling, communication, training, and all the other factors involved in getting the program up and running. Any program is a complex project; use project management techniques to get it right. If there is a lack of experience within the business, supplement it with outside help. Try to be consistent; too much wavering conveys a feeling of uncertainty and unease among the participants in the program. Expect that the implementation will be a multi-period (weeks or months) effort and adjust the pace accordingly. Keep in mind that the longer the program, the more challenging it will be to retain interest and enthusiasm. Above all, be patient.

■ *Evaluate and refine.* The evaluation of the program should assess both tangible and intangible criteria. Tangible items include the costs, savings, and benefits of the program. A number of intangible factors can also be considered, depending on the program and the needs of your organization. These can include such items as changes in the company image in the eyes of customers, competitors, and stakeholders; the morale of the workforce and their continued acceptance; and the compatibility of the program with other processes in the company. If actual results do not match expectations, take the necessary actions to correct or improve the situation.

■ *Assimilate the program into normal operations.* At some point after a successful implementation, it will be appropriate to plan for the end of the program as a separate and discrete project in the company. The evaluation phase should make it possible to select those program elements that should be assimilated into the normal operations of a business. Assimilation implies that the program elements that should be retained are continued on a long-term basis.

■ *Prepare for the next program.* After assimilating the selected features of the program, it is time to plan for the next program. New opportunities or threats may have emerged that will signal new areas for improvement. There is always a need to improve processes and that is what management programs can accomplish. If the previous program was successful, identify those factors you think contributed to its success. Likewise, mistakes should be noted so that the implementation of the next management program will be more seamless.

Future of Management Programs

What is the future for management programs? Ideally, all the workable components of the various management programs will be assimilated into the normal operations of a business. Managers will select those things that work and adapt them to their operations. However, new programs will continue to be spawned, implemented, adapted, and blended into management practice.

Management for the Twenty-First Century

As changes take place in society, changes also occur in the various key roles that individuals in society play. One rapidly changing role is that of the contemporary manager, especially the operations manager who must deal with employees on a day-to-day basis. Many of the routine activities that were formerly managerial tasks have been computerized. Examples of these tasks include the scheduling of employees and the purchasing of supplies and raw materials. Other tasks have been delegated to empowered employees and self-directed work teams. The time needed

to gather data to compile in reports has been shortened as well as the task of interpreting these reports, which are now often accompanied by notices of key variances
that are being violated. Relieved of the task of accumulating and interpreting this
information, managers must effectively communicate the findings to their employees. This change means that managers must be more highly skilled in interpersonal
communications and human relations to perform their roles effectively.

Just as business and management have undergone change, so too have the
employees managers lead. Many of these workers have come to expect and demand
much more out of their jobs and life in general. For instance, there is more emphasis on nonmonetary rewards, and employees are looking increasingly toward their
work as a source of fulfilling their needs for self-esteem, accomplishment, challenge, and involvement. That is why top-performing companies organize to meet
the needs of their people and do not just serve as a place to "work."

Organizing around the needs of people is not as easy as it sounds. It means
going against the traditions of what managers do. Instead of giving orders and
gathering and interpreting information (managing), managers must set direction
and provide resources and encouragement to their employees (leading). It also
means aligning the systems, structure, and culture so that the needs of employees
are linked with the needs of the business. The business environment is simultaneously faced with the challenges posed by customer service, total quality, continuous improvement, re-engineering, and assorted other hot topics of the month. As
a result, many top-performing companies are turning to work teams as the way to
link their employees' needs with company needs.

Why teams? Work teams provide the opportunity for employees to get involved
in the production processes on a more comprehensive level. In addition, teams
can draw on a broader mix of skills, experience, and know-how than any one person could offer. You do not have to look far to find proof of the stunning success
of teams within corporations. Motorola used teams to produce the world's smallest and highest-quality cellular phones. Ford relied on teams to create its popular
Taurus model. 3M credits work teams as the source of its incredible stream of
innovations.[22] Although the team approach to achieving results started primarily
in the manufacturing sector, it is now becoming more common in service organizations as well.

Another example of blending a technique originally developed in a specific
management program into normal operations is the use of statistical process control (SPC) methods. SPC techniques originated in quality programs such as TQM,
and more recently, Six Sigma. However, the basics of SPC can be applied in a
variety of other areas, such as the variation in supplier fulfillment rates or on-time
deliveries, customer ordering patterns, or wait times in a hospital emergency room.

There always seems to be a need to consider some new technique or concept.
Therefore, we need to expect that we will always have new management programs
to consider. Table 4.5 lists a few of the possibilities for the future.

Table 4.5   Future Program Emphases

Aspect of the Program	Possible Future Emphases
Paradigm	Mass customization
	Virtual corporation
	The Learning Organization
Area of focus	Manufacturing core competency (outsourcing)
	Strategic and collaborative alliances (trust)
	Design for adaptability (equipment/employee interface)
Global issues	Ethics/values equilibrium
	Environmental equilibrium
	Social equilibrium

Summary

In this chapter, we have described the concept of management programs, why companies use management programs to effect improvements, how management programs involve packaging basic management concepts and practices, how systems and contingency theories help to focus the program on the needs of a company, and how failure to adapt the program to the company's needs can lead to limited success. We also posed some of the possibilities for future programs. In Chapter 5, we describe management programs that originated in the manufacturing sector and have been successfully transferred to service businesses. In Chapter 6, we describe programs that have moved from services to manufacturing.

References

1a. Walker, W.T., *Supply Chain Architecture: A Blueprint for Networking the Flow of Material, Information and Cash*, CRC Press, Boca Raton, FL, 2005.

1. Fink, S., *Crisis Management: Planning for the Inevitable,* Backinprint.com, Rev. Ed., 2000.

1b. Crandall, W., Parnell, J.A., and Spillan, J.E., *Crisis Management in the New Strategy Landscape*, Sage, Thousand Oaks, CA, 2013.

2. Coombs, T., *Code Red in the Boardroom: Crisis Management as Organizational DNA*, Praeger Publishers, Westport, CT, 2006.

3. Childe, S.J., Maull, R.S., and Bennett, J., Frameworks for understanding business process re-engineering, *International Journal of Operations & Production Management*, 14, 12, 22, 1994, p. 34.

4. Luekue, R., *Managing Change and Transition*, Harvard Business School Press, Boston, 2003, p. 34.

5. Abrahamson, E., Management fashion, *Academy of Management Review*, 21, 1, 254, 1996.

6. Spell, C.S., Management fashions: Where do they come from, and are they old wine in new bottles? *Journal of Management Inquiry*, 10, 4, 358, 2001.

7. Ponzi, L. and Koenig, M., Knowledge management: Another management fad? *Information Research*, 8(1), paper no. 145, 2002. [Available at http://InformationR. net/ir/8-1/paper145.html]

8. Carson, P.P., Lanier, P.A., Carson, K.D., and Guidry, B.N., Clearing a path through the management fashion jungle: Some preliminary trailblazing, *Academy of Management Journal*, 43, 6, 1143, 2000.

9. Abrahamson, E., Management fashion, *Academy of Management Review*, 21, 1, 254, 1996.

10. Ettorre, B., What's the next business buzzword? *Management Review*, 86, 8, 33, 1997.

11. Fenn, J. and Linder, A., Gartner's Hype Cycle Report for 2005, August 5, 2005, http:// www.gartner.com.

12. Crandall, R.E. and Crandall, W.R., An analysis of management programs: Origins, life cycles, and disappearances, *Proceedings of the 2007 Southeast Decision Sciences Institute Conference*, Savannah, GA.

13. Juran, J.M., *Managerial Breakthrough, A New Concept of the Manager's Job*, McGraw-Hill, New York, 1964.

14. McMahon, D. and Carr, J.C., The contributions of Chester Barnard to strategic management theory, *Journal of Management History*, 5, 5, 228, 1999.

15. Bedeian, A.G., *Management*, The Dryden Press, Chicago, 1986.

16. Herzberg, F., One more time: How do you motivate employees? *Harvard Business Review*, 65, 5, 109, 1987.

17. Bedeian, A.G., *Management*, The Dryden Press, Chicago, 1986.

18. Bedeian, A.G., *Management*, The Dryden Press, Chicago, 1986.

19. Schoderbek, P.P., Schoderbek, C.G. and Kefalas, A.G., *Management Systems Conceptual Considerations* (4th ed.), BPI Irwin, Boston, 1990.

20. Wakefield, R.L., Identifying knowledge agents in a KM strategy: The use of the structural influence index, *Information & Management*, 42, 7, 935, 2005, p. 943.

21. Beer, M. and Nohria, N., Cracking the code of change, *Harvard Business Review*, 78, 3, 133, 2000.

22. Katzenbach, J.R. and Smith, D.K., *The Wisdom of Teams: Creating the High-Performance Organization*, Harper Collins, New York, 1993.

Chapter 5

Adapting Manufacturing Techniques to Services

In Chapter 4, we looked at how management improvement programs are developed. In this chapter, we look at how these programs are being applied. The manufacturing sector has developed an impressive body of knowledge to help businesses become more efficient and to improve product quality. As these techniques become more refined, they can be extended from manufacturing to service businesses.

In addition to using techniques developed in manufacturing, service businesses are expanding their scope of activities to move upstream along the supply chain toward the manufacturing process. In fact, some service businesses are becoming directly involved in manufacturing to secure a greater level of consistency and quality in their service offerings. We illustrate this by describing two companies—Amazon and United Parcel Service (UPS) at the end of this chapter. Both are successful service businesses that have recognized the value of implementing improvement programs developed first in manufacturing.

Introduction

Manufacturing companies have long been known for their emphasis on reducing costs and improving quality. As a result, they have developed programs such as JIT and lean production to reduce costs, and Total Quality Management (TQM) and Six Sigma to improve quality. Although there are a number of other examples, we will concentrate on these to illustrate how programs developed in manufacturing can be extended to service companies.

First, let us look at continuous improvement programs such as JIT, or its successor, lean production. In their research, David Bowen and William Youngdahl described how service businesses such as Taco Bell, Southwest Airlines, and Shouldice Hospital have applied lean service—the application of lean manufacturing techniques.[1] Similar studies have found applications of JIT in insurance firms, retailers, mail-order firms, hospitals, and finance companies.[2] Although the terminology may vary from manufacturing to services, the concept of JIT as a process-oriented waste eliminator applies to both areas.

Quality programs have also been very popular; examples include TQM and its successor, Six Sigma. Although service quality is more difficult to measure because of the intangibility of the service, some of the improvement concepts and tools used in manufacturing can apply equally well to services. Ron Snee and Roger Hoerl provide a number of examples of how Six Sigma has been extended beyond the factory floor to "the real economy," as they describe nonmanufacturing, or services.[3] Their examples cover financial services, e-commerce, health care, nonprofit organizations, and services within manufacturing organizations such as delivery, finance, and human resources. They stress that one should focus on processes, not functions or transactions. Once past that hurdle, it is as easy to apply Six Sigma concepts and techniques to service processes as it is to manufacturing processes.

We use Table 5.1 as the starting point for our discussion. The top row shows the most common classifications of manufacturing processes—continuous, repetitive, batch, and job shop. The left column lists the major characteristics of interest to managers and researchers. Table 5.1 also shows that there may be as much difference between types of manufacturing processes as there is between manufacturing and services. For example, online banking may be more like continuous manufacturing processes than job shops are.

Description of Manufacturing Process Types

In manufacturing, processes are usually classified as continuous, repetitive, batch, and job shop. It is not unusual for a company to have a combination of processes at the same location. For example, an automobile assembly plant may use a repetitive process to assemble the cars, a batch process to make components for the car, and a job shop to make replacement tooling for the assembly line.

The process classification depends on a number of factors, such as the following:

- *Amount of labor content.* This factor is usually expressed as a percentage of product cost. Is the labor content low, as in a continuous process; moderate, as in repetitive or batch processes; or high, as in a job shop?
- *Employee skill requirements, mental.* Continuous processes require a high level of mental skill to operate the specialized equipment. Repetitive batch

Table 5.1   Four Types of Manufacturing Processes

Characteristics	Continuous	Repetitive	Batch	Job Shop
Labor content	Low	Low	Moderate	High
Employee skill: mental	High	Low	Specialized	High, versatile
Employee skill: physical	Intermittent	Low	Low	High
Employee flexibility	Limited	Low	Low	High
Equipment intensity	High automation	Specialized automation	Individual machines	General equipment
Facilities type	Special purpose	Special purpose	General purpose	General purpose
Product standardization	Standardized	Standardized	Standardized families	Customized
Process flow	Continuous	Assembly	Sequential intermittent	Variable intermittent
Batch size	Large	Moderate	Moderate	Small
Volume of products	High	Moderate	Moderate	Low
Product orientation	MTS	MTS to ATO	MTS	MTO to ETO
Manufacturing examples	Chemicals, liquids	Automobiles, appliances	Clothing	Replacement windows
Service examples	E-commerce, online banking	Fast food, drive-through banking	Airlines, theme parks, bank check processing	Hospitals, investment banking

processes have relatively low mental skill requirements; job shops rely on a high level of mental skills among employees.

■ *Employee skill requirements, physical.* Job shops require the highest level of physical skills because of the variety of tasks performed. Repetitive and batch processes have low physical skill requirements as regards the variety of tasks but do require dexterity and endurance. Continuous processes require only intermittent physical skills, as the employees are primarily equipment attendants and maintainers.

- *Level of employee flexibility through cross-training.* Job shops have the greatest flexibility in scheduling work because of the need to have cross-trained employees. Cross-training is more limited in batch and line operations, although more of it is being done as part of JIT and lean implementations. Continuous processes tend to have reduced flexibility because of their specialized operations.
- *Level of equipment intensity.* Continuous processes require the highest level of equipment intensity, as in the form of chemical processing systems or baking ovens for bread. Repetitive processes may require substantial investment in assembly lines or robotics. Batch processes usually have multiples of individual pieces of equipment, such as sewing machines or lathes. Job shops usually have general-purpose equipment capable of handling a variety of products.
- *Type of facilities required.* Continuous processes require the most specialized facilities—for example, a chemical plant with connections between buildings. Repetitive processes usually require some specialized facilities, such as a painting department for automobiles. Batch processes do not require special facilities and can be set up in smaller units. Job shops are the least demanding insofar as facility requirements are concerned.
- *Level of product standardization.* Continuous processes require standard products and are not easily changed from one product to another. Repetitive processes can be changed between products in the same family but also require relatively standard products. Batch processes can handle a wide variety of products within the same general product category, such as clothing. Job shops generally make a wide variety of products with low levels of standardization.
- *Regularity of process flow.* Continuous processes require an almost continuous flow of product, and this is the origin of its name. Continuous processes, such as in steel mills, often operate continuously to avoid disruptions to their furnaces or rolling mills. Repetitive processes can be interrupted without dire consequences; however, they tend to operate with a high level of capacity utilization for economic purposes. Batch flow is intermittent and the flow requirements are more for efficiency than for process requirements. Job shops are intermittent with a variety of jobs moving in the shop at the same time.
- *Batch sizes.* Batch sizes range from very large in continuous processes, where the changeover times are high, to very small in job shops, where changeovers and setups are a way of life. Repetitive processes can be changed from one product to another if they are similar, such as from six-cylinder to four-cylinder Honda Accords. Batch processes can vary batch sizes between different sizes or colors as in women's dresses. However, the pressure to reduce batch sizes has been growing to reduce the amount of work-in-process inventories.
- *Volume of products.* Continuous processes are designed to make a high volume of products with a low unit cost. Job shops make a low volume of products

with a high unit cost. Repetitive and batch processes are in between continuous and job shops insofar as the volume of products produced is concerned.

■ *Product orientation.* Continuous processes have a make-to-stock (MTS) orientation. They must anticipate (forecast) demand and make both the variety and amount of products required. Repetitive processes have traditionally been MTS producers, although in recent years some have moved more to the assemble-to-order (ATO) mode to become more flexible to demand. Batch processes have the flexibility to be ATO and even make-to-order (MTO) as it becomes more difficult to forecast demand, especially the product mix. Job shops have typically been MTO or even engineer-to-order (ETO) producers.

Product–Process Relationship

Robert Hayes and Steven Wheelwright were among the first to emphasize the need to match the production process with the product characteristics. They pointed out that, as the product evolves over its life cycle, it tends to move from a high-variety, low-volume state to high-volume, low-variety requirements. To be responsive to the demands of the marketplace, the manufacturing process has to change.[4]

Although a generalization, this concept illustrates the need to match the process with the product. Not correctly matching the process with the product is expensive. Using a job-shop process to make high-volume products results in unnecessarily high product costs. On the other extreme, using a specialized (continuous) process to make low-volume, high-variety products results in higher than needed investment costs per unit of product.

In the early stages of the life cycle, a job-shop approach is often appropriate because the demand may be for a wide variety of low-volume orders. As demand grows, the most popular products and processes can be adapted to batch or repetitive flow. Should the product reach very high volumes and low variety, it may be economically feasible to move to a continuous process. This reasoning works when moving along the life-cycle curve until the product reaches the mature stage. Then, a dilemma occurs. What type of process is best in the decline stage of the product as volume decreases? This question does not have a single answer; rather, it has many answers, depending on the situation at each business.

Service Industry Classifications

The manufacturing processes described above have been used for a number of years and appear to fit the different types of manufacturing processes well. Unfortunately, service classifications are not as well established. We will look at two prominent service classification models and examine some of the characteristics of services that make it possible to apply techniques developed in manufacturing industries.

The Chase Model

One of the best-known service typologies is that developed by Chase, Jacobs, and Aquilano.[5] They proposed that services could be classified by the level of customer contact they had. With this model, the level of customer contact progresses from no direct contact, with standardized processing, to face-to-face contact, with customized processing. Sales opportunities move from low in the low-contact segment of the matrix to high in the high-contact segment. Concurrently, production efficiency moves from high in the low-contact segment of the matrix to low in the high-contact segment.

Schmenner Model

Another well-known model is that presented by Roger Schmenner.[6] Like Chase, he shows the level of customer contact on the horizontal axis, but only as discrete low and high categories. He shows degree of labor intensity versus relative throughput times, with two discrete categories of low and high. The result is a two-by-two matrix in which he classifies type of services. In the upper-left quadrant, he shows a service factory (low contact, low degree of labor intensity, and low relative throughput time). The upper right is designated as a service shop (high customer contact, low degree of labor intensity, and low relative throughput time). The lower-left section represents the mass service group (low customer contact, high degree of labor intensity, and high relative throughput time). The lower-right quadrant covers the professional service group (high customer contact, high degree of labor intensity, and high relative throughput time).

Comparison of Manufacturing and Services

How can the service models be compared with the manufacturing model? If service processes are similar to manufacturing processes, then it makes sense to assume that techniques that worked in manufacturing can also work in services, albeit with a little modification. This approach has worked in many instances, although as we will show later, some manufacturing techniques work better than others in services. We provide another perspective in Figure 5.1, which shows product and service characteristics along a horizontal axis. Process characteristics are shown along the vertical axis. At the top of the figure, moving from left to right, product and service volumes increase, variety decreases, and the orientation usually moves from MTO to MTS. In some cases, a single factory may focus on simultaneous strategies by having some product use a MTS approach while another product uses a MTO approach. The bottom of the figure shows that the length of the product or service life cycle increases when moving from left to right. In addition, the time to develop new products or services also increases.

Low	Level of capital intensity	High
High	Degree of process flexibility	Low
High	Relative throughput time	Low
High	Degree of labor intensity	Low
High	Degree of customer contact	Low

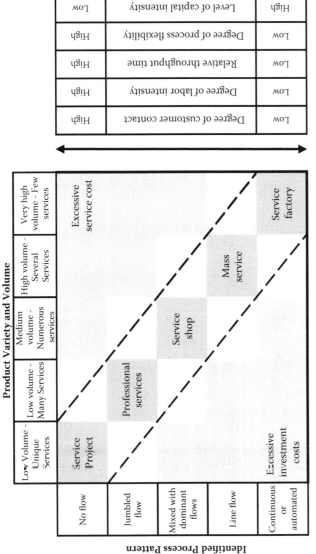

A process-focused positioning strategy is oriented toward low volume, customized products/services, such as in the upper-left corner of the matrix.

A process-focused positioning strategy is oriented toward high volume, standardized products/services, such as in the lower-right corner of the matrix.

Adapted to show Schmenner's service classification in a product–process matrix

Figure 5.1   Product–process matrix.

The left side of the figure shows variations in the process. Moving from top to bottom shows an increase in the importance of a structured flow, an increase in the level of capital intensity, the emphasis on low unit costs, and an emphasis on efficiency in both labor and equipment.

The right side of the figure also shows variations in the process. Here, moving from top to bottom shows a decrease in degree of customer contact, degree of labor intensity, relative throughput times, and degree of process flexibility.

The diagonal in the figure shows a movement from a project focus in the upper-left corner, where the project requirement dictates a special kind of process. As the product or service moves to the right, the process moves along the diagonal toward a flexible process. As product and service characteristics move even further to the right, the process must change to a more rigid process to achieve the high-volume, low-cost objectives. Arrows along the diagonal indicate that the process can change in either direction. Unfortunately, change is easier in moving down the diagonal toward the rigid process that entails heavier investment than it is in moving up the diagonal with emphasis on flexibility and the need for higher labor content.

Figure 5.1 shows that there is a way to relate service processes with manufacturing processes. However, if service processes are similar to manufacturing processes, then it makes sense to assume that techniques that worked in manufacturing also can work in services, albeit with a little modification.

Manufacturing Objectives

This section describes the types of programs that manufacturers have used to make improvements, which include efforts to reduce product costs, reduce inventories, increase resource utilization, improve quality, reduce response time, and reduce product development time.

Reduce Product Costs

One of the earliest efforts to reduce product costs was the specialization of labor. Instead of having one person do the entire job, such as the making of a musket, the job was divided into smaller tasks that could be completed by different workers. One of the earliest recorded examples of this approach was in the making of pins,[7] where the productivity of each operator increased dramatically. It also led to the need for interchangeable parts, although in the early stages of job specialization this created a need for filers and fitters to make the parts fit together.

Another major movement in reducing product costs was the use of equipment. David Hounshell offers a fascinating description of the early stages of this effort in his book *From the American System to Mass Production, 1800–1932*.[8] First, hand tools, such as hammers, saws, and augers, were used to assist the operators. Then power-assisted tools, such as lathes and drill presses, were provided. Finally, in an attempt to achieve the ultimate in product cost reduction, manufacturers moved to automation to replace the operator completely. This procedure worked well in some cases but not so well in others. The present thinking is to design a machine–worker combination that capitalizes on the best from both approaches.

Another major effort was in improving the methods used in making products. Beginning with the work of Taylor and others in the scientific management movement, there was an emphasis on eliminating unnecessary tasks, simplifying those tasks that remained by redesigning workplaces to reduce distances and weights handled, and combining tasks that were performed better together than separate. These efforts were one of the earlier efforts to reduce waste, particularly waste of motion and physical effort. Daniel Wren describes this period in detail in his book *The History of Management Thought*.[9]

Reduce Inventories

Accounting classifies inventory as an asset. At one time, an asset conveyed the impression of something good and a liability as something bad. However, over the years, inventory increasingly is being viewed as necessary, but not always desirable. If you think of inventory as stored expenses that will eventually show up in the income statement as a cost of goods sold, it is easier to understand that inventory should be subject to careful review. As products change more quickly today, it is important that inventory not contain obsolete products.

One of the major efforts to reduce inventory in the latter quarter of the twentieth century was the development of materials requirements planning (MRP). The concept of independent and dependent inventory originated with MRP. Dependent inventory items are the parts of an assembled product, the parent. If the final product is an automobile, then all the parts that go into an automobile, such as a fender or an engine, are required. Before MRP, most dependent demand components were managed as individual items without regard to the demand for the parent, or independent, demand item. As a result, it could be possible to have an inventory of dependent demand items even if there was no demand for the parent item. MRP attempted to correct this by calculating the demand for dependent demand items once the demand for the independent item was known. For example, if there was a demand for 50 automobiles, there was a corresponding demand for 50 engines, 200 wheels (four per automobile), 50 steering wheels, and so on. In theory, inventory for dependent demand items resulted only when there was demand for the independent demand item.

A more recent attempt to reduce inventory came from the movement to smaller lot, or batch, sizes. With smaller batches, it was possible to reduce inventories at all stages of completion—finished goods, work-in-process, and raw materials. However, to achieve smaller batch sizes, it was necessary to reduce setup and changeover costs by reducing the time required to make or purchase them. One of the concepts growing out of the JIT movement was SMED (single minute exchange of dies) where the emphasis was on making dramatic reductions in the time to effect a changeover from one product to the next. Many companies found that they could reduce their setup times by 80 to 90 percent.

A third approach to reducing inventories was the development of modular components. This is an extension of the interchangeable parts idea. Automobiles can again provide a good example. Traditionally, each model of automobile had uniquely designed parts, even down to the smallest part. There was little attempt to reduce the total number of parts by standardizing them throughout the product lines, such as with bolts, or even larger items, such as braking systems. Today however, standardization of parts and subassemblies is the norm in the automotive industry. By developing modular components and subassemblies, it is now possible to reduce both inventories and assembly times.

Although reducing inventories is now a popular initiative in many businesses, it is somewhat in conflict with another major improvement effort—increasing resource utilization.

Increase Resource Utilization

As the level of capital investment in the form of equipment and facilities increased, it became more imperative that this investment be fully utilized. A measure of this was called *equipment utilization*, which was a combination of both the efficiency in operating the equipment and the percentage of time that the equipment was operating. To achieve high utilization, several approaches were developed. These have been most applicable to the continuous process industries with secondary application to the repetitive industries.

One of the first approaches was to standardize the product. This often took the form of a reduced variety of final products. When coupled with specialization of labor, this movement produced a high volume of limited-variety products. When price was the primary concern, this was an ideal approach. Today, as customers become more selective in their choices, the need for product variety is becoming a greater concern.

Product standardization is closely coupled with long production runs. To achieve long production runs, it is necessary to forecast demand and operate in an MTS mode. When demand is predictable and relatively stable, this approach works well. For continuous process industries, such as chemicals, it is necessary because of the high cost of starting, stopping, and changeover of products, such as from automobile fuels to heating fuels. For repetitive products, long production runs are becoming a thing of the past and companies in this area are learning to adapt.

Capacity management is another way in which companies work to increase resource utilization. This approach includes careful analysis of the type of equipment best suited to the process needs, timing of the purchase with the capacity needs, careful monitoring of the equipment to ensure in-control operation, use of

Table 5.2 Areas of Potential Application of Improvement Programs: Group 1

Type of Improvement Program	Types of Processes and Level of Applicability in Each Type of Improvement Program			
	Continuous	Repetitive	Batch	Job Shop
Reduce Product Costs				
Worker specialization	Limited	High	High	Low
Automation	High	High	Limited	Limited
Methods analysis	Limited	High	High	Low
Reduce Inventories				
Dependent demand	Limited	High	Limited	Low
Reduce batch size	Limited	High	High	Low
Modular components ATO	Limited	High	High	Low
Increase Resource Utilization				
Product standardization	High	High	Limited	Low
Long runs MTS	High	High	Limited	Low
Capacity management	High	Limited	Limited	Low

preventive maintenance to avoid unplanned shutdowns, and planning for smooth transitions from one generation of equipment to the next. Table 5.2 offers a summary of the degree of applicability improvement programs can have on reducing product costs, reducing inventories, and increasing resource utilization.

Improve Quality

Although high quality of products has always been desirable, it became an imperative in the latter quarter of the twentieth century. This was especially notable in the automobile industry, where Japanese manufacturers made major inroads into the U.S. market, largely because they produced higher-quality automobiles. While U.S. manufacturers continued to rely on styling and price, the buying public was moving to more comfortable, reliable, and fuel-efficient automobiles. Repetitive industry businesses especially felt the need for this type of quality improvement program.

Process-control techniques were first developed at Western Electric in the 1920s, through the work by Walter Shewhart and others; disciples of Shewhart include Juran and Deming. Process control means that a consistent product is produced by a consistent process. Statistical techniques, especially the use of sampling techniques, were developed to monitor processes and to determine whether a process was out of control or moving toward an out-of-control situation. Over time, automated sensing systems have been developed to monitor time, temperature, pressure, and other variables of sophisticated continuous processes, such as the operation of nuclear reactors.

In addition to process-control techniques, a major movement has included empowering employees to be more responsible for product quality. The reasoning is that the equipment operator may be in the best position to influence product quality, through more awareness of the quality requirements and the assignment of responsibility to produce a quality product. This task requires a movement away from job specialization to job enlargement, and even further to employee empowerment. Although it increases labor costs, many companies have found it can also reduce the cost of poor quality, a cost that is largely hidden in most accounting systems.

Another approach to quality improvement is in product design. Companies have realized that no matter how carefully they control the process or how diligent the empowered employees, the product can never have a quality higher than in its inherent design. Consequently, they are now striving to design quality into the product. This effort requires better materials, closer fits, and improved fabricating methods.

Reduce Response Time

As the demands of the marketplace have increased, companies have had to address an increasing array of requirements (see Chapter 2 on Critical Success Factors) to be competitive. After reduced price and increased quality, reduced response time has been increasing in importance. Reduced response times not only reduce the level of inventory but also match production to demand more closely.

For years, both academics and practitioners developed scheduling algorithms that would make it possible to move jobs through the production process in an efficient (low cost) and effective (meet the required delivery date) manner. In the early days of their development, the techniques were manual with limited scope of application; however, in recent years, computerization has made them much more effective.

Responsiveness can be shortened with improved knowledge about demand. Better demand forecasting has been a paramount objective for several decades. This effect is characterized by both improved forecasting techniques and improved information processing. As more is known about demand patterns, especially about their basic features—stability, trends, seasonality, and events—a better forecast can be prepared. However, recent trends in shorter product life cycles and increased variability in demand are proving a difficult adjustment. Companies are finding that better forecasting techniques have inherent limitations; consequently, they are moving toward collaborative efforts with customers to enhance forecasts.

The use of modular components was discussed previously as a way to reduce inventories. It is also proving useful in responding to market demands. If the product can be carried in inventory as a generic model until the order comes in, it can then be finished quickly in a customized form and shipped to the customer. Dell has made this a hallmark of its business. Other companies are making use of this postponement technique to stock a smaller variety of "almost finished" products so that they have the ability to create a wide variety of finished products with the addition of customer-designated accessories.

Reduce Product Development Time

As product life cycles decrease, companies are finding it necessary to develop products faster, such as automobiles, airplanes, and especially software.

Concurrent engineering is a new approach to product development. Historically, companies used a sequential process to develop new products. Marketing came up with an idea for a new product and passed the idea to Product Engineering, who developed a prototype product. Product Engineering passed the idea along to Process Engineering, who developed a process for making the product. They passed the product and process along to Production, who adapted the recommended process to an existing process and began to make prototype or sample products. Along the way, Accounting got involved to consider product cost and the effect on the income statement. Finance also became interested because of the investment and balance sheet implications. Human Resources found out they had to hire more employees, some with skills not previously needed. Along the way, functional areas had questions that often required changes in the product design. Eventually, after iterations and revisions, a product emerged. In recent years, the process called "concurrent engineering" emerged, which required the use of cross-functional teams in the early stages to resolve questions, select the most promising alternatives, and eliminate dead ends in the product development. As a result, it has been possible to reduce the time to develop new products and improve the effectiveness of the new design.

Traditionally, the maker of the product took a make-and-sell approach in which they came up with a product idea, made it at low cost with their specifications for quality, and then tried to figure out how to sell it.[10] Today, businesses are attempting to figure out what the customer wants, often through participation from their customers. Then, they develop a product and the corresponding process that meets the customers' requirements. One technique that is becoming more popular is quality function deployment (QFD), in which customer requirements are related to company capabilities and competitor capabilities to decide on the attributes most suitable to the new product design.

Outsourcing is also being used to reduce product development time by requiring suppliers to participate in the product design process. For example, an automobile manufacturer may rely more heavily on a supplier to develop new braking systems or an engine manufacturer to develop the innovations in engines. This form of outsourcing requires a higher level of inter-organizational collaboration and relies heavily on mutual trust among businesses. This arrangement is proving to be a difficult transition for many companies. In fact, there are indications that some companies are finding that offshore outsourcing has many hidden costs and logistical problems that make it less attractive. The increase in risks incurred has generated a whole new area of supply chain management—risk management—that was not a major issue when suppliers were closer to their customers.

Most of the initiatives described here have been developed for application in the continuous process and repetitive manufacturing industries. Batch process manufacturers have tended to lag behind and many of the techniques are not as readily applicable to job shop types of operations. This problem is of special interest as we consider the transfer of knowledge to services because many service operations fit into the classification of batch flow and job shop operations. Table 5.3 offers a summary of the degree of applicability improvement programs can have

Table 5.3 Areas of Potential Application of Improvement Programs: Group 2

Type of Improvement Program	Types of Processes and Level of Applicability in Each Type of Improvement Program			
	Continuous	Repetitive	Batch	Job Shop
Improve Quality				
Process control	High	High	Limited	Low
Employee empowerment	High	High	Limited	High
Product design	Limited	High	Limited	Low
Reduce Response Time				
Job scheduling algorithms	High	Limited	High	High
Improved demand forecasting	High	High	High	Low
Modular components ATO	Limited	High	Limited	Low
Reduce Product Development Time				
Concurrent engineering	Low	High	Limited	Low
Quality function deployment	Low	High	Limited	Low
Outsourcing	Limited	High	Limited	Low

on improving quality, reducing response times, and reducing product development times.

Service Objectives

Service companies have some objectives that are similar to those of manufacturing organizations; however, they also have some that are different. We will examine those differences more fully in Chapter 6. Where service companies have objectives that are similar to those of manufacturing, the improvement programs that worked well in manufacturing may also work well in services. We explore some of those programs in the next section.

Programs That Work in Services

In an effort to achieve the goals described above, it became useful to develop programs that focused on specific objectives. We described the evolution of improvement programs in Chapter 4; in this section, we describe several programs that have been successfully implemented in services.

Programs work well in services when the service processes are similar to those for which the programs were developed in manufacturing. We will also describe some of the programs that worked in manufacturing but did not easily translate to the service side of business.

Theodore Levitt was one of the first to recommend the use of manufacturing techniques in services. He pointed out that even the so-called manufacturing industries contained a wide variety of service activities, such as systems planning, service support, software, repair, maintenance, delivery, collection, and bookkeeping. He emphasized the need for service industries to improve productivity and customer service to remain competitive and concluded by saying that the use of manufacturing techniques in service operations "can generate liberating new solutions to intractable old problems. It can bring to the increasingly service-dominated economies of the future the same kinds of vaulting advances in productivity and living standards as the newly created goods-producing factory economies brought to the world in the past."[11]

IBM has moved from a manufacturing company to a service company and provides services to purchasing, human relations, and customer relations programs. They have a growing interest in "services science." "The hybrid field seeks to use technology, management, mathematics, and engineering expertise to improve the performance of service businesses like transportation, retailing, and health care—as well as service functions like marketing, design, and customer service that are also crucial in manufacturing industries."[12] Services represent an "understudied field" that is attracting the interest of universities, corporations, and government agencies.

To provide a sampling of the progress that companies are making in adapting manufacturing programs to service applications, we will describe the successes for two types of programs. JIT and lean production represent programs to reduce costs and response times. TQM and Six Sigma have been used primarily to improve quality.

JIT and Lean

JIT is an outgrowth of the Toyota Production System (TPS), which to some extent was patterned after the work of Henry Ford at the River Rouge plant. It was designed to reduce inventories and eliminate waste. Lean production is a concept pioneered by Womack, Jones, and Roos following their study of automobile assembly plants throughout the world.[13] Lean production also stressed the need to eliminate waste and to smooth the flow of goods along the value chain. For practical purposes, it has become the successor to JIT in the minds of many writers.

One common theme throughout the lean production literature is the focus on manufacturing companies. Par Ahlstrom investigated why it is difficult to apply lean production to service companies. He found that lean production is applicable to service operations, although there are contingencies to the application. The contingencies stem from the characteristics of services. There are also instances where lean production is perhaps more applicable to services than manufacturing, again due to the nature of services. The conclusions indicate a need to make operations more general, avoiding making clear differentiation between manufacturing and services and instead focusing on the similarities between the two and thus the possibility to learn from each other.[14] In this light, it is interesting that one of the newsletters from the Lean Institute was about the opportunities to apply lean thinking to health care operations:[15]

- *Need for an integrated system.* Jeff Liker and James Morgan offer ideas on how to adapt the TPS, with its focus on the customer, to services.[16] Although they believe that the lean manufacturing concept can work in services, they also believe that many of the programs started in service sectors have been piecemeal or quick fixes to reduce lead times and costs or to increase quality. Consequently, these programs never create a true learning culture. They stress that it takes a true systems approach that "effectively integrates people, processes, and technology—one that must be adopted as a continual, comprehensive, and coordinated effort for change and learning across the organization."[16]
- *Consider service characteristics.* Some scholars provide a general approach to applying manufacturing techniques to services. One study points out that, when faced with the need to become more competitive, services are much like manufacturing in that they both use processes that add value to the basic inputs used to create the final product or service. Businesses identify the unique characteristics of services as inseparability of production and

consumption, intangibility, perishability, and heterogeneity. When integrated with the JIT themes of total visibility, synchronization and balance, respect for people, flexibility, continuous improvement, responsibility for the environment, simplicity, and holistic approach, they show that these themes can be applied in the service sector.[17]

■ *State government.* In a more specific study, Ron Bane reported on the successful application of lean principles to a state motor vehicle license bureau that was able to reduce waiting times by 70 percent.[18] The study also reports on other applications in the state of California, such as reduction of water main leaks, reduction of check processing times, and reduction in office space steps. The author cautions that successful application requires patience and top management support, and that continuous improvement is a never-ending process; however, he believes that there are "tremendous opportunities for nonmanufacturing organizations that embrace these approaches."[19]

■ *Bid quotation process.* Lean techniques have been applied to the bid quotation process, an especially important service function in job shops or MTO firms. Quotations convey two important bits of information—price and due date. Successful bids must be accurate and prompt, and the use of lean principles helps to achieve those objectives. One study found that application of lean techniques, coupled with the implementation of an ERP system, led to improvements in the following areas:
  - Reduction of paper use via electronic quoting
  - Reduction of waiting time through electronic reminders
  - Elimination of tasks in the quoting process
  - Coordinating with outside vendors
  - Collection of shop floor data for accurate labor costing.[20]

■ *Mass services.* Lean techniques can also be applied to *mass servicing*. Mass servicing is similar to mass production in that customer service employees are typically semiskilled representatives "trained a mile wide and an inch deep in handling a high volume of calls for all customer inquiries."[21] The traditional mass servicing model consists of a generically trained representative who rushes the customer off the phone to increase the number of calls handled per hour and reduces the unit costs; however, it almost guarantees errors and the need for callbacks. The proposed approach—customer-centric work cells—provides a simplified flow of the customer's order or request by grouping all knowledge and tasks associated with servicing the customer within the work cell. Employees are trained to perform multiple tasks (order entry, order fulfillment, service, and invoicing) so that the customer's request can be serviced without undue handoffs—"one call gets all."[22]

■ *Local government.* Sandra Furterer and Ahmad Elshennawy studied the use of lean practices in local government.[23] They looked at how the application of such lean concepts as kanban and visual control, waste identification and elimination, and one-piece flow can be used effectively to reduce the average processing times from 40 to 90 percent in processes such as payroll and pension reporting, purchasing/accounts payable, accounts receivable, and monthly reconciliation. They also found that a combination of lean with Six Sigma helped to increase both productivity and quality of financial services.

- *In association with Six Sigma.* Allen Pannell found that lean and Six Sigma are "powerful tools that complement—not compete with—each other." This has led to the use of the term *Lean Sigma* to describe improvement programs that use both lean and Six Sigma techniques. He explains that they are both management philosophies and sets of tools. Although they have many similarities, they also have some key differences in their primary focus. Both focus on reducing variation in the process. Although Six Sigma focuses on the use of statistical analysis to identify and reduce variation, lean takes more of a macro approach to reduce lead times and inventory. He cautions that "the challenge is not to look for a quick answer. Changing attitudes, thinking differently, and being willing to embrace the future takes leadership, time, and effort. It will pay off for companies that want to be in business 100 years from now."[24]
- *Customer service.* Siemens used lean methods to transform its culture to one that promotes active involvement, at all levels of the organization, in continuous improvement activities. The company made major improvements in customer service and won a ten-year contract to transform and manage the customer operations of National Savings & Investments, a major retail financial services organization.[25]

The health care industry is concerned with both reducing costs and improving quality of service. A small four-hospital group in Wisconsin looked to a nearby manufacturing company to learn how lean practices could help them. In addition to eliminating waste that reduced costs, they improved services. They reduced the "door-to-balloon" time—the minutes between a patient's arrival at the ER with chest pains and catheterization, a potentially lifesaving result. They also increased the percentage of time nurses could spend with patients by reducing the paperwork requirements.[25a]

Another novel use of lean in health care involved hospital facility design. While hospitals were designed with careful thought, they didn't always meet the needs of patients, staff, visitors, or care providers. Using lean techniques in designing hospitals can improve the work environment to create operational effectiveness, reduce the overall cost of the project, optimize patient flow and workflow to increase efficiencies, and provide a vision of the future.[25b]

These examples show that JIT and lean concepts have been applied to service operations successfully. However, there is no universal agreement that the transfer of manufacturing methods to service is always good. David Bowen and William Youngdahl caution that this industrialization of service is inappropriate because of the unpredictable nature of customer demand for services; however, these same writers advocate the transfer of service principles to manufacturing, a topic to be discussed more fully in Chapter 6.[26]

TQM and Six Sigma

Total Quality Management grew out of the statistical process-control work that originated in the Western Electric organization in the 1920s. Widely popular in the late 1980s and early 1990s, TQM programs were used extensively in both manufacturing and service companies. Although there were a number of successful programs, there were also a number of unsuccessful programs, often the result of inappropriate application or lack of focus by the implementers. In an attempt to revive the quality movement, Six Sigma was developed, originating with Motorola in the mid-1980s. It retained the primary focus of TQM but placed more requirements on the participants in the form of specific objectives (Six Sigma quality), an infrastructure (black and green belts), more training of participants, a formal approach to the improvement process including define, measure, analyze, improve, and control steps (DMAIC), and in general the need for a company to be serious about its implementation. Some of the service areas in which TQM or Six Sigma were successfully implemented include the following:

- *Universities.* Commenting on a 2001 Six Sigma survey in *Quality Digest,* Ron Bane reported that, although most of the respondents were in manufacturing, many were in nonmanufacturing areas such as document control, shipping, sales/marketing, customer service, and engineering. He also found Six Sigma being used in universities to reduce student scheduling errors.[27]
- *Health care.* Caldwell, Brexler, and Gillem examined why many Six Sigma efforts in health care fail. They found that although hospitals use processes, which is the necessary perspective heralded by Six Sigma authorities, those processes have a major difference when looking at the role of the physician. They studied both successful and unsuccessful implementations and found that the level of commitment by the physician was essential for success. They found that physicians resist participating in the program because the change places more burden on their processes—consuming more time, increasing complexity, and providing less service to them or their patients. They suggest that in order to engage physicians, health care providers should:
  - Seek to fully understand physician needs in general and within the specific process to be changed during the prework, define and improve phases of a DMAIC project, and learn the degree of support required.
  - Seek to build trust. This probably sounds simple, but trust between the physician and the hospital executive has dramatically eroded over the past decade.

- Educate physicians in all aspects of health care management, financial management, regulatory environment, and competitive pressures with an aim to establish a true visioning partnership about the future.
- Seek win-win projects. That is, find projects that will delight physicians, usually by improving the efficiency in physician–hospital interfacing processes.
- Negotiate a quid pro quo in which the organization provides a concession in some other area in exchange for physician agreement to embrace the desired change.
- Seek physician influences (referrers, physicians with high credibility) to lead the way, instead of hospital managers and executives.
- Integrate improvement work casually into existing physician committee and task force structures by discussion or nonaction items with action items.
- Consider incentives, but beware. Once offered, they become expected.
- Seek nonphysician caregivers to execute the change or influence physicians to embrace the change.

The authors caution that in health care situations, process innovation cannot be realized without physician engagement.[28]

- *Need for committed employees.* Members of the Gallup organization developed a concept they called Human Sigma, which they related to Six Sigma. They viewed Six Sigma processes as data driven, rational, and analytic and looked at their application to the employee–customer encounter at the retail level. They first found that emotionally satisfied customers were less likely to change to another store and bought more than rationally satisfied customers. They also found that those who were only rationally satisfied, as opposed to emotionally satisfied, were in some ways no different from dissatisfied customers. The Gallup studies also found that although an average performance measure may be satisfactory for a company, there was always considerable variability at the local level. This variability fit a normal curve, indicating an unmanaged situation. They found that employees with a higher level of commitment to the firm were more productive, had lower turnover, and were more effective than less committed employees in managing the customer encounter. To improve local performance, Gallup found that stores needed to look at both the level of customer engagement and employee engagement. These two attributes had a major effect on a store's financial performance. To implement rational programs such as Six Sigma in measuring and managing the employee–customer interface, companies need to take customers' emotions into account. Companies must also work to gain greater commitment from employees.[29]
- *Local government.* In addition to their study mentioned earlier, researchers Furterer and Elshennawy also looked at the effect of implementing lean Six Sigma programs in local government.[30] The authors applied lean Six Sigma

tools and principles to the financial administration processes in a 7000-citizen municipality. The financial processes included payroll, purchasing, and accounting functions of accounts payable, accounts receivable, and monthly reconciliations. The program achieved increased output, reduced processing times, and reduced errors.

■ *Hospital processes.* In their research study, James Harrington and Brett Trusko reported that there is no other industry more in need of reinventing itself than the health care industry.[31] They offer the following reasons:

1. It has been estimated that globally somewhere between 1,500,000 to 2,200,000 people die because of health care errors every year.
2. Health care errors kill up to 100,000 people per year in the United States alone.
3. One person in the United States dies every eight minutes because of nosocomial infection, and 95 percent of those deaths are preventable.
4. Few health care entities are ISO 9000–certified.

An ever-growing number of health care organizations are using Six Sigma to improve processes from admitting to discharge and all the administrative and clinical processes in between. This adoption is driven by several factors including the need to improve the organization's bottom line, eliminate medical errors, and position themselves for an imminent global consumer-centered health care revolution.

■ *TQM and change management.* Implementation of quality improvement programs such as TQM has major change-management implications. One multiyear study of twenty service companies found that the issues of unrealistic expectations of employee commitment, absence of process focus, lack of organization around information flow, holes in education and training, and failure to create a continuous improvement culture could contribute significantly to the lack of success of the programs. Unsuccessful companies tended to have a lack of commitment to the program; successful companies provided strong leadership based on strategic and tactical planning.[32]

■ *Intensive care.* Eva Lindberg and Urban Rosenqvist conducted a four-year study on the implementation of TQM in an intensive care unit (ICU).[33] They found that the TQM program was implemented during the introduction of two other major programs—a change in employee working hours and scheduling, and a change in the leadership of the ICU. The increased ambiguity of the situation made it difficult to evaluate the true impact of the TQM program. The authors, in speaking of the quality program, report that on the surface, the managers and staff are working in accordance with the quality system. However, they fear that the quality document has "promoted a mechanically functioning organization, similar to the attributes associated with mass production." They conclude that there is a need for a quality system that can avoid the negative effects by reducing the ambiguity in the organization.

- *Retail.* Fred Patton described the use of Six Sigma in a retail setting.[34] He believes that quality professionals understand how Six Sigma principles can be applied to any organization concerned with serving its customers as effectively as possible. The key is in convincing the service employees that there is a concept that was developed in manufacturing that can be adapted to a service setting. He states, "Getting service organization people to attend training is more painful for everyone involved than having your wisdom teeth extracted." Although there are obstacles, Patton is an advocate for applying Six Sigma concepts in service settings.
- *Similarity of manufacturing and service applications.* Some studies examined the difference between manufacturing and service firms with respect to the implementation of TQM practices and the relationship of these practices to quality performance. In a survey of 194 Australian firms that was equally divided between manufacturing and service firms, Prajogo found that there was no significant difference in the level of most of the TQM practices and quality performance between the two sectors. He believes that continuing fragmentation of industry sectors may be an area for further study and speculates that in manufacturing sectors, process management may be the key determinant in achieving a high-quality product, whereas in the service industry, people management may turn out to be the significant predictor.[35]
- *Sarbanes–Oxley.* One of the more innovative applications of Six Sigma was in developing a seven-step process to help "get and keep processes under control so management can comply with SOX 404 reporting requirements more cost effectively."[36] The authors caution that Six Sigma is only a tool and is most effective when used correctly.
- *Decision support systems.* David Rylander and Tina Provost studied how Six Sigma principles could be combined with technology, such as a decision support system (DSS), to enhance the customer service area.[37] Good frontline employees can reduce the negative effect of product problems; however, poor frontline employees can negate the positive effects of good products. They stress the need for employees to have good information to aid them in their customer interface dealings.

Programs More Difficult to Adapt to Service Operations

Programs such as JIT/lean and TQM/Six Sigma have found extensive application in service industries. Although each program required some adaptation, particularly in the employee acceptance area, the evidence supports the basic assumption that techniques from these programs can be beneficial in a variety of service applications.

In this section we examine some programs that have had more limited applicability in services. Although there are some successes, there is a general lack of widespread appeal. We identify some of the reasons these programs have been less successful in services than they were in manufacturing.

Product Costing

Stan Brignall studied the application of product costing techniques such as standard costs and activity-based costing (ABC) in service industries.[38] He also studied how performance measurement systems (PMS) developed in manufacturing could be extended to service sectors. As a reference, he used the service classification system developed by Rhian Silvestro and colleagues that included professional services, service shops, and mass services.[39] He observed that as services move from professional services to service shops to mass services, traceability of costs decreases and the level of overhead allocation increases. As a result, the concept of standard costs, with their corresponding variance analysis, may have some applicability in mass services but less in service shops and probably not at all in professional services. Performance measurement systems that tie shop-floor performance to overall company performance may be appropriate for many manufacturing companies (although less so for job shops); such systems are more difficult in service environments where there is more variability in the processes and quality of results.

Activity-Based Costing

Mostaque Hussain and Angappa Gunasekaran studied the applicability of ABC to service organizations.[40] Although they conclude that ABC can be useful in services, the results of their survey of banks indicate that ABC is not yet widely used. ABC was designed to more precisely determine the product costs in manufacturing. Products are easily identified and the unit costs do not have great variability for the same product. Services are more difficult to classify, making it more difficult to define a specific unit to be costed. The cost per unit can also vary considerably, because in many cases the customer is involved in the service and individual customers have different needs that can influence the time required to complete the unit of service. ABC can also be used for performance measurement in manufacturing; however, it is oriented toward financial measures, and service managers are finding that they need nonfinancial measures to measure performance best. The net result is that ABC has potential for services but needs considerable adaptation to be effective.

Materials Requirements Planning

Materials requirements planning (MRP) has been a fixture in the manufacturing toolkit for a half century. Because of its popularity and intricate structure, many researchers have tried to figure out how to apply it to services. There have been a number of clever applications. It could be used in surgical operations where the surgery to be performed is the independent demand item and the dependent items needed include the surgeons, the staff, the equipment, the supplies, and so on. MRP could also be used in a university to plan for the needs of students. The graduating student is the independent demand item and the courses needed to complete the degree are the dependent demand items. Although these and other applications sound interesting, MRP is not likely to be widely used in service applications, especially with the advent of lean production.

Enterprise Resources Planning

Enterprise resources planning (ERP) systems have been largely associated with manufacturing and have yet to be widely adopted in service organizations. Although there is a need, many service companies fail to consider how to select a system that provides what they need without buying all the extras that were designed into the manufacturing ERP systems. A study of professional firms reports that because an ERP system often is expensive and complex to deploy, it is important to understand the motivation for acquiring such a system. Programs that are project-oriented and have tools for tracking billable hours and producing what-if projections are the most practical in a service environment.

Performance Measures

The performance measures used in manufacturing often are based on specific units of output. It is not always easy to select a unit of output in the service industries because of the variability of the unit. For example, a unit of output may be a patient at a hospital. However, depending on the type of service and the characteristics of the patient, the time required per patient may or may not be a meaningful performance measure. In measuring quality, manufacturing companies use percent defective or parts per million, considering that the unit is either good or bad. This binary classification does not work as well in services, where the quality is more apt to vary along a continuum from very bad to very good, so another measure has to be adopted.

Automation

Automation is a common theme in manufacturing process improvement. It can provide both cost reductions and quality improvements. Automation is used extensively to reduce costs in service applications, such as in the banking industry with automatic teller machines (ATMs) and online banking, or in retail stores with self-checkout stations. However, quality in services is often associated with customer service, and the move toward automated services may not be compatible with the customer's perception of high quality. On the other hand, the use of online buying—books at Amazon.com where the website knows who you are if you have been there before—is considered to be a higher level of service than going to a bookstore with its limited selection of books and clerks who may not recognize you as a loyal customer.

Resource Utilization

Manufacturing managers have developed the utilization of resources—employees, equipment, facilities, and systems—to a real science. They work to utilize their assets effectively and efficiently. They have the luxury of building inventory ahead of demand to facilitate the use of level production plans, although this is changing somewhat because of the pressures to keep inventories low. Service companies cannot build inventory; therefore they have limited control over their utilization of equipment and facilities. They do have some discretion with respect to scheduling employees, and attempt to match capacity with demand whenever possible.

Keys to Extending Manufacturing Techniques to Services

A number of writers insist that the way to facilitate the movement of manufacturing techniques is to think in terms of processes. A manufacturing process involves the Input–Transformation–Output Model, in which materials (steel and wood) are transformed (manufactured) into outputs (automobiles and furniture). Using this model in services would make it possible to view an untreated patient as an input, the hospital surgery as the transformation process, and the treated patient as the output. Ron Snee and Roger Hoerl explain this in some detail in their book *Six Sigma beyond the Factory Floor.*[41]

There is also a need to recognize the differences and similarities between manufacturing and services. One way to approach this goal is to think in terms of subgroups within manufacturing and services. For example, there is a great deal of difference between continuous flow processes and job shops. However, job shops are more similar to professional service firms than to continuous flow manufacturing firms. Airlines share much in common with repetitive manufacturing processes. Online B2C systems like Amazon have commonalities with continuous flow systems. The point is that some service configurations are more similar to manufacturing units than they are to other service arrangements.

David Bowen and William Youngdahl remind us that manufacturing is also changing.[42] Businesses are learning to adapt some of the improvement programs that worked in stable, repetitive environments to rapidly changing, variable environments. As a result, program developers are moving closer to the type of programs needed in the more dynamic service sector. Robert Johnston suggests that the movement of manufacturing techniques to services has gone through several stages:

1. **Service awakening:** Initial interest in services
2. **Breaking free from product-based roots:** Awareness of the importance of services
3. **The service management era:** Cross-disciplinary research, especially among marketing, operations, and human resources
4. **Return to roots:** Refocus toward traditional operational issues and approaches

He believes that there is potential for applying the basic manufacturing techniques to service operations.[43]

Conclusions

This chapter provided a look at areas in which programs developed primarily in manufacturing have been extended to service industries. We described the types of manufacturing as job shop, batch, repetitive, and continuous. Many of the programs were developed in the repetitive and continuous flow processes where cost reduction and quality improvement were needed. JIT and lean manufacturing are programs that have been very successful in reducing manufacturing costs as well as providing other benefits that result from a smoother flow of products. TQM and Six Sigma have been successful as programs that improve quality. These programs are the ones most successfully applied to service operations.

Even though there have been numerous examples of successful applications of manufacturing programs to services, their success has been spotty. When programs such as lean manufacturing were used in manufacturing, they spread across industries, such as automobile and appliance manufacturing. This has not occurred in services, where the applications have been more company-oriented than industry-oriented. This lack of migration could be because it requires adaptation of these programs across industries while, in manufacturing, there are more similarities in processes across industries. It could also be that service industries have had other priorities and have developed their own programs to meet these needs, as we will see in the next chapter.

A characteristic of service industries is their concern with providing customer service. However, this strength can become a weakness. Some service industries have overshot their customer needs and are susceptible to new competitors who offer less service to customers at a lower cost.[44] Two examples stand out. Low-cost airlines, such as Southwest, have revolutionized the airline industry. Another example is Walmart in retailing. Their "Save money. Live better." approach has also been revolutionary, even though it has numerous critics. Low costs have a place in service operations, and many of the techniques developed in manufacturing to lower costs work in services.

However, other service industries, such as health care, need to both reduce costs and improve quality. Although there is a need to use the concepts developed in lean thinking and Six Sigma programs, these applications need to be applied with discretion. An additional area they need to address more closely is resource utilization, as in the deployment of expensive medical equipment, such as MRI scanners. Their approach to quality has been to add features, such as additional testing, to help diagnose and treat ailments; however, cost containment continues to be a problem.

We have developed two case studies of service companies—Amazon and UPS—that have used techniques developed in manufacturing to improve their operations. In addition, they have gone beyond improving their core functions; they have extended their scope of operations upstream along the supply chain to the point where they are performing some operations that include manufacturing. While we have not included a case for a health care company, there are several examples, such as the Mayo Clinic and the Cleveland Clinic, that use industrial engineering methodologies to both improve quality of care and reduce costs.

Manufacturing has been outstanding in improving internal operations. The need is to focus more on the customer, a strength of services industries, as we will see in Chapter 6.

References

1. Bowen, D.E. and Youngdahl, W.E., "Lean" service: In defense of a production-line approach, *International Journal of Service Industry Management*, 9, 3, 207, 1998.
2. Duclos, L.K., Siha, S.M. and Lummus, R.B., JIT in services: A review of current practices and future directions for research, *International Journal of Service Industry Management*, 6, 5, 36, 1995.
3. Snee, R.D. and Hoerl, R.W., *Six Sigma beyond the Factory Floor*, Pearson Prentice Hall, Upper Saddle River, NJ, 2005.
4. Hayes, R.H. and Wheelwright, S.C., Link manufacturing process and product life cycles, *Harvard Business Review*, 57, 1, 133, 1979.
5. Chase, R.B., Jacobs, F.R., and Aquilano, N.J., *Operations Management for Competitive Advantage* (10th ed.), McGraw Hill-Irwin, Boston, 2004.
6. Schmenner, R.W., *Plant and Service Tours in Operations Management* (5th ed.), Prentice Hall, Upper Saddle River, NJ, 1998.
7. Hounshell, D.A., *From the American System to Mass Production, 1800–1932*, The Johns Hopkins University Press, Baltimore, 1984.
8. Hounshell, D.A., *From the American System to Mass Production, 1800–1932*, The Johns Hopkins University Press, Baltimore, 1984.
9. Wren, D.A., *The History of Management Thought* (5th ed.), John Wiley & Sons, Hoboken, NJ, 2005.
10. Haeckel, S.H., *Adaptive Enterprise Creating and Leading Sense-and-Respond Organizations*, Harvard Business School Press, Boston, 1999.

11. Levitt, T., The industrialization of service, *Harvard Business Review*, 54, 5, 63, 1976, p. 74.

12. Lohr, S., Academia dissects the service sector, but is it a science? *New York Times*, April 18, 2006, p. C1.

13. Womack, J.P., Jones, D.T., and Roos, D., *The Machine That Changed the World, The Story of Lean Production*, Harper Perennial, New York, 1990.

14. Ahlstrom, P., Lean service operations: Translating lean production principles to service operations, *International Journal of Services Technology and Management*, 5, 5–6, 545, 2004.

15. Womack, J.P., Jones, D.T., and Roos, D., *The Machine That Changed the World, The Story of Lean Production*, Harper Perennial, New York, 1990.

16. Liker, J.K. and Morgan, J.M., The Toyota way in services: The case of lean product development, *The Academy of Management Perspectives*, 20, 2, 5, 2006.

17. Canel, C., Rosen, D., and Anderson, E.A., Just-in-Time is not just for manufacturing: A service perspective, *Industrial Management + Data Systems*, 100, 2, 51, 2000.

18. Bane, R., Leading edge quality approaches in non-manufacturing organizations, *Quality Congress. ASQ's Annual Quality Congress Proceedings*, 245, 2002.

19. Bane, R., Leading edge quality approaches in non-manufacturing organizations, *Quality Congress. ASQ's Annual Quality Congress Proceedings*, 245, 2002.

20. Buzby, C.M., Gerstenfeld, A., Voss, L.E., and Zeng, A.Z., Using lean principles to streamline the quotation process: A case study, *Industrial Management + Data Systems*, 102, 8/9, 513, 2002.

21. Erlich, B.H., Service with a smile, *Industrial Engineer*, 38, 8, 40, 2006.

22. Erlich, B.H., Service with a smile, *Industrial Engineer*, 38, 8, 40, 2006.

23. Furterer, S. and Elshennawy, A.K., Implementation of TQM and lean Six Sigma tools in local government: A framework and a case study, *Total Quality Management & Business Excellence*, 16, 10, 1179, 2005.

24. Pannell, A., Happy together, *Industrial Engineer*, 38, 3, 46, 2006.

25. Pollitt, D., Lean delivers savings for Siemens Business Services, *Training & Management Development Methods*, 19, 3, 223, 2005.

25a. Hyatt, J. Keen to be Lean, *CFO*, December, pp. 44–47, 2009.

25b. Meersereau, E. and Jimmerson, C. December 2011. *Improving Healthcare Delivery Using Lean Thinking for Facility Design*, http://www.leanhealthcarewest.com, accessed December 12, 2011.

26. Bowen, D.E. and Youngdahl, W.E., "Lean" service: In defense of a production-line approach, *International Journal of Service Industry Management*, 9, 3, 207, 1998.

27. Bane, R., Leading edge quality approaches in non-manufacturing organizations, *Quality Congress. ASQ's Annual Quality Congress Proceedings*, 245, 2002.

28. Caldwell, C., Brexler, J., and Gillem, T., Engaging physicians in lean Six Sigma, *Quality Progress*, 38, 11, 42, 2005.

29. Fleming, J.H., Coffman, C., and Harter, J.K., Manage your Human Sigma, *Harvard Business Review*, 83, 7, 106, 2005.

30. Furterer, S. and Elshennawy, A.K., Implementation of TQM and lean Six Sigma tools in local government: A framework and a case study, *Total Quality Management & Business Excellence*, 16, 10, 1179, 2005.

31. Harrington, H.J. and Trusko, B., Six Sigma: An aspirin for health care, *International Journal of Health Care Quality Assurance*, 18, 6/7, 487, 2005.

32. Hug, Z., Managing change: A barrier to TQM implementation in service industries, *Managing Service Quality*, 15, 5, 452, 2005.
33. Lindberg, E. and Rosenqvist, U., Implementing TQM in the health care service: A four-year following-up of production, organizational climate and staff wellbeing, *International Journal of Health Care Quality Assurance*, 18, 4/5, 370, 2005.
34. Patton, F., Does Six Sigma work in service industries? *Quality Progress*, 38, 9, 55, 2005.
35. Prajogo, D.I., The comparative analysis of TQM practices and quality performance between manufacturing and service firms, *International Journal of Service Industry Management*, 16, 3/4, 217, 2005.
36. Juras, P.E., Martin, D.R., and Aldhizer, G.R., Adapting Six Sigma to the SOX 404 compliance beast, *Strategic Finance*, 88, 9, 36, 2007.
37. Rylander, D.H. and Provost, T., Improving the odds: Combining Six Sigma and online market research for better customer service, *S.A.M. Advanced Management Journal*, 71, 1, 13, 2006.
38. Brignall, S., A contingent rationale for cost system design in services, *Management Accounting Research*, 8, 2, 325, 1997.
39. Silvestro, R., Fitzgerald, L., Johnston, R., and Voss, C., Towards a classification of service processes, *International Journal of Service Industry Management*, 3, 3, 62, 1992.
40. Hussain, M.M. and Gunasekaran, A., Activity-based cost management in financial services industry, *Managing Service Quality*, 11, 3, 211-13, 2001.
41. Snee, R.D. and Hoerl, R.W., *Six Sigma beyond the Factory Floor*, Pearson Prentice Hall, Upper Saddle River, NJ, 2005.
42. Bowen, D.E. and Youngdahl, W.E., "Lean" service: In defense of a production-line approach, *International Journal of Service Industry Management*, 9, 3, 207, 1998.
43. Johnston, R., Service operations management: Return to roots, *International Journal of Operations & Production Management*, 25, 12, 1278, 2005.
44. Christensen, C.M., Anthony, S.D., and Roth, E.A., *Seeing What's Next, Using the Theories of Innovation to Predict Industry Change*, Harvard Business School Press, Boston, 2004.

Appendix 5A: Amazon

Amazon started as an Internet bookseller. They were incorporated in 1994 and began operation in 1995. During the years since, they have expanded into a number of new businesses, including the following:

- Retailing
  - Internet reseller of books
  - Internet reseller of a wide variety of other merchandise produced by third parties
  - Internet reseller of used books, through third parties
  - Seller of electronic books for Kindle and other electronic readers
- Distribution (fulfillment)
  - Owner and manager of distribution (fulfillment) centers for books and other merchandise
  - Distributor of electronic books through downloads
- Manufacturing
  - Developer, manufacturer, and seller of Kindle
  - Publisher of books written by authors under contract with Amazon
- Services
  - Provider of marketing services for their parties
  - Provider of cloud software services.

They have aggressively built on their position as the leading Internet reseller to move into related businesses, at times sacrificing profits to achieve revenue growth and market position.

Overview

Amazon provides the following overview of the company in their 2011 Annual Report:

> *Our primary source of revenue is the sale of a wide range of products and services to customers.* The products offered on our consumer-facing websites primarily include merchandise and content we have purchased for resale from vendors and those offered by third-party sellers, and we also manufacture and sell Kindle devices. Generally, we recognize gross revenue from items we sell from our inventory as product sales and recognize our net share of revenue of items sold by other sellers as services sales. We also offer other services such as AWS, fulfillment, publishing, digital content subscriptions, miscellaneous marketing and promotional agreements, such as online advertising, and co-branded credit cards.
> *We seek to reduce our variable costs per unit and work to leverage our fixed costs.* Our variable costs include product and content costs, payment processing and related transaction costs, picking, packaging, and

preparing orders for shipment, transportation, customer service support, and a portion of our marketing costs. Our fixed costs include the costs necessary to run our technology infrastructure and AWS; to build, enhance, and add features to our websites, our Kindle devices, and digital offerings; and to build and optimize our fulfillment centers. Variable costs generally change directly with sales volume, while fixed costs generally increase depending on the timing of capacity needs, geographic expansion, category expansion, and other factors. To decrease our variable costs on a per unit basis and enable us to lower prices for customers, we seek to increase our direct sourcing, increase discounts available to us from suppliers, and reduce defects in our processes. To minimize growth in fixed costs, we seek to improve process efficiencies and maintain a lean culture.

Because of our model we are able to turn our inventory quickly and have a cash-generating operating cycle. On average our high inventory velocity means we generally collect from consumers before our payments to suppliers come due. Inventory turnover was 10, 11, and 12 for 2011, 2010, and 2009. We expect variability in inventory turnover over time since it is affected by several factors, including our product mix, the mix of sales by us and by other sellers, our continuing focus on in-stock inventory availability, our investment in new geographies and product lines, and the extent to which we choose to utilize outsource fulfillment providers. Accounts payable days were 74, 72, and 68 for 2011, 2010, and 2009. We expect some variability in accounts payable days over time since they are affected by several factors, including the mix of product sales, the mix of sales by other sellers, the mix of suppliers, seasonality, and changes in payment terms over time, including the effect of balancing pricing and timing of payment terms with suppliers.[1]

Some of the recent financial results for Amazon are shown in the following tables. Table A5a.1 includes the sales, income, and cash flows for the five years from 2007–2011. Sales have increased by over threefold during this period. Income also increased proportionately, except for 2011, when income dropped because of heavier than normal investment in new fulfillment centers and technology. This shows more clearly in Table A5a.2, which has a breakdown of operating expenses, both in dollars and as a percent of net sales.

Table A5a.3 shows the Net Sales by product category within geographic area. North America sales were more than 50 percent of the total for the three-year period shown. Merchandise sales is the largest and fastest-growing area, while media sales have declined over the past three years. Other sales represent the newest product category and include Amazon Web Services (AWS), which provides access to technology infrastructure that enables virtually any type of business to use the service. Amazon Web Services has grown to thirty different services with thousands of

Table A5a.1 Sales, Income, and Cash Flow for 2007–2011

Income Statement	2007	2008	2009	2010	2011
Net sales (millions of dollars)	14,835	19,166	24,509	34,204	48,077
Income from operations	655	842	1,129	1,406	862
Net income	476	645	902	1,152	631
Net income (% of sales)	3.2%	3.4%	3.7%	3.4%	1.3%
Cash Flow					
Net cash from operations	1,405	1,697	3,293	3,495	3,903
Purchase of fixed assets	224	333	373	979	1,811
Free cash flow	1,181	1,364	2,920	2,516	2,092

Source: Adapted from 2011 Amazon Annual Report.[1]

large and small businesses and individual developers as customers. One of the first AWS offerings, the Simple Storage Service, or S3, now holds over 900 billion data objects, with more than a billion new objects being added every day. S3 routinely handles more than 500,000 transactions per second and has peaked at close to a million transactions.[1]

As shown in Table A5a.3, sales in North America represented 56 percent of total revenues in 2011, with the remaining 44 percent in the rest of the world. With supply chains becoming more widely dispersed, being able to operate in various countries throughout the world will be a definite advantage for Amazon.

Retailing—Physical Products

Amazon had its beginning as an online retailer, and this continues to be the main portion of their business revenue. The company lists the following selection of merchandise for sale:

Audible Audiobooks
Automotive and Industrial Products
Books and Magazines
Electronics and Computers
Home, Garden, and Tools
Grocery, Health, and Beauty

Clothing, Shoes, and Jewelry
Sports and Outdoors
Movies, Music, and Games
Digital Games and Software
Toys, Kids, and Baby
Kindle Touch, Kindle Fire, and accessories[2]

In addition, they offer these additional electronic products and services, which are the latest additions to their product offerings:

Table A5a.2 Operating Expenses by Type for 2009–2011

Operating Expenses (Millions of Dollars)	2009	2010	2011
Net sales	24,509	34,204	48,077
Cost of sales	18,978	26,561	37,288
Fulfillment	2,052	2,898	4,576
Marketing	680	1,029	1,630
Technology and content	1,240	1,734	2,909
General and administrative	328	470	658
Other	102	106	154
Total Operating Expenses	23,380	32,798	47,215
Percent of Net Sales			
Cost of sales	77.4%	77.7%	77.6%
Fulfillment	8.4%	8.5%	9.5%
Marketing	2.8%	3.0%	3.4%
Technology and content	5.1%	5.1%	6.1%
General and administrative	1.3%	1.4%	1.4%
Other	0.4%	0.3%	0.3%
Total Operating Expenses	95.4%	95.9%	98.2%

Source: Adapted from 2011 Amazon Annual Report.[1]

Unlimited Instant Videos
MP3s and Cloud Player
Amazon Cloud Drive
Appstore for Android[2]

Amazon employs a multilevel e-commerce strategy. They started by focusing on business-to-consumer (B2C) relationships between themselves and their individual customers, while using business-to-business (B2B) relationships between themselves and their suppliers. By adding customer reviews as part of the product descriptions, the company developed customer-to-business (C2B) transactions. It now also facilitates customer-to-customer (C2C) business by using the Amazon marketplace as an intermediary to facilitate C2C transactions.[3]

Amazon places great emphasis on tracking the interests of individual customers or prospective customers. They greet returning customers by name (e.g., "Welcome,

Table A5a.3   Net Sales by Product Category for 2009–2011

Net Sales (Millions of Dollars)	2009	2010	2011
North America			
Media	5,964	6,881	7,959
Electronics, other merchandise	6,314	10,998	17,315
Other[a]	550	828	1,431
Total	12,828	18,707	26,705
International			
Media	6,810	8,007	9,820
Electronics, other merchandise	4,768	7,365	11,397
Other[a]	103	125	155
Total	11,681	15,497	21,372
Consolidated			
Media	12,774	14,888	17,779
Electronics, other merchandise	11,082	18,363	28,712
Other[a]	653	953	1,586
Total	24,509	34,204	48,077

Source: Adapted from 2011 Amazon Annual Report.[1]

[a] Includes nonretail activities, such as Amazon Web Services (AWS), miscellaneous marketing and promotional activities, other seller sites, and cobranded credit card agreements.

Richard") and do their best to recommend books their software determines may be of interest to the customer. Their system is streamlined to make ordering and paying easy, such as the use of the "1-Click Ordering" provision, which allows repeat customers to shortcut steps in the payment process.

Retailing—Electronic Products

Physical Products Downloads (Books)

Selling books in electronic form is being accelerated with the introduction of tablet readers such as Amazon's Kindle, Apple's iPad, and Barnes & Noble's Nook. Pricewaterhouse Coopers estimates e-books will represent 6 percent of consumer book sales in North America by 2013, up from 1.5 percent in 2009.[4] Amazon has been especially aggressive in selling e-books, largely to enhance the sales of their Kindle readers. Amazon has battled the print publishers over pricing. The publishers collaborated with Apple to reestablish "agency pricing," which is pricing where the publishers set the price. Amazon countered that this amounted to price fixing. In April 2012, the Department of Justice filed an antitrust lawsuit against Apple Inc., Hachette SA, HarperCollins, Macmillan, Penguin, and Simon & Schuster in New York district court, claiming collusion over e-book pricing.[5] No doubt other conflicts among online marketing companies will arise in the future as the technology and market structure continue to evolve.

Amazon Web Services

The Amazon website lists a number of web services available for subscribers. Needless to say, they are increasing their services almost on a daily basis. The list illustrates that Amazon is attempting to become a major player in cloud computing services.

Jeffrey Bezos, CEO of Amazon, Inc., has this to say about Amazon Web Services:

> Amazon Web Services has grown to have thirty different services and thousands of large and small businesses and individual developers as customers. All AWS services are pay-as-you-go and radically transform capital expense into a variable expense. AWS is self-service; you don't need to negotiate a contract or engage with a salesperson—you can just read the online documentation and get started. AWS services are elastic—they easily scale up and easily scale down.[6]

Distribution (Fulfillment Centers)

Amazon quickly discovered that sending individual customer orders to book publishers and expecting the publisher to ship individual books to consumers worked in theory but not in practice. As book sales increased during the 1996 Christmas season, publishers were overwhelmed and were unable to handle the large volume of orders. Amazon decided they must establish their own distribution centers (later named fulfillment centers) to stock some of the most popular selling books in order to meet customer demand. Although originally designed for books, the fulfillment centers quickly became a distribution center for all types of products that Amazon sold through its online website.

Once they established their fulfillment centers, Amazon has been trying to capitalize on these assets by expanding their use beyond Amazon's internal needs. Amazon provides distribution and shipping services for third parties. During the early part of 2012, these shipments represented over a third of the units shipped. Outside vendors that sell on the Amazon site pay Amazon a commission on sales, usually about 10 percent, as well as fees for storing and shipping the product in Amazon's fulfillment center. If an outside vendor doesn't sell on the Amazon website, they can still use Amazon's distribution service, designated as Fulfillment by Amazon, or FBA, where Amazon stores and ships the products to the third party's customers.[7]

Retailers who sell both through stores and online find the dual-outlet combination difficult; however, many are finding the distribution through both outlets even more difficult. That is why some are turning to Amazon to handle the distribution function at their fulfillment centers. "Integrating online and brick-and-mortar supply chains raises tricky questions, such as how to manage transportation between various origins and destinations, how to handle store returns for online-only merchandise, and how to supply online orders from store inventories without turning salespeople into order pickers and packers."[8]

Amazon tries to operate its fulfillment centers as efficiently as possible. They use the latest technology in storing and selecting items, where employees called stowers and pickers perform these functions. They use random storage, as opposed to having products occupy a certain space in the center, to maximize the use of storage space. They also use robots for some of the stowing and picking; they acquired Kiva Systems Inc., a robot manufacturer, in 2012 to have the latest in robotics technology.[9]

There are other technology applications. Automated scales check the weight of each box and compare it against the known weight of the items in the order. An IT system releases and routes orders in the best sequence to coordinate the picking operations with the schedule of outgoing trucks. Algorithms are run in Seattle to determine the best fulfillment center to handle any given order based on inventory availability, service requirements, and transportation cost. In addition to scaling each of the fulfillment centers through labor and shift adjustments to meet fluctuating demand, Amazon sets up fulfillment operations at supplier distribution centers to allow Amazon to ship directly from suppliers during peak seasons.[10]

Amazon considers its fulfillment centers an essential part of its total operations. As they point out: "If we do not adequately predict customer demand or otherwise optimize and operate our fulfillment centers successfully, it could result in excess or insufficient inventory or fulfillment capacity, result in increased costs, impairment charges, or both, or harm our business in other ways."[1] At the end of 2011, Amazon had sixty-nine fulfillment centers, about equally divided between the United States and other countries.[8]

> We rely on a limited number of shipping companies to deliver inventory to us and completed orders to our customers. If we are not able to negotiate acceptable terms with these companies or they experience performance problems or other difficulties, it could negatively impact our operating results and customer experience. In addition, our ability to receive inbound inventory efficiently and ship completed orders to customers also may be negatively affected by inclement weather, fire, flood, power loss, earthquakes, labor disputes, acts of war or terrorism, acts of God and similar factors.

Third parties either drop-ship or otherwise fulfill an increasing portion of our customers' orders, and we are increasingly reliant on the reliability, quality and future procurement of their services. Under some of our commercial agreements, we maintain the inventory of other companies, thereby increasing the complexity of tracking inventory and operating our fulfillment centers. Our failure to properly handle such inventory or the inability of these other companies to accurately forecast product demand would result in unexpected costs and other harm to our business and reputation (p. 16).[1]

In their annual report, Amazon cautions about the importance of managing their inventories and fulfillment centers. In addition to risks relating to fulfillment centers and inventory optimization, Amazon is exposed to significant inventory risks that may adversely affect their operating results because of seasonality, new product launches, rapid changes in product cycles and pricing, defective merchandise, changes in consumer demand and consumer spending patterns, and changes in consumer tastes with respect to their products. The company endeavors to accurately predict these trends and avoid overstocking or understocking products they manufacture and/or sell. However, demand for products can change significantly between the time inventory or components are ordered and the date of sale. In addition, when they begin selling or manufacturing a new product, it may be difficult to establish vendor relationships, determine appropriate product or component selection, and accurately forecast demand (p. 19).[1]

Manufacturing (Kindle)

Amazon introduced the first Kindle in 2007. It is a portable reader that wirelessly downloads books, magazines, newspapers, blogs, and personal documents to a high-resolution display. Kindle is now the best-selling product in the history of Amazon. Kindle Touch and Kindle Touch 3G are new additions that feature a touch screen that makes it easier to turn pages, search, shop, and take notes. Kindle Fire provides access to movies, TV shows, music, books, magazines, apps, games, and Web browsing with free storage in Amazon Cloud. It has a color touch screen and a dual core processor.[11]

The market for e-books is increasing rapidly. Forrester Research estimates there are 40 million e-readers and 65 million tablets in use in the United States. In the first quarter of 2012, e-books generated $282 million in sales, compared to $230 million for print, based on a study by the Association of American Publishers.[12]

While Amazon does not directly manufacture the Kindle line, they are "virtual" manufacturers in that they outsource the production process to contract manufacturers. They have suppliers in China (flex circuit connectors, injection-molded case, controller board, and lithium polymer battery), Taiwan (electrophoretic display), and South Korea (wireless card), according to Steve Denning.[7] In this way, Amazon is assuming greater control over the Kindle supply chain. It makes the product, markets it, and supplies books, both print and electronic, for Kindle users.

Origination (Book Authors)

Amazon Publishing

Amazon started by selling books online. When they found that publishers couldn't ensure prompt, consistent shipment of goods to customers, Amazon started their own distribution (fulfillment) centers. The fulfillment function has become a major activity, to the point where Amazon now provides distribution services for many of their partners who sell products, other than books, through Amazon.

Amazon began their move into digital books as early as 2005, when they acquired BookSurge (print-on-demand) and Mobipocket (e-books). As one writer puts it: "It was a clear message to both the digital and traditional book publishing industry that Amazon was investing in their vision of the future of book publishing."[13] Intrigued by the possibilities, Treanor[13] tracked a series of eighteen acquisitions by Amazon, beginning as early as 1998 through 2008. These acquisitions were investment in technology infrastructure to support their book business and other industries as well.

In 2009, Amazon began its move into the publishing business by establishing AmazonEncore to allow writers to digitally self-publish their work and to republish books that were no longer in print. This move was nonthreatening to the major book publishers, who also viewed Amazon as one of their major customers. In 2010, Amazon introduced AmazonCrossing to publish English translations of foreign-language books. It has since added several other imprints devoted to mysteries, thrillers, romance, and science fiction.[14]

Amazon brought their venture into publishing to a boil by hiring Laurence J. Kirshbaum to head up Amazon Publishing. Kirshbaum was once head of Time Warner Book Group and a well-accepted member of the publishing community. His movement to Amazon Publishing has, in the eyes of the major publishers, transformed him into a traitor and bitter enemy. In addition, Amazon encourages writers to publish their writing within Amazon and has recently announced the addition of several well-known authors to Amazon's stable of writers.[14]

It appears that Amazon is intent on becoming a major factor in the publishing world, moving them further up the supply chain, from retailer to distributor to manufacturer and even to originator, as they encourage authors to publish for Amazon. The goal is to connect the customer with what they want, regardless of the medium used to move the content to them.[15]

Summary

Figure A5a.1 is a simplified model of Amazon's business segments. The horizontal axis show the major sections of a supply chain as products and services move from

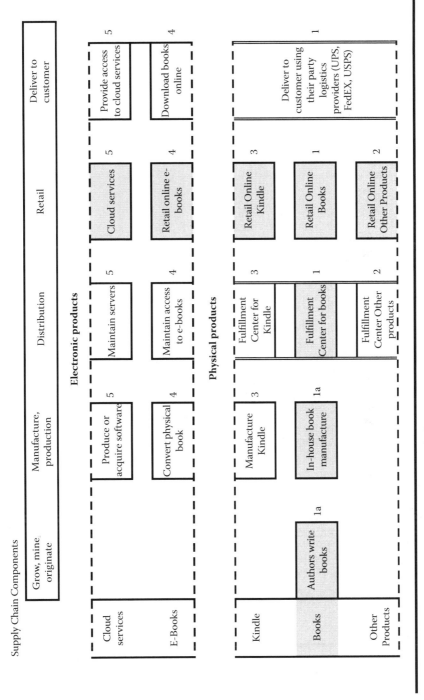

Figure A5a.1 Evolution of Amazon supply chains.

the point of origination to the manufacturer, to the distributor, to the retailer, and finally to the ultimate consumer.

The vertical axis shows the multiple business segments. They are numbered to indicate the major business segments that include the following:

1. **Books.** This is the main business in which Amazon started and is still the backbone of their company. As described earlier, they moved quickly from being purely an online seller of books to developing their own distribution (fulfillment) function. The 1a distinction is to show that the book publishing business has been developed in recent years but is a natural extension of the book supply chain.
2. **Other merchandise.** It soon became apparent that the Amazon website that sold books could also sell other merchandise, especially those products that were well known to consumers, such as clothes and small appliances. As consumers became more knowledgeable and more trusting in Amazon, the variety of merchandise expanded rapidly. As with books, Amazon was able to use its fulfillment centers to stock and deliver the merchandise to consumers quickly and reliably.
3. **Kindle.** Once the books and merchandise supply chains were operating well, the company recognized that physical books would eventually be replaced, at least in part, by electronic books, and that readers would need a device on which to download and store books for reading as they desired. The Kindle was first introduced as a reader and then was further developed into a tablet computer—the Kindle Fire. As described earlier, this also enabled Amazon to enter the manufacturing part of the book supply chain. Adding in-house authors moved Amazon back into the originating portion of the supply chain.
4. **E-books.** Acquiring and delivering physical products has enabled Amazon to grow into an international business with markets and facilities throughout the world. Their knowledge of electronic communications also enabled them to quickly become a major supplier of e-books to readers. Electronic downloads were also something that consumers were becoming more comfortable with, especially with the growth in music downloads. It was a small step to downloading text materials and Amazon was quick to grasp the potential in this area.
5. **Cloud computing services.** Ever aware of potential new businesses, Amazon didn't miss the interest in cloud computing services. The company is gearing up its portfolio of services and is expecting to become a major player in the game.

Amazon demonstrates both of the characteristics we describe for UPS—they use techniques developed in manufacturing to improve their operations, and they are expanding their business segments upstream along the supply chain to gain greater control over each facet of the business.

Amazon is expanding horizontally as a retailer and has become a competitor in a wide variety of industries. One of their more recent entries is in the home improvement field, where they are rapidly becoming a competitor to Home Depot, Lowe's, and Sears as a seller of tools and appliances.

Amazon is also extending their retail prowess back along the supply chain in books. It appears likely they could integrate backward in other product areas to secure a more dependable supply chain. To date, they have been successful in pursuing these aggressive expansion strategies.

References

1. Amazon, Annual report, 2011.
2. Amazon, Earth's Biggest Selection (full line of products and services), http://www.amazon.com/gp/site-directory/ref=sa_menu_fullstore, 2012 (accessed July 27, 2012).
3. Wikipedia 2012.
4. The future of publishing: E-publish or perish, *The Economist*, 395 (April 3, 2010), retrieved from http://www.economist.com/node/15819008.
5. Catan, T., Trachtenberg, J.A., and Bray, C., U.S. alleges e-book scheme, *Wall Street Journal*, http://online.wsj.com/article/SB100014240527023044446045773375730546151152.html, April 11, 2012 (accessed July 30, 2012).
6. Bezos, J. P. 2012. Letter to the Shareholders, 2011 Amazon Annual Report.
7. Denning, S., Why Amazon can't make a Kindle in the USA, *Forbes.com*, http://www.forbes.com/sites/stevedenning/2011/08/20/does-it-really-matter-that-amazon-cant-manufacture-a-kindle-in-the-usa/, Aug. 12, 2011 (accessed July 27, 2012).
8. Bonney, J., Amazon's supply chain: Delivering clicks and bricks. *Journal of Commerce*, Jan. 30, 2012.
9. Kucers, D., Amazon wrings profit from fulfillment as speaking soars, *Tech, Bloomberg Business Week*, http://www.bloomberg.com/news/2012-03-21/amazon-wrings-profit-from-fulfillment-as-spending-soars-tech.html, May 21, 2012 (accessed July 25, 2012).
10. Johnson, M.E., Virtualizing fulfillment operations—Physical clouds allow Amazon to scale, *Center for Digital Strategies* Blog, http://www.tuck.dartmouth.edu/digital/about/blog/detail/virtualizing-fulfillment-operations, 2011 (accessed June 29, 2012).
11. Overview of Amazon 2012. http://phx.corporate-ir.net/phoenix.zhtml?c=176060&p=irol-mediaKit (accessed June 29, 2012).
12. Alter, A., Your e-book is reading you; digital-book publishers and retailers now know more about their readers than ever before. How that's changing the experience of reading, *Wall Street Journal (Online)*, June 29, 2012.
13. Treanor, T., Amazon: Love them? Hate them? Let's follow the money. *Publishing Research Quarterly*, 26(2), 119–128, 2010, doi: 10.1007/s12109-010-9162-7.
14. Stone, B., Amazon's hit man. *Business Week*, 1 (Jan. 30, 2012). Retrieved from http://www.businessweek.com/magazine/amazons-hit-man-01252012.html.
15. Welly, K. Amazon knows the medium doesn't matter. *EContent*, 34(8), 8-9, 12, 2011. Retrieved from http://www.econtentmag.com/Articles/News/News-Feature/Amazon-Knows-the-Medium-Doesnt-Matter-78026.htm.

Appendix 5B: United Parcel Service (UPS)

> Over the past 100 years, UPS has become an expert in transformation,
> growing from a small messenger company to a leading provider of air,
> ocean, ground, and electronic services. Ever true to its humble origins,
> the company maintains its reputation for integrity, reliability, employee
> ownership, and customer service. For UPS, the future promises even
> more accomplishments as the next chapter in the company's history
> is written.[1]

UPS is a company that has used manufacturing techniques, especially those associated with industrial engineering, to improve its service processes. This strategy will be described in the first part of this review of the company.

In addition, UPS has taken advantage of its position as a third-party logistics provider in global supply chains to extend its service segments along the supply chain in a planned and rational manner. This will be described in the latter part of this study.

A Brief History of the Company

UPS started as the American Messenger Company in Seattle, Washington, in 1907.[1] Jim Casey and his partner Claude Ryan ran the service from a small office located under the sidewalk. Messengers ran errands and delivered packages on foot or used bicycles for longer trips. From the beginning, the messengers were well groomed, honest, and polite, characteristics that Casey insisted on.

As improvements in automobiles and telephones reduced the demand for messenger service, the company shifted its focus to package delivery for retail stores. The company acquired its first delivery vehicle, a Model T Ford, with its new name on the side—Merchants Parcel Delivery—to reflect its new mission. In 1916 Charlie Soderstrom joined the company, bringing added expertise, and he is credited with settling on the color brown for the growing fleet of delivery vehicles.

By 1919, the company expanded outside of Seattle to Oakland, California, and assumed its present name, United Parcel Service. They also added common carrier services that enabled it to establish rates that were comparable to those of parcel post. This facilitated their move into the first of many technology innovations that would become the hallmark of the company—a conveyor belt system for handling packages. Jim Casey decided early that industrial engineering techniques developed in manufacturing industries could be adapted to improve package delivery services.

In 1930, UPS entered the East Coast market by consolidating the deliveries of several large department stores in New York City and Newark, New Jersey. In the 1940s and 1950s, the U.S. population began to move to the suburbs and shop at large new shopping centers with ample parking for automobiles. This caused a decline in package deliveries from retail stores to consumers and UPS made another major shift in its emphasis—to that of a common carrier delivering packages between all customers, both private and commercial.

Over the period from about 1950 through the 1970s, UPS fought many battles with the Interstate Commerce Commission to obtain the rights to deliver packages in all states and between states. In 1975, they became the first package delivery company to serve every address in the forty-eight contiguous United States. During this period, they also resumed air delivery service by shipping packages in the cargo sections of regularly scheduled airlines.

By 1981, the demand for air parcel delivery increased, both domestic and foreign. In 1985, UPS began providing air package and document service throughout the United States and six European nations. In 1988, UPS received authorization from the Federal Aviation Administration (FAA) to start its own airline. UPS Airlines became one of the fastest-growing airlines in FAA history and today is one of the largest airlines in the United States. As with their ground delivery systems, the airline employs some of the latest technology for planning, scheduling, and load handling.

Today, UPS operates an international small package and document network in more than 200 countries and territories, spanning both the Atlantic and Pacific oceans. They continue Casey's early interest in continuous improvement by adopting technology throughout their operation. They introduced the handheld Delivery Information Acquisition Device (DIAD) in 1992 to track all ground packages and subsequently to allow customers to track their package status.

By the late 1990s, UPS entered another transition phase. They began to branch out into a new line of expanded supply chain services. They built on their expertise in global deliveries to become a third-party logistics provider to companies that were extending their supply chains through outsourcing and entering foreign markets. In 1995, UPS formed the UPS Logistics Group to provide global supply chain management solutions and consulting services. In 1998, UPS Capital was started to provide integrated financial products and services to help companies expand their business.

The 2000s saw the extension into UPS Supply Chain Solutions to provide an even greater array of global services. In 2001, UPS acquired Mail Boxes Etc. Inc., the world's largest franchisor of retail shipping, postal, and business service centers. UPS continues to expand worldwide, and in 2005 launched the first nonstop delivery service between the United States and Guangzhou, China. UPS, through its joint venture partner in China, now services twenty-three cities and more than 80 percent of the country's international trade.

In addition to internal growth, UPS has used acquisitions to expand into express and air cargo carrier (1999 acquisition of Challenge Air), heavy air freight (acquisition of Menlo Worldwide Forwarding in 2004), and ground freight services in North America (acquisition of Overnite in 2005).

Ever mindful of its responsibility to society and the environment, UPS operates the world's largest fleet of compressed natural gas (CNG) vehicles. Although UPS has long practiced environmentally conscious innovations, it recognized the need to document its practices. In 2003, UPS issued its first Corporate Sustainability report, now an annual report.[1]

UPS is a service company, but today its services extend well beyond having courteous drivers, driving brown trucks, and delivering packages on time.

The Crusade for Continual Improvement

UPS's success involved the use of techniques imported from manufacturing. We will expand on them in the following sections.

Long History of Embedded Culture of Efficiency

Jim Casey, one of the original founders of UPS, and for a long time its CEO (1907–1962), instilled a culture focused on customer service. He stressed honesty, fast and accurate deliveries, and courtesy by the messengers. He believed in efficiency in every phase of the operation, and this led to the extensive use of industrial engineering techniques to plan, monitor, and measure every step in the delivery process. Casey was relentless in his drive to produce a high-quality service at the lowest cost.[2]

Casey was also relentless in his pursuit of new business and was never reluctant to try new ventures. When deliveries from department stores faded with the post-World War II movement to the suburbs, he turned to delivering packages to department stores. He also spent a great deal of time in battling the U.S. Post Office, and later the U.S. Postal Service, to gain access to the delivery of packages throughout the United States.

Ever alert to new technology delivery methods, he moved from delivering on foot to bicycles, to motorcycles, to automobiles, to trucks, and to airplanes. This was all done in the pursuit of better delivery methods and expanded geographic markets.

Above all, he instilled in the company a culture of frugality and hard work. While he wanted employees to work hard, he also rewarded them with above-average wages, and UPS was one of the first companies to introduce employee stock purchase plans.[2]

Heavy Use of Industrial Engineering Techniques

Casey was always interested in finding ways to improve the operation of UPS. In most cases, there was no model for the company to follow in designing its systems and infrastructure, at least within the delivery industry. So Casey turned to other industries and eventually settled on the need to use the industrial engineering pioneered by Frederick Taylor. He hired Russel Havighorst in 1923 as the company's first industrial engineer and the use of industrial engineering techniques, most developed in manufacturing industries, have been very successfully applied to the services areas of UPS. That industrial engineers enjoy widespread acceptance is evidenced by the election of Mike Eskew as CEO, originally hired as an industrial engineer. Casey coined the term *constructive dissatisfaction* to encourage managers to be constantly on the lookout for improvement opportunities and to be prepared to change the direction of the company when needed.[2]

Progressive Use of Technology

UPS is at the leading edge in the use of technology to help move packages from sender to receiver. The following are some of the areas in which their technology is at the leading edge:

Sorting and loading packages onto trucks. At the Worldport air hub in Louisville, Kentucky, they process 16 million packages a day through the use of automated conveyor lines that move packages through at blinding speed and shunt them into delivery chutes for loading onto trucks. The trucks are loaded in a LIFO sequence, so that items come off the truck in the correct order for delivery. The trucks are also loaded to maximize the space available.[2] They have recently introduced the use of portable scanners worn on a finger and a small terminal worn on the wrist or hip, which enables the employee to more quickly read one- and two-dimensional bar codes and to accelerate the transfer of package tracking and processing.[3]

Routing trucks. Delivery routes are planned and scheduled to minimize distances traveled and to reach the customer consistently on time. Routes are planned to avoid left-hand turns to reduce wait times and reduce the chance for accidents.[2]

Fleet planning. Industrial engineers spend time in deciding the types of vehicles best suited for the needed operations. Some are conventional "brown" vans and others are semis (feeders) for larger deliveries. They also recognize the need for sustainability and have been among the first to use hybrid, all electric, natural gas, and even hydrogen cell fuels for experimental vans.[2]

Package tracking. The introduction of the DIAD ushered in a new phase of customer service. Now customers can track their package throughout its delivery cycle. In 2012, UPS announced the fifth generation of their Delivery Information Acquisition Device (DIAD V), based on Gobi radio technology, which allows instant switching of cellular carriers if one carrier's signal is lost, ensuring the device will stay connected to the UPS network at all times. The device includes a color camera that can be used for proof of delivery, and when not in use, as a training device for 90,000 drivers worldwide. In addition to facilitating tracking of 32 million online tracking requests daily, it enables UPS operators to forward customer requests to change package delivery instructions to the UPS driver while on the road.[4]

Airplane scheduling and operation. UPS has one of the largest airlines in the world. They plan and schedule air flights as carefully as they do trucks.

Always striving for faster and safer flights, they have developed a landing system that will enable them to land planes closer together. Initially competitive with conventional landing systems, UPS is gaining acceptance of it from the FAA. There is also extreme attention to getting flights loaded and off to meet the delivery time commitments.[2]

Other Keys to Success

While the constant drive to improve efficiency and effectiveness has been a hallmark of UPS, there are other unique qualities that have also been instrumental in the company's success.

Stability at the CEO Level (Promote from Within)

One of the strengths of UPS is its stability at the CEO level. All CEOs have been promoted from within. This not only sustains the culture but also minimizes the discontinuities in management decision making. UPS has had nine CEOs in the fifty years since Mr. Casey resigned as CEO. They include:

1962–1972	George D. Smith
1972–1972	Paul Oberkotter
1973–1980	Harold Oberkotter
1980–1984	George Lamb
1984–1990	John W. Rogers
1990–1997	Kent C. "Oz" Nelson
1997–2001	James P. Kelly
2002–2007	Michael L. "Mike" Eskew
2008–present	D. Scott Davis

Tenures as CEO of five to seven years meant that each was more than a caretaker and could add developments of his own creation. It also meant that each recognized that his time as CEO was limited and that part of his responsibility was to help select and prepare his successor for promotion.

Emphasis on Drivers as the Key Customer Interface

The driver of the UPS trucks is a key interface with the customer. The company stresses efficiency; however, drivers are expected to never skimp on service to the customer. They are dependable and persistent in getting the package delivered to the correct person or representative, and are to be pleasant and helpful in doing it.

Few, if any, drivers start out as drivers. They go through extensive training and work at other jobs, such as sorting and loading, before being appointed to drive. One of the newer training approaches is to use simulated conditions to teach proper lifting methods and other safety procedures subsequently encountered in making deliveries. Drivers are expected to operate safely and are recognized for having accident-free service. Drivers work long hours until all deliveries are made.[5]

Concern for Employee Welfare

UPS is a company with low employee turnover. Jim Casey was demanding, but he always was very much aware of the need to select, train, and nurture employees. Two practices that stand out are promoting from within and making stock ownership available to all employees. All CEOs have come from within the company.

Managers have long had an ownership position in the company because part of their compensation is in stock. All employees have had the opportunity to buy stock since the company had its initial IPO in 1999, the largest public offering of stock at the time.[2]

Relationship with Unions (Teamsters)

Another interesting feature of the driver relationship is that the Teamsters have long been the union representative for UPS. Except for one extended strike in 1997, UPS has maintained an effective relationship with one of the most noted unions since 1939. One of the reasons for the good working relationship with the union is that most members of management had been members of the Teamsters.

The strike in 1997 was over the use of part-time employees and pensions. The concern in the 1990s was that many companies were shifting full-time jobs to contractors or part-time employees. The Teamsters wanted to shift from a multi-employer pension fund to a newly administered, joint UPS-Teamster pension fund. The strike was settled and UPS quickly regained its lost business.[2]

Sustainability

UPS places a great deal of emphasis on their sustainability efforts. It is at the forefront in the use of low-emission delivery vehicles. They reduce fuel consumption by designing delivery routes that minimize distances traveled and the need for backtracking on missed deliveries. This practice applies not only for ground delivery vehicles but also for aircraft.

UPS continues its expansion and innovations today, driven by the values established by its founder, Jim Casey. In their search for the answer to UPS's success, Brewster and Dalzell provide the following insight:

> So you go to UPS corporate headquarters in Atlanta. Mike Eskew, UPS's chairman and chief executive officer, sits in the company cafeteria and sips from his mug as he talks about postal codes in Germany, time studies in Bloomington, and overflow volumes in Pittsburgh.

"You know, going all the way back to our founder, Jim Casey, we have this extraordinary culture, this culture of constructive dissatisfaction," Eskew says. "Jim Casey is still a strong presence here, because the things he said are true even today." Now Eskew is talking about—ghosts?

No. You realize that he is saying that maybe UPS's century-long success is about something more than driving trucks and sorting packages and wiring mainframes and flying airplanes. Maybe it's also about the simple values that Casey embraced, about a singular culture of never being satisfied, about the discipline to execute, about the willingness of tens of thousands of like-minded people to pull together and transform an organization. Maybe it's about—surprise!—managing large-scale change.[2]

This heavy emphasis on change, both incremental and strategic, is part of what drives UPS to aggressively enter new types of services, although they are careful to make sure that there are solid connections between their core business and the new service applications. The next section describes the evolution of UPS from a package deliverer to a supply chain integrator.

Evolution along the Supply Chain

UPS has long been an integral part of supply chains. They provide key links in moving products along the supply chain from points of origin to the ultimate consumer. This strategic position has enabled them to expand their services along the supply chain upstream toward the source of products and downstream toward the customer.

Heavy Focus on Customer

From the beginning, Jim Casey placed great importance on providing the utmost service to the company's customers. This included prompt and careful delivery of packages by honest, courteous, and well-groomed delivery persons.

This focus on the customer continues today as UPS continually stresses reliable service, coupled with constant feedback to the customer about its package location and expected time of arrival. UPS is also looking for ways to customize their services to add value for their customers.

Movement Upstream in Supply Chain

In 2004, UPS began repairing Toshiba laptops. UPS would pick up the laptop to be repaired and move it to the central UPS hub in Louisville. There, it would be repaired and returned to the customer by UPS. This eliminated the need to return the laptop to the Toshiba factory and reduced the repair time from weeks to days. UPS had already been servicing printers for Lexmark and Hewlett-Packard. This service falls under the Supply Chain Solutions segment of UPS which works with clients to manage inventory, ordering, customs processes, and disposal of unwanted electronics.[6]

Movement Downstream in Supply Chain—Retail Stores

In 2001, UPS entered the retail business by acquiring Mail Boxes Etc., Inc., the world's largest franchisor of retail shipping, postal, and business service centers. By 2003, there were over 3,000 locations throughout the United States.[1]

Provides Integrated Logistics Services—Domestic and Global

As supply chains have become more global in recent years, especially as a result of the outsourcing movement, the logistics function has increased in importance. Logistics has not been a major consideration for all companies, either in manufacturing or in retail. Consequently, they are turning to third-party logistics providers (3PL), such as UPS, to either assume the delivery responsibility or assist in the management of the logistics function within the company. Supply chains in the future will likely have greater need for "maestros," or third parties to coordinate all, or parts, of the supply chain.[7]

In addition to knowing how to plan and schedule, UPS is becoming more important with their understanding of how to do business in over 200 countries and territories throughout the world. They know how to get products through customs and on their way to the customer with a minimum of delay and interruption.

Business Segments

UPS has three major service segments: U.S. Domestic Package, International Package, and Supply Chain Solutions. Supply Chain Solutions is further divided into the following components: freight forwarding, customs brokerage, logistics and distribution, UPS Freight, and UPS Capital. Table A5b.1 shows the most recent five-year revenues and operating income for each of these service segments. Table A5b.2 provides a breakdown of the revenues for each segment for the year of 2011.

Global Small Package

UPS provides domestic delivery services within fifty-six countries and export services to more than 220 countries and territories around the world. They offer same-day pickups at over 150,000 domestic and international entry points including 40,000 drop boxes, 1,000 customer centers, 4,700 independently owned and operated locations of the UPS Store worldwide, 13,000 authorized shipping outlets

Table A5b.1   Summary of Revenues and Income by Business Segment

Business Summary

United Parcel Service is engaged in the field of transportation services, primarily domestic and international letter and package delivery. Through its Supply Chain & Freight subsidiaries, Co. is also a provider of transportation, logistics, and financial services. Co. serves the global market for logistics services, which include transportation, distribution, forwarding, ground, ocean and air freight, brokerage, and financing. Co. has three reportable segments: U.S. Domestic Package; International Package; and Supply Chain & Freight. Co. provides domestic delivery services within 56 countries and export services to more than 220 countries and territories around the world.

	2007	2008	2009	2010	2011
Business Segments	*Revenues (Millions of Dollars)*				
U.S. Domestic Package	30,985	31,278	28,158	29,742	31,717
International Package	10,281	11,293	9,699	11,133	12,249
Supply Chain & Freight	8,426	8,915	7,440	8,670	9,139
Total	49,692	51,486	45,297	49,545	53,105
Business Segments	*Operating Income*				
U.S. Domestic Package	(1,531)	3,907	2,138	3,373	3,764
International Package	1,831	1,580	1,367	1,904	1,709
Supply Chain & Freight	278	(105)	296	597	607
Total	578	5,382	3,801	5,874	6,080
Geographic Analysis	*Revenues (Millions of Dollars)*				
United States	37,741	38,553	34,375	36,795	39,347
International	11,951	12,933	10,922	12,750	13,758
Total	49,692	51,486	45,297	49,545	53,105

and commercial counters, 6,300 alliance locations, and 86,300 UPS drivers who can accept packages provided to them.

The growth of online shopping has generated a higher rate of returned goods. To address this need, UPS offers a returns service that is available in over 100 countries. UPS drivers can simplify product exchanges by delivering a replacement item at the same time they are collecting the item to be returned.

Table A5b.2   UPS Revenues for 2011 by Business Segment

Segment	Revenues (in Millions of $)	% of Total
U.S. Domestic Package		
Ground	22,189	42%
Next day air	6,229	12%
Deferred	3,299	6%
Total U.S.	$31,717	60%
International Package		
Export	9,056	17%
Domestic	2,628	5%
Cargo	565	1%
Total Other Countries	$12,249	23%
Forwarding and Logistics	$6,103	11%
Freight	$2,563	5%
Other	$473	1%
Total	$53,105	100%

Source: From United Parcel Service, Quick report, *Hoovers*, http://subscriber.hoovers.com/H/company360/quickReport.html?companyId=40483000000000, 2012 (accessed July 18, 2012).

UPS operates a global ground fleet of over 100,000 vehicles and a global air fleet of over 500 aircraft, making them one of the largest airlines in the world. Their central hub in the United States is in Louisville, Kentucky, with regional U.S. air hubs in Hartford, Connecticut; Ontario, California; Philadelphia, Pennsylvania; and Rockford, Illinois. The Louisville hub has a capacity of 416,000 packages per hour.[8]

Domestically, UPS delivers over 11 million ground packages on time every day. Customers can choose same-day, next-day, two-day, and three-day delivery alternatives. Many of these services offer a definite cutoff time for delivery (e.g., by 8:30 a.m.).[8]

UPS provides small package operations in Europe, Asia, Canada, and Latin America and offers similar service to that in the United States. The largest international air hub is in Cologne, Germany, with other regional international hubs in Miami, Florida; Canada; Hong Kong; Singapore; Taiwan; and China. The Cologne hub is being expanded to a capacity of 190,000 packages per hour, with a scheduled completion slated for 2013.[8]

Supply Chain Solutions

In 2001, Mike Eskew, newly appointed CEO, described the strategy that spawned SCS this way:

> Our strategy is focused on complementing and growing our core package delivery business with new, innovative supply-chain solutions. We'll accomplish that by inserting our unmatched capabilities in managing the flow of goods, information and funds across our customers' supply chains. Over the past year, our strategy has been manifested by the launch or acquisition of new logistics, freight forwarding, customs brokerage, financial services, mail, and consulting businesses that expand the scope and power of our distribution and supply chain solutions.[5]

UPS manages supply chains in over 200 countries and territories, with approximately 35 million square feet of distribution space worldwide.[9] The Supply Chain Solutions (SCS) group works with companies to manage inventory, ordering, and disposing of unwanted electronics.[6] The role of the SCS group is to assist customers with logistics, global freight, mail services, consulting, and financial services. As global competition and outsourcing caused supply chains to be extended throughout the world, companies needed help in linking the pieces together in a rational, effective, and efficient way; that is the goal of UPS's SCS.[5]

The services include the following:

- Freight forwarding. UPS is the second-largest U.S. domestic air freight carrier and among the top six international air freight forwarders globally.
- Customs Brokerage. UPS is among the world's largest customs brokers by both the number of shipments processed annually and the number of dedicated brokerage employees worldwide.
- Distribution Services: Comprehensive distribution services are provided through a global network of distribution centers that manage the flow of goods from receiving to storage and order processing to shipment, allowing companies to save time and money by minimizing their capital investment and positioning products closer to their customers.
- Post Sales: Services support goods after they have been delivered or installed in the field. The four core service offerings include (1) Critical Parts Fulfillment; (2) Reverse Logistics; (3) Test, Repair, and Refurbish; and (4) Network and Parts Planning. There are over 600 stocking locations to ensure the right type and quantity of customers' stock to meet the needs of their customers.
- UPS Mail Innovations: Offers an efficient, cost-effective method for sending lightweight parcels and flat mail to global addresses from the United States.
- UPS Freight: Offers regional, inter-regional and long-haul less-than-truckload ("LTL") services, as well as full truckload services, in all 50 states, Canada, Puerto Rico, Guam, the U.S. Virgin Islands, and Mexico.[9]

UPS CrossBorder Connect, a ground freight service between the United States and Mexico, is designed to address heavyweight freight supply chain challenges for companies investing in cross-border trade. This is in response to the nearsourcing movement in which companies are deciding that faster response times and reduced product movement risks can result from moving production closer to consumption points.[10]

UPS operates pharmaceutical distribution centers, which involves temperature control, regulation, and liability issues. It also capitalizes on UPS's ability to track and redirect shipments, when necessary.[11] They continue to expand this business by investing in megawarehouses to service multiple pharmaceutical companies with freezers for medicines and high-security vaults for controlled substances.[12]

One of the new services added during 2011 for the fast-growing business-to-consumer (B2C) market include UPS My Choice[SM], which allows the receivers of UPS shipments to control aspects of their deliveries.[9]

In another extension of the reverse logistics service, UPS introduced UPS Returns® Exchange and UPS Returns® Pack and Collect. These two services provide retailers in North America and Europe with new and cost-efficient ways to process returns.

UPS provides a variety of supply chain solutions to some of the major businesses in the world, including the following:

- Ford Motor Company—UPS redesigned the entire delivery network to optimize the flow of vehicles from factory to dealer.[5]
- Harley-Davidson—To improve the distribution of parts and accessories to dealers, UPS helped optimize the entire transportation operation by having everything sent to one UPS cross-docking facility and shipped out immediately.[5]
- Royal Philips Electronics—UPS redesigned the service parts logistics for the medical systems.
- Nikon—UPS coordinated Nikon's manufacturing centers in Japan, Indonesia, and Korea with air and ocean freight and customs brokerage. The products are centralized at the UPS Supply Chain Solutions Logistics Center in Louisville where UPS employees prepare Nikon kits, adding accessories like batteries and chargers, then repackage them and deliver them throughout the United States.[5]
- Toshiba—UPS collects Toshiba laptops to be repaired from retail return centers and transports them to Louisville, where they are repaired by UPS employees and returned to the customer in days instead of weeks.[5]
- NASCAR—Agreement, first signed in 2007 and renewed in 2012, to be the sport's official logistics partner, providing critical pickup and delivery services for race teams, vendors, and suppliers.[13]
- Live Nation Entertainment—UPS helps the 250 artists represented by Live Nation entertainment to find more environmentally responsible ways to manage their concert tours. UPS first measures the CO_2 emissions for their existing

transportation methods and develops a customized plan to reduce emissions and waste while meeting the speed and reliability needed for the tour.[14]

■ 2012 London Olympics—UPS was the official logistics partner for the Olympics. They secured 80,000 square meters of warehousing space to meet the demands of moving 30 million items into the warehouse where they are unloaded and x-rayed for security purposes before sending them on to the thirty-seven competition venues. After the competitions, the items are returned to the warehouses for disposal.[15]

Supply Chain Capital

In business financing services, industrial manufacturers can have financial needs that are almost as complex as a modern assembly plant. They can rely on UPS Capital for products and services that can help them improve cash flow, protect goods, and accelerate and protect payments.

■ Flexible Parcel Insurance. Safeguard time-sensitive goods and other hard-to-value items in the event of loss, damage, or delay.
■ Receivables Management Solutions. Get cash for your invoices, reduce collection costs and hassles, and protect your business from customer credit defaults.
■ Cargo Insurance. Protect your freight shipments against loss or damage from the time they leave your facility until your client receives them.
■ UPS Capital Cargo Finance. Increase liquidity and secure working capital earlier in your supply chain by using your in-transit inventory as collateral.
■ Global Asset-Based Lending. Borrow operating capital against goods warehoused abroad or moving through your supply chain within the UPS network.
■ Export Credit Agency Financing. Expand overseas sales by extending credit to prospective buyers in emerging markets at rates and terms typically unavailable in their local markets.[16]

Key Financial Results

Has UPS been successful? Tables A5b.3 and A5b.4 show the financial results; they demonstrate that UPS is a financial success.

Summary

UPS has evolved from a deliverer of packages to a supply chain integrator. Figure A5b.1 shows the progression UPS has taken to extend its products and services along the supply chain. Their evolution has included the following major phases:

Table A5b.3   Revenues and Income for 2007–2011

As Reported Annual Income Statement (Millions of Dollars)					
Report Date (Year Ending December 31)	2007	2008	2009	2010	2011
Revenue	49,692	51,486	45,297	49,545	53,105
Total operating expense	49,114	46,104	41,496	43,671	47,025
Operating profit	578	5,382	3,801	5,874	6,080
Income before income taxes	431	5,015	3,366	5,523	5,776
Net income (loss)	382	3,003	2,152	3,488	3,804
Total number of employees	425,300	426,000	408,000	400,600	398,000

Source: Adapted from Mergent Online (2012).

Table A5b.4   Profitability Ratios for 2007–2011

Profitability Ratios	2007	2008	2009	2010	2011
ROA% (net)	1.06	8.45	6.75	10.65	11.14
ROE% (net)	2.76	31.59	29.87	44.69	50.67
ROI% (operating)	2.70	26.94	22.49	32.66	32.88
EBITDA margin%	1.84	14.12	12.27	15.48	14.89
Calculated tax rate%	11.37	40.12	36.07	36.85	34.14
Revenue per employee	$116,840	$120,529	$111,022	$123,677	$133,430

1. They started by delivering packages from retail stores to customers.
2. When that business declined, they added delivering packages from suppliers to retail stores.
3. By expanding to common carrier stature, they were able to deliver packages from business to business, or from business to consumer, thus completing linking together businesses along the supply chain.
4. As supply chains became longer and more complex, they began to offer distribution services to reduce handling, manage inventory, and facilitate prompt delivery to customers.
5. They initiated collection of returns from customers or retail stores and, in the case of Toshiba, they provided repair services to reduce waiting times for customers.

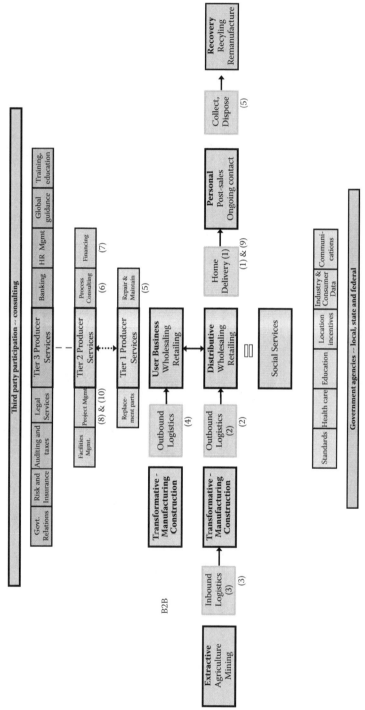

Figure A5b.1 Hierarchy of services provided to businesses.

6. Building on their supply chain expertise, they provide consulting services to design improvements in company supply chains through their Supply Chain Solutions unit.
7. To help businesses grow, they provide financial services.
8. They provide complete project management capability, as in designing and managing the logistics for the 2012 Olympics.
9. With the advent of online selling, UPS returned to its roots as a deliverer of purchases from online retailers to consumers.
10. UPS provides the logistics services to some unique businesses, such as moving bands from concert to concert in the most environmentally friendly way.

Each of these steps in the company's evolution is highlighted in Figure A5b.1. Table A5b.5 provides a comprehensive list of services developed during the company's history.

UPS is generally not competing with their customers when they enter new supply chain businesses; they are complementing their customers by helping them do their jobs better. Their success has resulted from their careful building of new services on already successful parts of their business.

UPS is building its logistics services throughout the supply chain. It is also using its experience and corporate knowledge to extend their services into Tier 2 services such as consulting, financing, and project management. Up to this point, they have been aggressive in building their logistics services along both B2B and B2C networks. As supply chains become more complex and more geographically dispersed, more manufacturing and retail businesses are outsourcing this function to third-party logistics providers (3PLs).

At this point, they have not entered the manufacturing business. This probably is because they anticipate that they can generate greater growth through their logistics services expansion. It also precludes them from competing with some of their customers.

UPS's success is built on a culture of moving quickly to service customer needs and adapting to changing conditions. The strong value system instilled under Jim Casey has been carried forward by CEOs who have all been longtime employees of UPS.

Table A5b.5   UPS Timeline of Services

Years	Services Provided	Added Observations
1907–1912	Messenger service—ran errands, delivered packages, and carried notes, baggage, and trays of food from restaurants.	James E. (Jim) Casey borrowed $100 from a friend and started the American Messenger Company of Seattle, making deliveries on foot or bicycle.
1913–1918	Focused on package deliveries for retail stores, as improvements such as the automobile and the telephone were causing a decline in the messenger business. Changed name to Merchants Parcel Delivery.	Began to use motorcycles for some deliveries. For about two years, the company's largest client was the United States Post Office. Began using consolidated deliveries—combining packages for the same area. Acquired its first delivery car, a Ford Model T.
1919–1930	The differentiating features of common carrier service included automatic daily pickup calls, acceptance of checks made out to the shipper in payment of CODs, additional delivery attempts, automatic return of undeliverables, and streamlined documentation with weekly billing.	In 1919, made its first expansion beyond Seattle to Oakland, California, and adopted its present name, United Parcel Service. In 1924, debuted one of the technological innovations that would shape its future: the first conveyor belt system for handling packages.
1930–1952	Expanded to East Coast; began retail store delivery operations in New York and Newark, NJ. WWII reduced business because of fuel and rubber shortages. Expanded common carrier rights between all customers, in competition with the U.S. Postal Service.	After WWII, the population began extending to the suburbs with malls and contract services when retail stores became limited. Therefore the company began to look for new opportunities to expand its core business. In 1953, Chicago became the first city outside of California with common carrier services.
1953–1974	Resurrected air service, originally started in 1929 but abandoned during the Great Depression. Offered two-day service to major cities on the East and West Coasts.	UPS packages flew in the cargo holds of regularly scheduled airlines. Called Blue Label Air, the service grew, until by 1978 it was available in every state, including Alaska and Hawaii.

Table A5b.5 (continued)   UPS Timeline of Services

Years	Services Provided	Added Observations
1975–1980	Forged "Golden Link," became the first package delivery company to service every address in the 48 contiguous United States. Blue Label services expanded to all 50 U.S. states.	In the 1950s, UPS was restricted from operating in many parts of the country by the Interstate Commerce Commission. Finally, in 1975, UPS was authorized to operate its delivery service throughout the United States.
1981–1988	UPS Airlines. Demand for air parcel delivery increased and federal deregulation of the airline industry created new opportunities for UPS. UPS Airlines became the fastest-growing airline in FAA history. Started international air service between U.S. and six European countries.	In 1981, purchased first aircraft for use in air delivery service. In 1988, received authorization from the FAA to operate its own aircraft, thereby officially becoming an airline. Began operations from the Louisville air hub. Featured some of the most advanced information systems in the world, like the Computerized Operations Monitoring, Planning and Scheduling System (COMPASS).
1988–1990	Entered the international shipping market in earnest, establishing a presence in a growing number of countries and territories in the Americas, Eastern and Western Europe, the Middle East, Africa, and the Pacific Rim.	Operated an international small package and document network in more than 185 countries and territories. In 1990, first scheduled flight to Asia on UPS aircraft.
1991–1994	Embracing Technology. By 1993, UPS was delivering 11.5 million packages and documents a day. Technology spanned from small handheld devices to specially designed package delivery vehicles, to global computer and communications systems. In 1993, established UPS Logistics Group to provide supply chain management solutions.	The handheld Delivery Information Acquisition Device (DAID) tracks packages, allowed drivers to stay in constant contact with their headquarters, keeping abreast of changing pickup schedules, traffic patterns, and other important messages. In 1992, electronic tracking of packages began. In 1994 UPS.com went live.

continued

Table A5b.5 (continued)   UPS Timeline of Services

Years	Services Provided	Added Observations
1994–1999	Expanded Services. UPS began to branch out and focus on a new channel, services. Expanded its company's expertise in shipping and tracking to become an enabler of global commerce and a facilitator of the three flows that make up commerce: goods, information, and capital.	Began strategically acquiring existing companies and creating new kinds of companies that didn't previously exist. By focusing on unique supply-chain solutions, UPS allowed its customers to better service their own customers and focus on core competencies. In 1988, founded UPS Capital to provide a comprehensive menu of integrated financial products and services that enable companies to grow their business.
2000–2007	Global Commerce and Transformation. UPS Supply Chain Solutions provides logistics, global freight, financial, and mail services to enhance customers' business performance. In 2001, acquired Mail Boxes Etc., Inc., the world's largest franchiser of retail shipping, postal, and business service centers. In 2001, launched direct flights to China with China Express. In 2002, began operations at new intra-hub located in Philippines.	Expanded Worldport, the air hub in Louisville, Kentucky, as well as the European air hub in Cologne. In 2000, added capability to calculate rates and find transit times for shipments on any digital wireless device in the U.S. In 2003, issued the first Corporate Sustainability Report, highlighting the importance of balancing economic, social, and environmental objectives. Purchasing Menio Worldwide Forwarding in 2004 added heavy air freight capability, while the acquisition of Overnite in 2005 expanded the company's ground freight services in North America.

Source: From http://www.ups.com/content/corp/abouthistory/1929.html, accessed January 18, 2012.

References

1. UPS company history, *UPS.com*, http://www.ups.com/content/corp/abouthistory/ 1929.html, 2012 (accessed January 18, 2012).
2. Brewster, M. and Dalzell, F., *Driving Change, The UPS Approach to Business*, Hyperion, New York, 2007.
3. UPS pressroom, *UPS.com*, http://pressroom.ups.com/Press+Releases/Homepage+Press+ Releases/UPS+Deploys+New+Scanning+Device+in+its+Package+Sorting+Operation, 2012 (accessed September 7, 2012).
4. UPS equips drivers with networked handheld devices, *Material Handling & Logistics*, http://mhlnews.com/archive/ups-equips-drivers-networked-handheld-devices, Feb. 29, 2012.
5. Niemann, G., *Big Brown: The Untold Story of UPS*, Jossey-Bass, John Wiley & Sons, San Francisco, 2007.
6. James, G., The next delivery? Computer repairs by UPS, *CNN.com*, http://mondy.cnn. com/magazines/business2/business2_archive/2004/07/01/374848/index.htm, 2004 (accessed January 18, 2012).
7. Bitran, G.R., Gurumurthi, S., and Shiou, L.S., The need for third-party coordination in supply chain governance. *MIT Sloan Management Review*, 48(3), 30–37, 2007.
8. United Parcel Service, Quick report, *Hoovers*, http://subscriber.hoovers.com/H/ company360/quickReport.html?companyId=40483000000000, 2012 (accessed July 18, 2012).
9. Davis, D.S., Letter to the shareholders, *UPS 2011 Annual Report*, 2012.
10. UPS introduces cross-border services into Mexico, *Material Handling & Logistics*, http://mhlnews.com/archive/ups-introduces-cross-border-services-mexico, 2012 (accessed April 16, 2012).
11. Adamson, L., UPS poised to profit from recovery, *Institutional Investor*, Oct. 2009.
12. Levitz, J, and Martin, T. 2012. Business technology: UPS, other big shippers carve health-care niches. *Wall Street Journal*, Jun 28, p. B.4.
13. United Parcel Service, Inc., *MarketLine*, http://advantage.marketline.com/Product?pid =233ADDC2-32B5-4702-AD8D-SECA699536AA&View=History, 2012 (accessed July 7, 2012).
14. Dickens, M., Live concerts get greener with help from UPS and Live Nation Entertainment. *UPSide, the UPS Blog*, http://blog.ups.com/2011/05/24/live-concerts-get-greener-with-help-from-ups-and-live-nation-entertainment/, 2012 (accessed July 24, 2012).
15. Howells, R., The logistics of the Olympics is a marathon, not a sprint, Forbes, http:// www.forbes.com/sites/sap/2012/07/16/the-logistics-of-the-olympics-is-a-marathon-not-a-sprint/, 2012 (accessed July 16, 2012).
16. Industrial manufacturing, UPS.com, http://www.ups.com/content/us/en/bussol/ browse/industries/indust_manuf.html (accessed May 15, 2012).

Chapter 6

Extending Service Techniques to Manufacturing

Introduction

In Chapter 5, we described the movement of continuous improvement programs from the manufacturing sector to the services sector. In this chapter we describe the movement of continuous improvement programs from services to manufacturing. For a long time, services were considered a necessary by-product of the more important agriculture and manufacturing sectors of the economy. Many authors even claimed that true wealth could be created only through the production of goods. In recent years, however, there has been a change in this thinking. Today, most business authorities agree that services are an integral and increasingly important part of the global economy, at least in industrialized nations. In other words, services have become "the tail that wags the dog." Service businesses have been dependent on the manufacturing sector for its improvement ideas for a long time. However, a number of new concepts and techniques have been developed in the services sector and are being extended back into the manufacturing sector. Before we look at some of these improvement programs, we review the reasons that services have achieved their increased level of dominance.

The other main theme of this chapter is that manufacturing companies are adding services as part of their product portfolios. This strategy has been most successful when the services are a natural extension of the product, and less successful when manufacturing companies add services simply because it makes them look like a more profitable business. The successful application of services to manufacturing will be elaborated on in the case studies of General Electric and Hewlett-Packard at the end of this chapter.

The Rise of Services as a Part of the Economy

Services are sectors of the economy whose time has come. The movement from agriculture to manufacturing and then from manufacturing to services is a natural progression for most countries. In its simplest form, the progression moves from providing survival (food) to providing an increased standard of living (manufactured goods) to increased enjoyment (services). Automation and improved processes leading to significant increases in productivity reduced the need for workers in agriculture and provided the opportunity for increased manufacturing of goods that could further increase productivity (tools and machines) or provide a higher standard of living (enhanced transportation and appliances). As increased productivity reduced the need for manufacturing employees, many moved into the services area, some because it was an attractive change and others because they had no other alternative.

The Swing of Power from Manufacturing to Retail

The growth of services has also brought a shift in economic power. Until the latter part of the twentieth century, manufacturing companies tended to have leverage over service companies because of their size. Although manufacturers were the dominant businesses in the past, retailers (e.g., Walmart and Home Depot), software developers (e.g., Microsoft), telecommunications companies (e.g., AT&T), information sources (e.g., Google), and a host of social media sites have become more significant participants in the business world. Online retailers such as Amazon are rapidly gaining market share. Most large retailers have added websites to their traditional bricks-and-mortar locations. As retailers have increased in size and purchasing power, they have imposed service requirements on manufacturers. Retail clothing stores want the manufacturer to make the apparel "floor ready" by sorting by category and attaching prices and security tags. Grocery stores want service merchandisers in such areas as bakery goods or health and beauty aids to keep the store's inventory stocked with the right products and an attractive appearance. Restaurant chains want frequent and reliable merchandise deliveries to keep their inventories low and fresh. Home improvement chains want attractive packaging and liberal return policies, especially for seasonal goods.

In the United States, services account for over 75 percent of the workforce and 90 percent of the new jobs being created. A similar pattern exists in other industrialized countries. As a result, the focus of most companies, whether manufacturing or service, is on the customer. Improvement programs, even those designed to reduce costs or improve product quality, must consider the implications for the customer. As this occurs, some programs can help manufacturing companies provide better customer service; however, they may also increase the cost of providing the product/service bundle.

Maturity of the Customer as a Shopper

In the past, shoppers were less informed about the product or service offered by a company. As a result, they were often satisfied with a basic product with little in the way of a complementary service package. Shoppers today are more knowledgeable about most products thanks to the Internet, where they can make product and price comparisons. As they find an added feature on one product and another feature on a second product, they become more demanding of the product they buy. To enhance their products, manufacturers must often add services. As described in Chapter 3, customers not only are the recipients of goods and services but also may be involved in the design of the goods and services they buy.

Increasing Complexity of the Marketplace

The marketplace is changing. Customers have become more discriminating and desire greater variety in the products and services from which they can choose. Manufacturers have created greater variety in products; however, so have their competitors. To gain a competitive advantage, many companies are turning to a greater level of services. This adds complexity, causing the marketplace to take on a different configuration than in the past. Markets no longer distribute themselves in a normal curve; they tend to have at least two major segments—one with customers who continue to want low-cost products and services, and a second market segment with customers who want customized products and are willing to pay and wait for that customization. This is described later in this chapter.

Need for Manufacturing Companies to Add Services

Manufacturing companies are finding they need to add services to complement the goods they produce. Customers are demanding services to accompany the products they buy. For personal computers, customers want help desks and security software downloads. When restaurants order after-dinner mints, they want the manufacturer to put their names on the package. For home assembly of furniture, buyers expect the directions will be complete and easily understood. James Brian Quinn was one of the early writers to predict that successful companies in the future would gain their competitive advantage from having service-based "core competencies" to complement their superior products. He further predicted that physical facilities—manufacturing plants and other capital-intensive facilities—would not likely be enough to provide the competitive edge companies would need. Quinn believed that companies would have to become intelligent enterprises to derive sustainable advantage from knowledge and service-based activities that would leverage their intellectual assets.[1]

One way manufacturers can provide an almost seamless link between their product and the accompanying services is by embedding the service as part of an electronic network that can detect potential problems, and in some cases take automatic action to prevent or mitigate the problem. Several strategies can be employed in this approach:

- *Embedded innovator.* This is the most product-centric model. Customers will still perceive the physical product as the source of primary value, and they will expect to continue receiving the support services they have in the past (installation, warranties, and maintenance contracts). Example: Heidelberg, a maker of high-end printing presses, is an embedded innovator.
- *Solutionist.* A single product is still the dominant gateway to a business opportunity, but the scope of high-value activities with the product is broader. Example: GE with its MRI scanner illustrates the solutionist.
- *Aggregator.* A number of entities are involved in providing a collection of services, and the aggregator is the firm that controls the actual data collection and central-processing power. Aggregators are primarily product companies that do not integrate all aspects of the services related to their product. Example: Gardner Denver makes air compressors and monitors the compressed air processes to provide information to manufacturers of air coolers, filters, and dryers.
- *Synergists.* Firms that participate in the system by providing a specific device or service to the system. Example: Phillips provides lighting ballasts and controls.[2]

All of these added features require manufacturers to consider their customers and how best to please them. These are concerns that service companies have always dealt with. Although this extension of products and services is innovative, it also increases the complexity of a business, perhaps to the point where it is difficult to separate innovation from complexity.[3]

Jay Galbraith cautions that blending products and services is not easy. He proposes that companies need to change from offering a product to providing solutions. When they do attempt this change, they may have trouble because they have not restructured their organizations to become customer-centric.[4]

Another possible complication in adding services to a goods-producing company is in the decision-making process. Whereas decisions for a goods-producing company can usually be based on quantitative measures, providing customer service often requires a trade-off between increased quantitative costs and less tangible benefits.[5]

Move from Product-Centric to Customer-Centric

Another factor moving service industry thinking back to manufacturers is the change in orientation from a product-centric to a customer-centric environment.

This change is more than just a need to beef up the product with selected services or to involve the customer in some of the discussions on product design. It is a new way of thinking about the goals and strategies of a manufacturing company. Barbara Bund describes this as an "outside-in" way of looking at the business, meaning through the eyes of the customer.[6] Stephen Haeckel describes it as a "sense and respond" approach rather than a "make and sell" approach.[7]

Table 6.1 compares the characteristics of product-centric companies with customer-centric companies. Many companies are experiencing difficulty in moving through this transition. It is both an organizational problem and a people problem. How do you begin? How do you measure progress in becoming customer-centric? Some researchers warn that many companies become complacent and underestimate the long-term need for building a more customer-centric business.[8] They point out that most established companies have the means to renew growth effectively

Table 6.1 Comparison of Product-Centric and Customer-Centric Strategies

Organizational Factors	Company Strategies	
	Product-Centric	*Customer-Centric*
Goal of the company	Make the best product for the customer	Provide the best solution for the customer
Mental process	Divergent thinking: How many possible uses of this product?	Convergent thinking: What combination of products is best for the customer?
Organizational structure	Based on product segments	Based on customer segments
Most important process	New product development	Customer relationship management
Performance measures	Tangible: Number of new products, percent of market share	Intangible: Customer satisfaction, lifetime value of a customer, retention
Organizational culture	Open to new product and process ideas; internally focused	Searching for more ways to satisfy customers; externally focused
Main offerings	Specific and cutting-edge products	Personalized package of services
Human resource philosophy	Power to employees who develop products	Power to employees with in-depth knowledge of customers' business

Source: Adapted from Galbraith, J.R., Organizing to deliver solutions, *Organizational Dynamics*, 31, 2, 194, 2002.

by creating by-products from the company's core business. Others caution that it is important for companies to match their organizational structure and behavioral norms when switching to a customer-centric business strategy.[9] Historically, businesses have focused on the lifetime value of a product and sometimes failed to measure the lifetime value of a customer.[10] Managerial Comment 6.1 describes steps in becoming a customer-engaged organization.

Managerial Comment 6.1   Steps to a Customer-Engaged Organization

1. Reframe the task at hand. Move from "creating a marketing program" to "creating an organization where everyone is focused on and energized about the mainstream of the business—delivering value to customers." In other words, you are working in the world of organizational and cultural change.

2. Mobilize the leadership team. Have your leadership team work together to develop a set of "customer value hypotheses"—real issues, propositions, concerns, and other customer-related matters that speak to what makes customers choose you rather than your competition. Then, have the leadership team charter a functionally and hierarchically diverse team—a diagonal slice of the organization whose task is to conduct face-to-face value conversations with your customers.

3. Engage customers in value conversations. *Conversations* is exactly the right word here. It suggests the open discussion that enables your people to understand more profoundly—both intellectually and viscerally—what is important about your customers' businesses. These kinds of probing, open-ended questions are certainly nothing new to marketing professionals. But remember who it is that is engaging customers in these value conversations: representatives of all functions throughout your organization. It will be an energizing revelation to them.

4. Create a customer value guide. The customer value guide (CVG) is a synthesis of all the value conversations. Its purpose is to provide a "North Star" by which everyone can navigate through the maze of decisions that are a part of work. The CVG is a series of brief, compelling statements

that capture and convey the essence of what represents value to your customers.

5. Conduct customer-engagement workshops with all employees. These workshops serve as the visceral link that connects everyone in the organization with customers. Ultimately, they are intended to ensure that all employees can answer two fundamental questions:

   What makes our customers decide to buy from us rather than from our competition?
   How does the way I do my job each day affect the likelihood of customers choosing us over the competition?

6. Integrate customer value insights into the organization. Customer engagement has to become part of "the way we work around here." This has significant implications for activities such as business planning; organization design; performance appraisal; performance improvement infrastructure (e.g., TQM, Six Sigma, re-engineering, CRM [customer relationship management]); processes, systems, and structures; and keeping the customer connections "alive" for all employees.[11]

Services as a Separate New Business Segment

Manufacturers may find that the movement of the economy toward services will provide an opportunity to view services as new "product" opportunities. If manufacturers view the life cycle of their products more comprehensively, they may find potential opportunities.

There are several service possibilities to complement a company's existing product line:

- Financing the purchase of their product
- Installing the product
- Modifying other products or processes to work with the product
- Maintaining the product and selling replacement parts
- Selling information to support product use
- Training personnel to use the product
- Disposing of product waste
- Disposing of the product.

Glen Allmendinger and Ralph Lombreglia state that any industrial manufacturer that has not awakened to the fact that it must become a service business is in serious peril today. Soon it will not be enough for a company to offer services; it will have to provide "smart services." Smart services are services that help a customer to know ahead of time that a machine is about to fail or that a customer's supply of consumables is about to be depleted. It is a way to help the customer use the product more effectively and efficiently, usually by providing a connected device that sends feedback to the customer for manual intervention, or senses a problem and automatically takes a corrective or preventive action.[12]

Manufacturers add services to generate additional revenue. IBM was originally a manufacturing company with practically all of its revenue originating from the computers it produced. Today, it is known more as a service company because the majority of its revenue originates from service lines of business.

Economic Growth and the Need for Added Services

As economic conditions improved in the industrialized countries, consumers had time and money to pursue added interests, especially in leisure activities. Travel and tourism were available to an increasing percentage of the world's population. Sports activities, such as skiing and golf, grew rapidly. Professional sports—football, baseball, and tennis—attracted huge crowds. In other words, businesses had to develop new ways (services) for individuals to spend their money.

The Movement from Make-to-Stock to Make-to-Order

As customers become more discriminating, they desire more choices not only in the products they buy but also in the services that accompany those products. One recent movement is toward customization of orders. If manufacturers can reduce their response times sufficiently, they can wait until they get the order before they make the product. This requires greater communication and coordination between supplier and customer, both service-type activities.

June Ma and her colleagues studied the effects of make-to-stock (MTS) versus make-to-order (MTO) programs on a supply chain. They found both benefits and problems in making the transition. Specifically, they found that introducing the volatility of MTO demand could cause problems in production planning and execution. They concluded that businesses should take an integrated view of demand management, production planning, transportation service selection, and information flow across the whole supply chain. For best results, businesses should carefully combine their goods-producing processes with their service-providing processes.[13]

The Movement toward Mass Customization

Closely aligned with the MTO movement is the concept of mass customization. The idea behind mass customization is to provide a product or service almost as fast and at a price almost as low as standard products and services that were provided in the past.

B. Joseph Pine popularized the movement toward mass customization. He described the approach as one in which manufacturers could customize their products by making their processes more flexible and capable of transforming their products.[14] Some manufacturers have succeeded in making their processes more flexible and are able to keep inventory in a subassembly stage until they receive the customer order, after which they quickly customize the finished product to meet the customer's requirements. They are essentially moving from a make-to-stock strategy to an assemble-to-order strategy. This product design strategy, which shifts product differentiation closer to the consumer by delaying identity changes (such as final assembly and packaging) to the last possible supply chain location, is known as postponement.[15] Dell is a prominent user of this strategy.

However, many companies have tried to address the need for mass customization by simply making a wider variety of products. This approach offers more options to smaller market segments, but it is not the same as customizing the product for an individual customer, whether it is a business or a consumer. Some consider the increased variety as both costly to the manufacturer and stressful to the consumer.[16]

Mass customization increases the complexity and costs of making products. As an offset, mass customization provides a benefit when the customer is involved in the co-design of the product because this simplifies and focuses the entire product line and increases customer loyalty.[17]

These factors have contributed to the growth in services. But what are services? What are some of the continuous improvement programs that were developed in services and are applicable to manufacturing? We examine both of these questions in the following sections.

What Are Services?

The word *services* has many meanings. Roger Schmenner suggests that services are better defined by what they are not rather than what they are. He explains that "the U.S. government defines service employment as a residual, namely, as nonfarming, nonmanufacturing employment."[18] Heskett also concludes that services can be classified as any non-goods-producing business. The goods-producing industries are mining, agriculture, and manufacturing; all else are services.[19] In a more recent book, the authors explain, "Economically, the term services is often defined not by what it is, but by what it is not. Historically, economic reports identify activities as 'service producing' that are not 'goods producing,' which includes manufacturing and construction, and are not 'extraction,' such as agriculture, forestry, fishing, and mining."[20] In Chapter 3, we described the Input–Transformation–Output (ITO) Model and explained that the model could be used to describe either goods-producing or service-producing businesses.

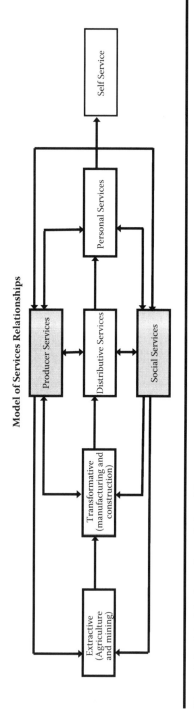

Figure 6.1   Model of service relationships.

Figure 6.1 shows the relationships among various types of goods and service producers. We will also identify some of the industries that fall within each of these service types. In Chapter 11, we describe more fully the relationship among these business areas. In this chapter we focus primarily on the improvement programs developed in services that are now spreading into the manufacturing and extractive industries.

Managerial Comment 6.2 reviews the basic attributes of service.

Managerial Comment 6.2 Service

In economics and marketing, a service is the nonmaterial equivalent of a good. Service provision has been defined as an economic activity that does not result in ownership, and this is what differentiates it from providing physical goods. It is claimed to be a process that creates benefits by facilitating a change in customers, a change in their physical possessions, or a change in their intangible assets.

By supplying some level of skill, ingenuity, and experience, providers of a service participate in an economy without the restrictions of carrying stock (inventory) or the need to concern themselves with bulky raw materials. On the other hand, their investment in expertise does require marketing and upgrading in the face of competition that has equally few physical restrictions. Providers of services make up the tertiary sector of industry.

Services can be described in terms of their main attributes:

- *Intangibility.* They cannot be seen, handled, smelled, etc. There is no need for storage. Because services are difficult to conceptualize, marketing them requires creative visualization to evoke effectively a concrete image in the customer's mind. From the customer's point of view, this attribute makes it difficult to evaluate or compare services prior to experiencing the service.
- *Perishability.* Unsold service time is "lost," that is, it cannot be regained. It is a lost economic opportunity. For example, a doctor who is booked for only two hours a day cannot work those hours at a later time—she has lost her economic opportunity. Other service examples are airplane seats (once the plane departs, those empty seats cannot be sold) and theater seats (sales end at a certain point).
- *Lack of transportability.* Services tend to be consumed at the point of "production" (although this does not apply to outsourced business services).
- *Lack of homogeneity.* Services are typically modified for each client or each new situation (customized). Mass

production of services is very difficult. This can be seen as a problem of inconsistent quality. Both inputs and outputs to the processes involved in providing services are highly variable, as are the relationships between these processes, making it difficult to maintain consistent quality.

- *Labor intensity.* Services usually involve considerable human activity, rather than a precisely determined process. Human resource management is important. The human factor is often the key success factor in service industries. It is difficult to achieve economies of scale or gain dominant market share.

- *Demand fluctuations.* It can be difficult to forecast demand (which is also true of many goods). Demand can vary by season, time of day, business cycle, etc.

- *Buyer involvement.* Most service provision requires a high degree of interaction between client and service provider.

- *Client-based relationships.* Based on creating long-term business relationships. Accountants, attorneys, and financial advisers maintain long-term relationships with their clients for decades. These repeat consumers refer friends and family, helping to create a client-based relationship.[20a]

Figure 6.1 shows the relationships among types of industries. At the left, we see the extractive industries of agriculture and mining. These industries provide the raw materials for the transformative industries of manufacturing and construction. Everything else in the diagram can be considered as services. Transformative industries depend on distributive services to move the goods from the factory to the retail stores or directly to the consumer, the personal services area. This horizontal flow completes the supply chain (described in Chapter 3). In addition, we have two other major service areas: producer or business services (the services provided to the goods-producing businesses by other businesses) and social services (provided by government or other nonprofit entities). We explain each of these service areas briefly below and more fully in Chapter 11.

Distributive Services

There are two principal kinds of services in this category—transportation, as in the movement of goods, and interim storage, as in regional distribution centers. The functions are becoming more costly as supply chains lengthen, and more critical as customer service demands increase. Distributive services also include retail outlets that are primarily concerned with making products such as food, clothing, and appliances available to consumers. In Chapter 5 we described how UPS is a major player in this area.

Personal Services

There are two main areas of personal service. One is the retail outlets that provide some products but are primarily thought of as service outlets, such as restaurants, hotels, movie theaters, and theme parks. The other is represented by one-to-one service, such as that provided by doctors, lawyers, teachers, automobile mechanics, hair stylists, travel agents, stockbrokers, and physical fitness trainers.

Self-Service

A special kind of service is increasing in popularity—self-service, where the consumer provides part of the service package by becoming an active participant in the service process, as in pumping gas at a service station, using the self-checkout station at a grocery store, or buying products online. Technology provides the capability to make self-service possible and the pressure to reduce costs makes it attractive. In addition, busy lifestyles encourage consumers to participate willingly, even eagerly.

Producer (Business) Services

Producer services are those services provided to businesses by other businesses. Some examples of this type of service include banks, law firms, insurance companies, consultants, public accounting firms, and architectural firms. In addition, other third-party businesses may provide services that were previously handled internally by the manufacturing companies such as food service, janitorial services, and information processing.

Social Services

Governments or nonprofit agencies provide social services. These are services that are not purchased directly, such as producer services, but which may be of great value to a business. Such services include employee training, economic benefits to attract industry, business activity information, and regional, such as roads and utilities access. They may be beneficial; however, they may also impose constraints on a business if they involve mandates to reduce hazardous emissions or increase financial reporting requirements.

Knowledge Transfer from Services to Manufacturing

In Chapter 5 we described how service companies could improve by using improvement programs that had been developed in manufacturing. In this chapter we describe programs that have been developed in services and can be used in manufacturing. In some cases, manufacturing companies will want to use these concepts because they will provide a benefit. In other cases, manufacturers will have to adopt the programs to remain competitive.

Areas of Manufacturing Expertise

Manufacturers designed and implemented programs focused on cost reduction, product quality, and reliability improvement. They tended to be isolated or decoupled from direct customer contact, both physically and systemically. Physically, most of the customer contact resided in the sales and marketing departments. Systemically, manufacturers tended to produce standard products to a demand forecast in a make-to-stock mode. To utilize their costly investment in plant and equipment efficiently, they used a level production strategy. When demand was less than production, they built inventory, and they worked down the inventory when demand exceeded production. They designed production jobs to be specialized to improve productivity. They managed the supply side of the demand/supply relationship. To a large extent, they operated in a closed-system environment, shielding their technical core from outside interference. In doing this, they also built a collection of service-type operations—payroll, accounting, engineering, human resources—that were primarily concerned with supporting the manufacturing operations and only incidentally concerned with servicing the customer. Although this promoted efficiency, it left a lot to be desired when it came to customer service.

On the other hand, companies in service industries designed and implemented programs that were primarily focused on improving customer service. Because they did not have the luxury of building inventory to carry them through the peaks and valleys of demand, they had to develop the capability to handle demand variability by designing volume-flexible capacities. Because employees came in direct contact with customers, service companies had to empower employees to handle a variety of situations. In so doing, these companies found that the traditional vertical hierarchical organization structure lacked flexibility; consequently, service industries began to develop alternate forms of organizations that were flatter and facilitated horizontal information flow and decision making. Of necessity, service companies operated in an open-system environment. They managed the demand side of the supply/demand relationship.

Areas of Service Expertise

Although service businesses have not been as effective or efficient as manufacturing businesses, they have always had a customer orientation, a quality sometimes lacking in manufacturing. Many of the continuous improvement programs that developed in the service sector provided improved customer service. The following are some of the areas in which service businesses have special skills:

- Customer relationship management (CRM)
- Adaptive organizations
- Response time reduction
- Quality as customer perception
- Qualitative goals and performance measures
- Supply chain management
- Flexibility and agility to adapt to changing needs—customization
- Knowledge workers.

Services have been concerned with customer service, quality as a customer perception, and flexibility of demand management.[21] Although manufacturing companies respond or react to demand by using inventory as a buffer against demand variability, service companies are more inclined to manage demand.

Examples of Programs Developed in Services

It is easy to find articles about how manufacturing has provided concepts and techniques that can be used in service businesses. It is more difficult to find the reverse—how service concepts can be used in manufacturing.

Some of the programs that have originated in services are described here. A number of the concepts that originated in services involved the participation of manufacturing as a partner in the process, such as in Quick Response programs. Other programs, such as the use of the Internet to sell direct to consumers, have been adopted by manufacturing companies. Figure 6.2 shows the approximate relationship of programs, both as to origin and with respect to time. The location of the programs is representative, not absolute, because it is impossible to separate the programs into neat categories. The figure suggests that, contrary to the general movement of improvement programs from manufacturing to services, some programs have originated in services.

Customer Relationship Management

Customer relationship management (CRM) has evolved from a data-collection technique to a relationship-building strategy. It is a program with value for both the services and manufacturing sectors.[22] CRM builds on the concept that it is better—more profitable—for a company to retain customers than it is to keep trying to find new customers. There is a widespread feeling that loyal customers buy more and may cost less to service than occasional customers do.[23] CRM programs help to

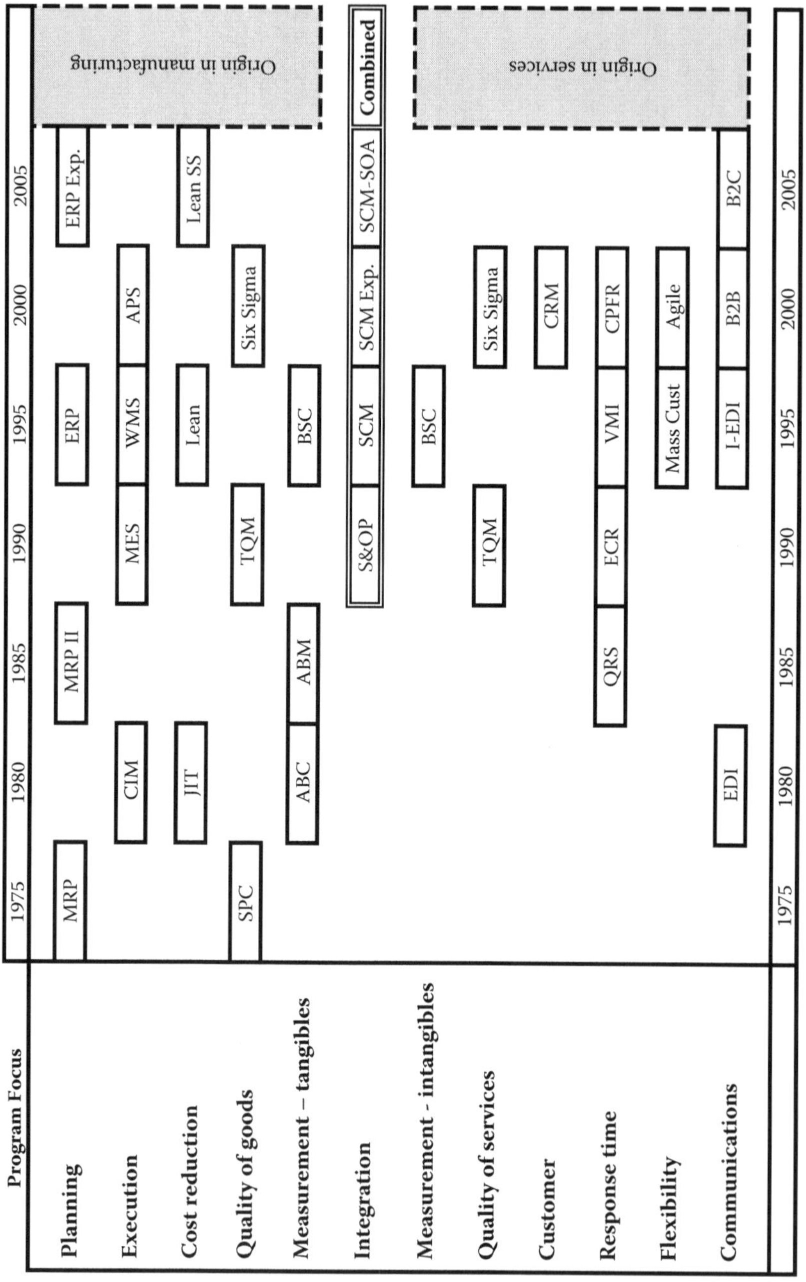

Program Focus	1975	1980	1985	1990	1995	2000	2005
Planning	MRP		MRP II		ERP		ERP Exp.
Execution		CIM		MES	WMS	APS	
Cost reduction		JIT			Lean		Lean SS
Quality of goods	SPC			TQM		Six Sigma	
Measurement – tangibles		ABC	ABM		BSC		
Integration				S&OP	SCM	SCM Exp.	SCM-SOA
Measurement - intangibles					BSC		
Quality of services				TQM		Six Sigma	
Customer						CRM	
Response time			QRS	ECR	VMI	CPFR	
Flexibility					Mass Cust	Agile	
Communications		EDI			I-EDI	B2B	B2C
	1975	1980	1985	1990	1995	2000	2005

Origin in manufacturing · Combined · Origin in services

Figure 6.2 Evolution of continuous improvement programs—manufacturing and services oriented.

develop relationships with customers that cause the customer to want to continue to do business with the company. These programs provide a framework for developing business processes and their supporting infrastructure to improve service delivery. Initially, they focus on automating internal functions, such as sales, marketing, and customer service. However, the hardware and software have to be supplemented by employees and business processes to implement the CRM system successfully.[24]

CRM programs initially focused on obtaining information about customers, such as location, income levels, and buying habits. Companies used this information to plan their product and service strategies. Satisfied that there was value in finding out more about their customers, some businesses went even further by asking customers directly what they wanted or needed. Over time, suppliers built a relationship with their customers that made it possible to anticipate demand patterns more accurately. With this knowledge, suppliers were able to provide a higher level of customer service.

Definitions

One study lists twelve different definitions for CRM, ranging from data-driven marketing to an e-commerce application, with numerous variations and extensions in between.[27] Another study found descriptions of CRM as a process, a strategy, a philosophy, a capability, and a technology.[28] A study of acronyms reports that CRM is guilty of "straying acronym syndrome" (SAS) in which the words used vary over time. Agnes Nairn tracked CRM from customer relationship marketing to customer relationship management until 1998; since 1999, PRM, for partnership relationship management, has been in vogue, and even more recently e-CRM, for electronic or e-business, has become popular.[29] A final reference describes Operational CRM (OCRM), Analytic CRM (ACRM), Conceptual CRM (CCRM), and Technical CRM (TCRM) and provides an Enterprise Information Roadmap (EIR) to show the relationships among the different versions.[30]

We will use the *APICS Dictionary* definition for CRM:

> CRM is a marketing philosophy based on putting the customer first. The collection and analysis of information designed for sales and marketing decision support (as contrasted to enterprise resources planning information) to understand and support existing and potential customer needs. It includes account management, catalog and order entry, payment processing, credits and adjustments, and other functions. *Syn:* customer relations management.[31]

For the most part, users have not been ecstatic about their results of CRM implementations. Many have failed to realize the "perfect order" status heralded by CRM proponents. A Meta Group study found a 55 to 75 percent implementation failure rate.[32] Another study reported that 55 percent of all customer relationship management (software solutions) projects do not produce results.[33] In a survey of 451 senior executives, 25 percent reported that these software tools failed to deliver profitable growth and in many cases damaged long-standing customer relationships.[34]

Should we conclude that CRM is ready to join a host of other "management fads" and fade away? Just the opposite. From its origin in the mid-1990s, CRM had a burst of popularity until 2000, when it fell into decline because of limited success and tighter IT budgets. After a low point about 2004, it is suddenly in vogue again. There should be renewed growth, particularly in nontraditional service areas, such as government and education.[35] One reason for the renewed interest is that many supporters conclude that CRM is not just a marketing program; it is an essential part of the modern supply chain.

Background

How did CRM get started? One driver was global competition, which forced many companies to become more customer-oriented as a means of securing a competitive advantage. Progressive businesses want to become "customer-centric." Many expressed this as simply formalizing their ever-present "caring for the customer" position; however, a number of people viewed it as a blending of capabilities not heretofore available. A host of IT technologies made it possible to collect and analyze data about customers and then to translate that knowledge into meaningful marketing strategies.

Another driver was the need for ERP vendors to find new products to sell. One of the golden opportunities came along with Y2K. As many readers will remember, Y2K was the alarm bell for many companies anticipating what their computer systems might do when the date rolled over from December 31, 1999, to January 1, 2000. The concern was for those systems with two-digit years, such as 99 for 1999. How would the computer interpret the year 00? Would it be 2000, 1900, or something else? Many ERP vendors proudly claimed their systems had four-digit years, so there was no need to worry if a business installed their system. After the dizzying burst of ERP implementations leading up to Y2K, it became apparent that the next wave of software implementations was in supply chain management (SCM). SCM offered a bonanza of applications that included CRM, along with other processes that needed software and implementation consulting help. Although specialist vendors such as Siebel developed CRM software, the major ERP vendors (SAP and Oracle) are selling extensions of their ERP packages that include CRM software. Salesforce.com, one of the pioneers, continues to be a strong participant. Microsoft also continues to be a participant in the market.[36,36a]

What Does It Do?

What should the CRM program do? Shaun Doyle provided a detailed checklist of activities in CRM.[37] Raymond Ling and David Yen developed an analysis framework and implementation strategies for CRM. They also described the evolution from the direct sales of a bygone era to mass marketing in the 1960s to target marketing in the mid-1980s to CRM in the 1990s.[38] Adrian Payne and Pennie Frow provided a conceptual framework for CRM strategy consisting of the following processes: strategic development, value creation, multi-channel integration, information management, and performance assessment.[39] CRM programs have the following major components:

- CRM collects information about customers, primarily from sales transactions, and also from a number of other "touch points," such as complaints or inquiries. E-CRM is popular as a means of recording website contacts, searches, and other nonsales activities. For example, Amazon.com recommends books based on your previous searches and requests that you provide reviews of books that you have purchased.
- From the collected data, a CRM program organizes the customer base into segments or groups of similar customers. The groups may organize around age, income level, location, books searched, or whatever the marketing group deems useful. The ultimate objective is to arrive at a group of one (an individual customer), if that is practical.
- The marketing department designs a program to appeal to the groups described above. Although the primary emphasis of a CRM program is customer retention, marketers are not above designing sales programs that will attract new customers. One example describes typical programs as customer service, frequency/loyalty programs, customization, rewards programs, and community building.[40]
- Marketing implements the programs designed to enhance the relationship with existing customers. Sometimes sales people may feel they are asked to change from a "hunt and kill" mode to a "tend the farm" mode.[41] This change means that progress in these programs will be monitored closely and results will be measured.
- The marketing literature does not say much about "how" the customer is better served. It stresses the need for customer service but does not always expand on its content. Sometimes, "inventory" slips into a diagram, but there is not much space devoted to its role. Operations has a key role, which will be described later in the chapter.
- The CRM program develops a set of metrics to measure the results and to revise, modify, discontinue, and reverse the marketing initiatives that have been introduced.

Benefits

The primary benefit of a CRM program is increased customer retention. There is some indication that customer retention has greater value than customer acquisition and that businesses should expect to see improved financial results from their CRM program.[42–44]

In addition to these tangible benefits, there are intangible benefits. The relationships with customers should be more open and effective. Hopefully, things will proceed smoothly; however, if there are incidents concerning customers, the closer relationships with them should help in their resolution.

Internally, the need to develop cross-functional programs should increase the collaboration among internal functional departments. Even without considering the operational functions, such as purchasing, production, distribution, and inventory management; there is a need to get sales, marketing, and IT people more comfortable with one another. Agnes Nairn highlighted the difficulty in communication between the "emotions-driven sales force and the clinical binary-driven IT expert" as difficult unless intelligently managed.[45]

Problems

Why have CRM programs not been more successful? It would be easy to blame the software vendors for overhyping their product, but that does not excuse the companies that bought the software from doing their part. One book describes 53 myths attributed to customer loyalty programs. These myths include a number of promises that have not been met satisfactorily in practice.[46]

One of the leading problems has been the myopic view of CRM in considering it an IT technology and not a strategic process. It is not sufficient to create a database of customers, no matter how cleverly designed. Another limited-scope problem has been to consider CRM as only a marketing program. Although marketing is the driver, cross-functional support is needed from the rest of the organization.

Sometimes businesses design the software around what they think they can do (their capability) rather than around the needs of the customer. As a result, the program may be efficient at performing tasks that the customer does not care about or respond to. Some companies fail to obtain the support within their organization for the CRM program. Marketing personnel may not like the closer monitoring, finance may not agree with the deployment of resources, operations may feel left out, and IT may resent the cavalier attitude toward the innovative systems design. Companies may attempt to undertake too much too soon. Most researchers advocate a selective approach to implementation of CRM.

Finally, customers may be "turned off" instead of "turned on." Most customers want some but not excessive attention. Although subtlety is not generally associated with marketing efforts, it may be a trait to cultivate in CRM.

Relation to the Supply Chain

Although the idea may be in conflict with the marketing world, Douglas Lambert and Terrance Pohlen, in the logistics field, have proposed that CRM is only one of eight processes in the arena of supply chain management. Other processes are customer service management, demand management, order fulfillment, manufacturing flow management, supplier relationship management, product development and commercialization, and returns management.[47]

CRM's Future

CRM research supports the conclusion that it is a management "fashion" and not a "fad." It should have great appeal in service industries, such as banks,[48] public accounting,[49] retailing,[50] insurance,[51] and increased use of Web services.[52] However, CRM is a long way from achieving its potential. Some companies reap substantial benefits by assimilating the principles of CRM into their normal business practices. Others will go through the motions, stall, and move into a newer program that promises to be easier to implement and provide even greater benefits. Unfortunately, still others will never know that CRM was available.

Response Time Reduction

A variation of the CRM program involves programs that provide the customer with products and services faster. For example, a retail store sells stock items. When the items are out of stock, the customer may select a competitor to purchase the same product. Unfortunately, the customer may like the competitor and return there for future purchases. In response to this type of dilemma, programs such as Efficient Consumer Response (ECR) in the grocery industry and Collaborative Planning Forecasting and Replenishment (CPFR) quickly replenish stock at the retail store. These programs also help to reduce excess inventory of slow or nonselling items that require markdowns to move them.[53]

Quick Response Systems[54]

In the early 1980s, a group of manufacturers and retailers in the textile and apparel industries initiated a program that would reduce the number of stockouts by providing a closer matching of demand and supply. The program was to provide a way to reduce lead times for stock replenishment orders from manufacturer to retailer and to reduce the lead times for introducing new products. There were a number of components to the program, but the essential ingredients were a willingness and a capability to communicate actual demand data more quickly from the retailer to the supplier who could have more time to prepare for orders from their downstream customers. Another goal was to reduce the bullwhip effect by placing orders on a more regular and predictable schedule.

The potential was high, so how did it do? Alan Hunter reported the expected benefits as reductions in pipeline inventories, greater probability of garment designs and colors being acceptable to the consumer, ability to re-estimate stock keeping unit (SKU) demand at retail, and greater competitiveness for domestic producers facing increased levels of imports.[55]

Hunter also described some of the problems that have delayed the widespread adoption of Quick Response Systems (QRS):

- Naivety—participants did not realize the magnitude of the task
- Difficulty in creating "partnerships"—the retailers received the benefits while the suppliers incurred the costs
- Structural issues, such as the staggering number of unique SKUs (1.2 to 1.4 million at a department store every four months), overwhelming effect of fashion (shelf lives are decreasing), and the makeup of the pipeline (retailers and textile companies dominate, apparel manufacturers are small)
- Technical problems, including inadequate accuracy of bar codes, inadequate storage, manipulation of inventory and sales data, and lack of standards in information transmission or electronic data interchange (EDI).

Finally, Hunter outlined the factors necessary for growth of QRS including the need for UPC/EAN compliance and standardization, the clarification and acceptance of the role of value-added networks (VANs), and the need to find a way to extend EDI to smaller manufacturers.

Kurt Salmon Associates expanded the scope of QRS to include product development and product sourcing as well as product distribution. Despite its slow beginning, QRS programs provided a model for other industries to follow and for subsequent programs to build upon.[56]

Continuous Replenishment Programs

Closely allied with QRS, continuous replenishment programs (CRP) encouraged automatic replenishment ordering so that customers would automatically place an order when their inventory management system indicated a need for a reorder.[57]

Efficient Consumer Response

Encouraged by the positive results of the QRS programs, in 1992 several grocery executives formed a voluntary group, known as the Efficient Consumer Response Working Group, and commissioned a study by Kurt Salmon Associates to identify opportunities for more efficient, improved practices in the grocery industry. The consultants returned in early 1993, claiming that the industry could reduce inventory costs by 10 percent, or $30 billion.[58] In addition to efficient replenishment, this group added the requirement of category management, consisting of efficient new product introduction, efficient store assortment, and efficient promotion. The program included collection of demand (sales) data with point-of-sales (POS) terminals, and feedback of this data to suppliers with EDI. Suppliers could then avail themselves of a variety of techniques, such as cross-docking, to move the product more quickly to the customer.

Vendor-Managed Inventory

Some retailers decided that inasmuch as the suppliers had the consumer demand data, they (suppliers) could assume the responsibility of managing their (retailers') inventory. Although the idea of suppliers managing a retailer's inventory was not new—rack jobbers and service merchandisers had performed this function in the health and beauty aids categories years before—it did have the added element of rapid feedback of demand information. One study by Andres Angulo and colleagues revealed the information technology challenges, especially the effects of information delay and accuracy.[59] Another study has concluded that even for products with stable demand, a partial improvement of demand visibility can increase production and inventory control efficiency.[60]

Sales and Operations Planning

This program has been around for at least thirty years and was originally intended to encourage marketing and manufacturing personnel to collaborate on a production schedule within their company. Sales and operations planning (S&OP) represents a way to get companies to talk with one another and smooth the flow of goods along the supply chain. In one survey, it was identified as the No. 2 initiative of global companies, following strategic sourcing of direct materials.[61]

Collaborative Planning, Forecasting, and Replenishment

In 1997, voluntary inter-industry commerce standards (VICS) created a subcommittee to develop collaborative planning, forecasting, and replenishment (CPFR) as an industry standard. The following year, VICS issued the first document on CPFR: "VICS CPFR Guidelines," which has been constantly updated since then.[62] Although QRS and ECR provided the flow of demand information from the retailer upstream to suppliers, it was the responsibility of the supplier to anticipate demand and the retailer (except in vendor-managed inventory) to actually do the ordering. CPFR attempted to eliminate this disconnect by advocating that both the retailer and the supplier collaborate to plan a joint demand forecast and replenishment schedule. Several issues addressed for the first time with CPFR were:

- The influence of promotions in the creation of the sales forecast (and its influence on inventory management policy)
- The influence of changing demand patterns in the creation of the sales forecast (and its influence on inventory management policy)

- The common practice of holding high inventory levels to guarantee product availability on the shelves
- The lack of coordination between the store, the purchasing process, and logistics planning for retailers
- The lack of general synchronization (or coordination) in the manufacturer's functional departments (sales/commercial, distribution, and production planning)
- The multiple forecasts developed within the same company (marketing, financing, purchasing, and logistics).

The report also provided an excellent summary of benefits, barriers to implementation, and enablers of the CPFR process, and concluded that trust and good information technology must be present for success.[63]

Another study reports that although most companies think of CPFR as primarily a technology, it is the business process supported by the internal culture that makes CPFR successful.[65]

Supply Chain Management

A supply chain management (SCM) program includes all the features of the programs described above plus other attributes that are described more fully in Chapter 10. An integrated supply chain is the ultimate goal in linking entities through communications and collaboration.

Present Status

How are companies doing in their quest to reduce response times? A simplistic answer is that the programs work but few companies are getting the full benefit of the programs because of incomplete implementation. There are three key components to any program of this type: technology, infrastructure, and change management. The technology, primarily in the form of electronic data collection—with POS terminals—and data transmission—with EDI or the Internet—is good and getting better. The infrastructures—organization, systems, and functional relationships—are inconsistent among companies and even industries, but improving. Managing change is a continuing problem, particularly in the area of company culture and trust. How much do you trust? Which customers/suppliers do you trust? Almost every study finds that trust is not only essential for success but also often lacking in many programs.

Coleen Crum and George Palmatier list the following reasons that demand collaboration programs have not realized their potential:

- The pace of adopting new ways of doing business is slow.
- Demand information supplied by customers is not put to use in trading partners' own demand, supply, logistics, and corporate planning in an integrated manner.

- Demand management and supply management processes are not integrated, and sales and operations planning is not utilized to synchronize demand and supply.
- There is a lack of trust among trading partners to share pertinent information and collaborate on decision making.
- There is a desire to partner but not to commit to executing the communicated plans.
- A common view exists that demand collaboration is a technology solution and that the current technology is too complex.[64]

Computer Sciences Corp., in conjunction with *Supply Chain Management Review,* concluded that some companies understand the advantage of leveraging their buying across a more strategic supply base, and others are content to pursue more limited, tactical improvements. Despite the progress needed in some areas, most companies pursuing the various supply chain initiatives were generally happy with the results. However, only 28 percent said that an awareness of the need to increase customer satisfaction ratings was a main factor in the success of their initiatives.[66]

One seemingly simple problem is the need for consistency in product data identification and transmission. UCCnet, a nonprofit unit of the Uniform Code Council standards organization, established a global online registry that requires product data with as many as 151 attributes, or descriptors, about 40 of which are mandatory. One well-known company found that it was transmitting information by phone, e-mail, fax, CDs, EDI, PDFs, spreadsheets, websites, and printed price pages. The same survey found that the percentage of companies that felt they have the business processes in place to take full advantage of real-time information varied from 20 percent in the construction and engineering industry to slightly over 70 percent in the logistics and transportation industry, with the overall average of all industries around 45 percent.[67]

Another nagging question is the relative benefits between retailers and suppliers. Corsten and Kumar studied the question of whether collaborative relationships with large retailers benefit suppliers. They found that although suppliers benefit in the economic sense and in capability learning, they believe that suppliers bear more of the burden and receive less of the benefits they deserve.[68]

Some activities are increasing the response times, most notably the offshore outsourcing movement. The route to cheaper supplies from overseas sources may be attractive, but supplier lead times are getting longer, a step backward from the improvements in domestic operations from years of lean implementations. Some companies may find the payoff worth it. Others may find themselves with cheaper raw materials and components but fewer customers. The problem is that it just takes longer to get products from another country. The costs may be lower when outsourcing, but the response times and uncertainty of supply availability are higher.

Balancing the trade-offs between supplier responsiveness and order costs requires a long-term strategic outlook. In addition, the extended supply chains increase the risk of disruption in the flow of goods. These disruptions increase the difficulties of maintaining consistent short lead times to customers.

Future

What does the future hold? No doubt, the pressure to clean out excess inventories in supply chains and gain a better matching of supply with demand will continue. Individual programs such as those described above are losing their focus as they become a part of more general programs such as supply chain management and integrated demand-driven collaboration systems. A survey of over sixty retailers in the United States, Europe, and Asia found that the persistent problem of being out-of-stock was their biggest inventory challenge.[69] Another study suggested that retail exchanges, stemming from improved IT, may provide part of the answer.[70]

Whatever the program is called, companies must integrate their communication systems and develop sufficient trust with one another to collaborate effectively and to gain the benefits of their efforts. If they do, they can hope to succeed. If they do not, they face an uncertain future.

Flexibility

The service sector has been progressive in increasing its flexibility in several areas—product and service offerings, organization structure, demand management, facility location, and employee versatility.

Product and Service Flexibility: Mass Customization

Although the idea of mass customization is still developing, it is becoming a more common theme in many companies. Just as in product or service design, the customer becomes the focal point in that mass customization means building a product or providing a service exactly the way the customer wants it. Service companies can achieve some level of mass customization by empowering employees to make the modifications needed such as "hold the mustard" in a deli, to "a little more off the side" in a barber shop, to "put more emphasis on growth stocks" in a stockbroker's office, to "I want to carry fifteen hours this semester" at a university. Manufacturers often must make major changes in their production processes to achieve mass customization; however, they are being driven to do this by the service industries. B. Joseph Pine outlines the future of mass customization in Managerial Comment 6.3. (See also Figure 6.3.)

High	Containing	Mitigating
	Workforce flexibility Information technology Efficiency measures	Restructuring Risk pooling Outsourcing
	Absorbing	Shielding
	Time buffers Slack capacity buffers	Pricing and rationing Demand management models Managed care controls

Range of Flexibility (vertical axis, High to Low)

Low   High

Demand Uncertainty

Figure 6.3 Mass customization alternatives.

Managerial Comment 6.3   Mass Customization

Mass customization is a new way of viewing business competition, one that makes the identification and fulfillment of the wants and needs of individual customers paramount without sacrificing efficiency, effectiveness, and low costs. It is a new mental model of how business success can be achieved, one that subsumes many of the "silver bullets" of prevailing management advice such as time-based competition, lean production, and micromarketing. Further, the development of Mass Customization as a paradigm of management explains why product (and service) life cycles are decreasing, why development and production cycle times must follow, why businesses are re-engineering into networked organizations.[71]

Table 6.2 summarizes some of the factors that influence operational effectiveness in mass customization.

Horizontal Communication and Organization Structure

The traditional vertical hierarchical organization structure—the command and control concept developed by the military—was good in maintaining structure and discipline from top to bottom of an organization. However, it was slow in moving information and decisions in a horizontal direction—toward the customer. This type of organization was effective in managing the normal day-to-day operations

Table 6.2 Operational Effectiveness in Mass Customization

Mass Customization Practice	Strategic Priorities Affected	Mechanisms of Enablers of Strategic Advantage
Modular design	Customization, cost, speed	Complexity reduction, ease of planning, economies of scale, demand variance reduction, and lead time reduction through delayed differentiation
Delayed design, fabrication, assembly (postponement)	Cost, speed, customization	Reduction in demand variability, safety stocks, cycle times, and lead times
Cellular or flexible processes	Cost, quality, speed	Cell efficiency, recurrent problem-solving skills, process stability and flexibility, reduced setup costs and times, quick learning efficiencies, complexity reduction, planning efficiencies
Customer co-design	Customers' satisfaction	Product fit, customer perception of design ownership, reduced cost of warranty, returns, product liability, reduced risk of obsolescence

Source: Copyright © 2007. Reprinted with permission of the Institute of Industrial Engineers from September 2006, *Industrial Engineer.* All rights reserved.

of a stable demand business. However, it is deficient when managing a business in a more dynamic industry. It is also not very effective in planning and managing improvement programs or projects. Service organizations have been active in developing more flexible organizational forms than manufacturing because services have several unique needs:

■ They often have smaller operating units that are widely dispersed—chain restaurants, insurance agents, and stockbrokers. They need decentralized authority to handle day-to-day operations.
■ They are closer to the customer and need empowered employees to address a variety of transactions and problems.
■ The smaller units cannot afford to support a specialized staff; therefore, staff support is located at the central home office and unit operations are self-supporting with general-purpose employees.
■ Service operations do not have the tradition of functional staffs that became prominent in manufacturing during the scientific management era and continue today.

One of the most recognized authorities on both service operations and their flexible organization forms is James Brian Quinn. In Managerial Comment 6.4, he describes some of the types of organizations and their common characteristics and problems. Quinn provides detailed descriptions and diagrams for each of the organization forms in his book *Intelligent Enterprise: A Knowledge and Service Based Paradigm for Industry*,[72] and with others in a separate journal article.[73]

Managerial Comment 6.4
Flexible Organization Forms

Service and service technologies have opened a wide variety of new organizational options for managing intellect on a much more disaggregated basis. The infinitely flat, spider's web, and inverted forms embrace the most common structures. But there are many other configurations, the most significant being "starburst," "cluster," and "voluntary" organizations. All these radical organizations enjoy certain common characteristics and similar problems:

- All tend to be flatter than their hierarchical predecessors. All assume a constant dynamic rather than stasis in their structures. And all support greater empowerment of people closer to the customer contact level.
- All develop around a maintainable set of core service competencies at their center. These competencies—typically consisting of special depth in some unique technologies, knowledge bases, skills, or organizational–motivational systems—are then radiated outward toward customers by information technologies or through organizational dynamics that enable mixed clusters of people, with greater knowledge access, to work directly with customers.
- All recognize intellect, motivation, and knowledge bases as the most highly leverageable assets of a company. They use the new organization forms and their supporting systems as improved ways to attract, keep, and leverage key people as knowledge resources.
- Successful companies use their new organizations as both aggressive and defensive strategies. Most companies look to these forms initially to lower costs, decrease cycle times, and ensure more precise implementation of strategy. At the same time, they allow the companies to bring more energy, focus, and responsiveness to bear on customers than ever before. But companies have also found that these organizations make the firm a more exciting, personally challenging, and satisfying place to work.

Like earlier decentralization or SBU concepts, some of these new organization modes have often been touted as cures for almost any management ill. They are not. Each form is useful in certain situations, and not in others. But more importantly, each requires a carefully developed infrastructure of culture, measurements, style, and rewards to support it. When properly installed, these disaggregated organizations can be awesomely effective in harnessing intellectual resources for certain purposes. When improperly supported or adapted, they can be less effective than old-fashioned hierarchies.[74]

Demand Management

Service industries have learned to live with variable demand patterns, whether daily or weekly at a restaurant, or monthly in a variety of seasonal businesses. Consequently, they have learned to vary their capacity, especially in the form of employees, by developing scheduling techniques to best match the number of employees to the demand highs and lows. In some cases, where the employees represent a scarce and costly resource, such as among physicians, lawyers, and other professional people, they use an appointment system to control the demand to fit the availability of their capacity. These and other demand management techniques can be extended to a number of manufacturing applications.

A number of approaches have been suggested to match supply better with demand. Christopher Lovelock and Robert Young were among the first to suggest that customers could help reduce the variation in demand by participating in the service process. "Service managers, utilizing demand management, can use marketing tools to encourage consumers to modify their behavior so that services can be delivered in a more productive and economically efficient manner."[75]

Kenneth Klassen and Thomas Rohleder provided a thorough review of the demand management literature. They described the desirability of simultaneously planning demand and capacity by involving both marketing and operations in the planning process, especially in a service environment where it is not possible to build inventory as a buffer against demand fluctuations.[76]

Demand management involves sales forecasting, usually a responsibility of sales and marketing functions. However, a company's position in the supply chain affects their role in sales forecasting. The business closest to the consumer—the retailer—can forecast independent demand. Once the independent demand is determined, all demand upstream in the supply chain is dependent on that demand forecast. "Recognizing the differences between independent, dependent, and derived demand, recognizing which type of demand affects a company, and developing techniques, systems, and processes to deal with that company's type of demand can have a profound impact on supply chain costs and customer service levels. To realize the benefits that are possible from managing demand in a supply chain, communication among the enterprise must be open, coordination in the form of sales and operations planning processes must be implemented, and a culture of collaboration must be established."[77]

Demand management extends beyond day-to-day operations. When combined with logistics and new product development, demand management can contribute to improving the financial performance of a firm.[78]

The health care industry is one that could benefit from better demand management. One study in an acute care hospital found that a demand management program better utilized bed and clinic capacities, reduced patient length of stay and treatment costs, and provided relevant and timely input for the nursing staff.[79]

Shielding strategies work best when demand uncertainty is high and flexibility is low. Pricing and resource rationing changes offer a strategy available under shielding. Absorbing strategies are recommended when both demand uncertainty and flexibility are low. Using time buffers and slack capacity buffers is recommended with this strategy. Containing strategies are offered when demand uncertainty is low and flexibility is high. Focusing on workforce flexibility and information technology is recommended. Finally, mitigating strategies are recommended when both demand uncertainty and flexibility are high.

Location near the Market versus Lowest Cost

Manufacturing companies build large plants to be low cost and efficient. Service companies have to build small outlets to be close to their markets, such as retail grocery stores, restaurants, hospitals, and branch banks. Consequently, service companies have learned how to manage smaller units, often through process designs that are modular and scaled to fit the situation. Manufacturers need to be more responsive to the customer in terms of both faster response and more appropriate products. To achieve these objectives, they have to locate closer to their markets and use smaller facilities.

Closely allied with the concept of smaller facilities is the need to manage multiple locations or branch operations within the total company. Some of this can be done with standardized processes for use at the branch, such as customer contact services, with centralized processes maintained at the corporate office, such as procurement. In a retail specialty store, such as The Gap, the in-store personnel handle the direct sales to the customer; the corporate office handles stock replenishment functions, based on sales information collected at the point-of-sale registers. Even with specific guidelines, empowered employees are needed to carry on the branch operations.

Evolution from Job Specialization to Self-Directed Teams

Services have generally been more oriented toward broader job designs, as opposed to the specialization of manufacturing. This stems from the variability of services, the need for lower-level decision making, and the smaller size of the operating units. As a result, the human resources function in service businesses has been responsible for designing programs that move from training employees to do simple tasks, to education programs that enhance the capabilities of the employees to perform multiple tasks and to work effectively in cross-functional relationships. Because of the increased capability of its workforce, a business can gain a competitive advantage through improved utilization of its human resources. Although operations management can provide contextual insights to human resources, human resources can provide behavioral insights to operations management.[80]

Replacement of Inventory with Information

For years, the availability of inventory has covered a multitude of sins in manufacturing. Today, largely through the development of information, inventory is being replaced. In services, money, as a medium of exchange, has transformed from cash to checks to online payments to electronic funds transfer. With each step in the progression, there is less inventory but more information. In manufacturing, better demand information has made it possible to reduce the amount of inventory needed to satisfy demand. An even more tangible example is the replacement of newspapers with electronic news transmissions. The computer download capability could mean that some time in the future electronic storage and accessibility will replace books, although this carries with it the need for readers to adapt from reading books to reading computer screens.

Inter-Organizational Communications

Another area in which service companies have taken the lead is inter-organizational communications (IOC). Although EDI originated almost three decades ago, the use of the Internet for IOC is increasing rapidly and will probably become the primary tool in the future.

Traditional EDI

Traditional electronic data interchange is the paperless (electronic) exchange of trading documents—such as purchase orders, shipment authorizations, advanced shipment notices, and invoices—using standardized document formats. It became available to companies in the early 1970s as a means of transferring information electronically from one business to another. One of the more popular applications was in order processing. A business could place an order with a supplier, who would acknowledge the order and ship the order with an invoice. The receiving business would prepare a receiving report, match it with their purchase order and the invoice, and authorize payment through their bank, which would send the payment to the supplier's bank. This process eliminated paperwork because all the transactions and accompanying documents described above were electronic. EDI reduced costs of order processing by as much as 90 percent, reduced errors, speeded up deliveries, and reduced the time required of employees to identify errors or track orders. The biggest problem was that it was expensive to implement and operate. As a result, only a small fraction (less than 5 percent in 1995) of the businesses adopted EDI.[81] Many of those that used EDI did so reluctantly because their major customers demanded it. Because it was a one-to-one kind of communication, it was secure. However, it required that each customer–supplier relationship be set up individually.

Large companies implemented EDI and found it worth the investment. Some of the major applications included global communications, financial funds transfers, health care claims processing, and manufacturing and retailing.[82] Small companies could not achieve the volume necessary to make it a worthwhile investment. If they had to use EDI to conduct business, they usually did it through value-added networks (VANs), entities that facilitated the transfer of information between customer and supplier. Using a VAN eliminated a portion of the initial investment cost but did not change the high transaction costs. Sawabini reported, "Where EDI fails, it is because (1) it's cost prohibitive and too complex for smaller suppliers, and (2) it offers few bottom-line benefits for suppliers."[83] In the early days, users viewed EDI as an inter-organizational system (IOS). In recent years, as the concept of an IOS has enlarged, writers now view EDI as an element of an IOS.

In summary, EDI was good for a few companies but a burden for many others. Today, competitive pressures make the need for rapid information transfer imperative. Is the Internet a viable alternative? Many say yes; however, a few are more cautious.

Internet EDI

Internet EDI looks attractive because of the lower costs potential. The initial investment is lower and the transaction costs are lower.[84] In addition to the lower costs, other reasons the use of the Internet is attractive include the following:

- Publicly accessible network with few geographical constraints
- Offers potential to reach the widest possible number of trading partners
- Tools facilitating inter-organizational systems are becoming available
- Consistent with the interest of business in increasing electronic services

- Can complement or replace current EDI strategies
- Leads to an electronic commerce strategy.[85]

The Internet also offers close to real-time transactions because there is no longer the need to go through VANs that use the batch-and-forward method of transmitting information. But how do companies ensure the secure transmission of information? If a satisfactory solution to this issue becomes available, Internet EDI could become a major growth application in the next few years.

Research tells us to expect that Internet EDI will become more widely used than traditional EDI because of its lower costs. At the same time, traditional EDI will also become more widely used because of its advantage over paper processing. Both forms of electronic data interchange will begin to realize their potential in the development of inter-organizational systems. With the ever-expanding global business community, it is no longer possible to use hard copy documents in most business transactions. The speed and efficiency of electronic information is compelling businesses to make the transition to paperless systems. Although Internet EDI may become the standard at some point in the future, traditional EDI will remain a significant factor for several years because it is a proven method and many companies already have an investment in their present systems.

Like many innovations, the use of the Internet as a replacement for traditional EDI has many supporters, some of whom have a stake in seeing it succeed. Technology opportunities abound. For those who understand the technology, it is like being a kid in a candy store. For those who do not understand, it is a jungle with danger lurking behind every acronym. A number of articles and websites provide more complete descriptions of the technology involved.[86–88] Although the future looks promising, Internet EDI is not without some hurdles, some of which have to do with the evolution of changing standards.

Additional actions become necessary to provide greater security for data transmission over the Internet. One example is the use of firewalls at both the sender and the receiver, and a virtual private network (VPN) between the two. VPNs use the Internet and its standardized protocols to communicate information by combining the global connectivity of the Internet with the security of a closed, private network.[93]

Internet EDI must overcome the problem that restricted the growth of traditional EDI—dominance by a few large companies in a supply chain. When there is an imbalance among supply chain members, not all participants share equally. Smaller firms may have to adopt innovations to maintain supplier relationships with large customers, but without realizing any of the savings provided by them.[94]

Any IOS is more than technology. It also includes organization and people. Trust between participants is necessary if the relationship is to be an effective one, supporting collaboration among entities. Even if the cost and security issues can be resolved, the trust issue may remain. Company culture and trust are significant issues when it comes to the use of EDI in either the traditional or Internet format. To move from coordination to collaboration, trust is required.[95]

EDI, either traditional or Internet, cannot stand alone. To be completely effective, a company should integrate it with its mainstream information systems. This requires the capability to interface two disparate systems or to modify the EDI system to allow integration of the data into the main systems, such as Enterprise Resource Planning (ERP), Supplier Relationship Management (SRM), or Customer Relationship Management (CRM).

Traditional EDI is a proven method of transmitting data between entities. Where there is sufficient volume, it is of significant benefit. Unfortunately, only a limited number of companies have the requisite volume. A promising supplement is Internet EDI, when the level of security can be raised to acceptable levels. Other approaches include the use of portals—access points (or front doors) through which a business partner accesses secured, proprietary information from an organization. These portals may be distribution portals (single supplier, multiple buyers), procurement portals (single customer, multiple suppliers), or trading exchanges (balance between suppliers and buyers).[96] Until some approach dominates, companies may use a hybrid approach, continuing to use traditional EDI for the more sensitive data applications and moving more deliberately into Internet EDI with less sensitive data, using trading portals where they offer benefits. As with any worthwhile innovation, the widespread diffusion of electronic data exchange will take time and will likely involve changes in direction or emphasis during its life cycle.

Other Service Developments

In addition to the major areas described above, services have also pioneered in a number of other ways, as described below.

Working in an Open-System Environment

Service industries are exposed to an open-system environment more than manufacturing industries. Service industries such as retail, insurance, health care, banking, and hospitality are highly visible to the consumer and other elements in the economic world. Consequently, they are less able to shield their operations from the effect of external entities.

Product Development as a Result of Customer Inputs

Historically, product ideas have been passed along from sales and marketing, who may have interpreted what they thought was the voice of the customer, to engineering and operations, which designed and made the product. Often there were a series of modifications to the design, so that the resulting product was not what the customer wanted. Some companies have found that direct customer involvement in the design of products and services is beneficial and increases the likelihood that the product will be a successful entry in the marketplace.

Quality as Customers' Perceptions, Not Just Conformance to Specification

Manufacturers have long measured product quality by comparing the end product with the product specification. If the product met the specification, it was considered to be of good quality, whether or not it was suitable for the customer. Although Juran and Gryna introduced the concept of "fitness for use,"[97] the quality measures still tended to be quantitative and measurable without regard for the customer. Services have changed this. The service industries must consider customer perceptions when assessing their service quality. The hamburger that was made to specification may not please the customer because the order taker was not courteous or there were food scraps on the floor. Designing products to meet customer perceptions, where each customer has slightly different perceptions, is a new way of life for manufacturers.

Managing the Customer Encounter

Customer encounter implies a face-to-face meeting between employee and customer. It is a common occurrence in services. Much of the success of such an encounter depends on the knowledge, ability, and motivation of the employee. Customers look for a combination of service content, quality of service, and humane treatment. To achieve this combination of outcomes, companies can use an approach that weaves together a consistent method for assessing the employee–customer encounter and a disciplined process for managing and improving it.[98]

Nonquantitative Performance Measurement

Manufacturing has always been thought of as tangible, but services have been assigned the attribute of intangibility. Whether or not this differentiation is valid, services do have the need to use some nonquantitative measures, whether it has to do with customer perceptions of quality or the decision-making capability of empowered employees. Manufacturing is facing the need to become more adept in measuring nonqualitative variables. For example, what is the risk cost of outsourcing to an offshore supplier? What is the impact of a packaging change on the customer's propensity to buy that product? What is the cost of a back order? Although services may not have developed foolproof ways to handle these dilemmas, they are forcing manufacturers to address the issue.[99]

The Use of Business Intelligence

Business intelligence is a term that is usually associated with the increased understanding of the available information, developed from collected data, about a company's business. It tends to focus on the demand side of the business, rather than the supply side, so we associate it with external factors that affect the business. It probably originated with point-of-sales data collection and has been enhanced with the collection of additional data from surveys and websites. The additional information processing capabilities make it feasible to begin to analyze, interpret, and act on this information. An Aberdeen report by Greg Belkin examined the status of business intelligence in retailing. He found that the top factors driving the implementation of business intelligence included the need for more rapid response to consumer demand, the need to become more operationally efficient, and the need to manage demand across multiple channels. He also determined that some of the most pressing challenges for retailers involved the quality of their data— for example, the data is not good enough, it is not well organized, and it is not customer-specific enough to generate valid analysis and actions.[100]

Summary

The boundary between manufacturing and services is vanishing.[101] Progressive companies are building a bridge over the chasm that separated "hard goods" from "soft services." For years, we have read how programs that originated in the manufacturing area are being adapted for use in service applications. Authors suggest that MRP has application in academic institutions, JIT works in retail, banks use TQM, lean manufacturing revolutionizes distribution, and the health care industry benefits from Six Sigma.

Although many writings describe using manufacturing techniques in services, an increasing number report how concepts and techniques that originated in service industries are adopted in manufacturing. While manufacturing industries concentrated on reducing costs and improving quality, service industries focused on improving customer relationships and service. Certainly, service businesses want to reduce costs and improve quality; that is why they adapt manufacturing techniques to fit their needs. Beyond that, they understand that businesses exist because of their ability to find customers and satisfy their needs. As Peter Drucker described it, "With respect to the definition of business purpose and business mission, there is only one such focus, one starting point. It is the customer. The customer defines the business. Businesses exist for one purpose only—to find customers and provide service to them."[102] Companies must look at their business from the outside in—from the perspective of the customer. Barbara Bund builds on Drucker's basic ideas and includes a number of case histories of businesses that have succeeded, in part because they have taken the "outside-in" look at their companies. She uses service companies such as FedEx, eBay, and Costco to illustrate the customer-centric thinking. She also considers GE and Dell as outside-in thinkers, but only because they think of themselves as service businesses.[103] Low cost is not enough for survival in the future; manufacturing companies must integrate service strategies in their planning.

Does this mean that manufacturing companies are not interested in their customers? Of course not; however, most manufacturing customers are other businesses, not individual consumers. The service industries deal with other businesses, but they also focus on individual consumers. This understanding requires that they have a more diverse customer base and deal with a greater variety of needs and wants. There are no mass markets anymore. Although we have not yet reached single-customer markets, we are certainly into niche markets and the niches are getting smaller.

Differences between Manufacturing and Services

A key difference between manufacturing and services is that manufacturing companies provide a tangible product and service companies provide a less tangible service. Almost all companies provide a combination of goods and services; however, a manufacturing company provides a product as its focus, such as an automobile, and surrounds the product with a service package that may include such things as post-sales service or financing. Service companies provide a service—a complete dining experience—that includes tangible goods such as food. The difference is in the primary focus. There are "defensible differences" between manufacturing and services. There are differences in production strategies, production processes, and performance measures.[104] However, the fact that manufacturing and services are different does not mean that they are incompatible. On the contrary, they are an important complement to each other. As Chase and Garvin put it, "Today's flexible factories will be tomorrow's service factories."[105] They emphasize that manufacturing companies must bundle services with their products; in addition, they must anticipate and respond to an increasing range of customer needs. Bund refers to the "augmented product"—the basic product plus the support, brand name, documentation, and all the other attributes that customers think matter. In this section we show how knowledge transfer between manufacturing and service businesses is helping each to be more competitive.[106]

From Services to Manufacturing

What have the service industries been doing that is of value to manufacturing? Many of these ideas or concepts are not neatly packaged like JIT or Six Sigma, so they are not always obvious. It is easy to find articles about how manufacturing has provided concepts and techniques that can be used in service businesses. It is difficult to find the reverse—how service concepts can be used in manufacturing. A few of the ideas that originated in services are described here.

Service businesses have not been viewed to be as effective or efficient as manufacturing businesses. However, they have always had a customer orientation, a quality sometimes lacking in manufacturing. One of the programs being actively pursued by a number of companies is CRM. As CRM becomes more fully defined, understood, and applied, this concept—which heavily depends on knowledge workers—will be assimilated more effectively into manufacturing businesses.

Historically, quality in manufacturing has been measured by how well the product conformed to the specifications. Although Juran and Gryna pointed out "quality is customer satisfaction" or "fitness for use," an idea that encouraged manufacturers to consider customer needs, it has been convenient to measure product performance against tangible and quantifiable specifications.[107] On the other hand, services have always been somewhat at a disadvantage because customers do not always agree on what is good quality. How do you measure such desired attributes as reliability, responsiveness, competence, access, courtesy, communication, credibility, security, and understanding the customer?[108] As a result, services have always needed to recognize that quality is more than conformance to specifications; it includes customers' perceptions. As manufacturers become more involved in providing complementary services, they must also consider individual customer perceptions of quality.

As a corollary to the intangibility of quality, there is a need to develop qualitative goals and performance measures. How do we "improve customer service" or "enhance the acceptance of our product in the marketplace"? Sometimes the answers can be meaningful, even if not precise. Feedback information comes from surveys, focus groups, extended interviews, complaint letters, and personal observation. As a result, service businesses search for meaning in a labyrinth of ambiguity. Although manufacturers may be uncomfortable with this uncertainty, they have to learn how to deal with it in a meaningful way. The number of customer complaints may not sound as tangible as number of defective units, yet it is a measure of performance. Bowen and Ford advise that "assessing organizational effectiveness and efficiency for an intangible product relies on subjective assessment by the customer; traditional objective measurements need to be supplemented with subjective measures for assessing service experiences."[109]

Manufacturers have subscribed to the notion that bigger is better in building facilities. They locate plants to minimize costs and worry about how to get the product to the market later. Service businesses have no choice; they must locate near their market. This has forced them to learn how to manage smaller units. As manufacturing companies focus on reducing lead times, they will have to consider moving closer to the market and learn how to manage smaller, more flexible units.

Organization structures are also changing—from hierarchical to horizontal or flat. Quinn describes how "service firms—because of their communications–information intensity and localized delivery needs—have generally pioneered" some of the more innovative organization forms such as the matrix, spider web, or inverted pyramid forms.[110] New organization structures enable empowered employees to respond to the customer faster and more effectively.

Collaborative Efforts

Some concepts and programs are difficult to classify as manufacturing or services, such as mass customization. Both manufacturing and service businesses are active in this area. Mass customization represents the convergence of basic manufacturing and service principles, rather than a transfer from one to another. It combines the individual customization associated with service and the efficient volume associated with manufacturing. Mass customization is service-oriented because it is responsive to the customer. However, manufacturing makes it happen.

Quick Response programs also seem to be a collaborative effort. Retail organizations recognized that if they were to have the right goods available at the right time in the right quantities, they had to recognize that accurate forecasts were not always achievable and rely more on rapid replenishment of goods from the manufacturer. Services had the need; manufacturing provided part of the answer.

Manufacturers have worked for decades to eliminate the worker through automation. To date, we have not achieved that ultimate state. On the other hand, the labor-intensive services sector has been learning how to capitalize on the versatility and knowledge of the worker.[111] We are nearing the time when the effort will be on how to optimize the worker–equipment interface. How do we design equipment to aid the workers, not to replace them?

Conclusion

Does it matter where the ideas originate? Most improvement programs require a cross-functional approach. Whether in manufacturing or services, sales and marketing provide the customer contact; operations and finance provide the analysis and execution. The world is changing, and competing successfully is becoming more difficult. Businesses need to take advantage of good ideas regardless of the source. Manufacturing and services can learn from each other; many companies have already started.

We include two case studies on the following pages to provide examples of how manufacturing companies have moved into the services areas. GE has actually been providing services to their customers almost since their beginning, but it has only been in the past half-century that services have become a substantial portion of their business. Services grew to about 45 percent of revenues during the last years of Jack Welch's tenure as CEO but has been reduced to a little over one-third of revenues in recent years under CEO Jeffrey Immelt. As successful as GE has been in adding services to their product line, they have found that some of their ventures were too far afield even for a diversified company. For example, GE Capital was successful in providing financing to customers but ran into difficulty in financing real estate ventures, as described in the GE case included at the end of this chapter.

HP is the second company that has added a significant line of services to their product segments. As we show, they have not been as successful as GE, especially in the past two decades. When HP added services primarily to augment their products, the result was favorable. However, when they made a major move in shifting their product lines from lab equipment for other manufacturers (B2B) to computers for consumers (B2C), their attempt to add significant services became a struggle. HP's strategy regarding products and services has been thwarted by their unparalleled turnover in CEOs, after years of stability with William Hewlett and David Packard serving as CEOs for the first forty years of the company's existence.

References

1. Quinn, J.B., *Intelligent Enterprise: A Knowledge and Service-Based Paradigm for Industry*, The Free Press, New York, 1992.
2. Allmendinger, G. and Lombreglia, R., Four strategies for the age of smart services, *Harvard Business Review*, 83, 10, 131, 2005.
3. Gottfredson, M. and Aspinall, K., Innovation vs. complexity: What is too much of a good thing? *Harvard Business Review*, 83, 11, 62, 2005.
4. Galbraith, J.R., Organizing to deliver solutions, *Organizational Dynamics*, 31, 2, 194, 2002.
5. White, R.E. and Pearson, J.N., JIT, system integration and customer service, *Organizational Dynamics, International Journal of Physical Distribution & Logistics Management*, 31, 5, 313, 2001.
6. Bund, B., *The Outside-In Corporation: How to Build a Customer-Centric Organization for Breakthrough Results*, McGraw Hill, New York, 2006.
7. Haeckel, S.H., *Adaptive Enterprise Creating and Leading Sense-and-Respond Organizations*, Harvard Business School Press, Boston, 1999.
8. Slywotzky, A. and Wise, R., The dangers of product-driven success: What's the next growth act? *The Journal of Business Strategy*, 24, 2, 16, 2003.
9. Olson, E.M., Slater, S.F., and Hult, G.T.M., The importance of structure and process to strategy implementation, *Business Horizons*, 48, 1, 47, 2005.
10. Jain, D. and Singh, S.S., Customer lifetime value research in marketing: A review and future directions, *Journal of Interactive Marketing*, 16, 2, 34, 2002.
11. Band, W. and Guaspari, J., Creating the customer-engaged organization, *Marketing Management*, 12, 4, 34, 2003.
12. Allmendinger, G. and Lombreglia, R., Four strategies for the age of smart services, *Harvard Business Review*, 83, 10, 131, 2005.
13. Ma, J., Nozick, L.K., Tew, J.D., and Truss, L.T., Modeling the effect of custom and stock orders on supply-chain performance, *Production Planning & Control*, 15, 3, 282, 2004.
14. Pine, B.J., II, *Mass Customization, The New Frontier in Business Competition*, Harvard Business School Press, Boston, 1993.
15. Blackstone, J.H., *APICS Dictionary* (13th ed.), 2010, Chicago, IL, APICS—The Educational Society for Resource Management.

16. Schwartz, B., *The Paradox of Choice: Why More Is Less,* Harper Collins, New York, 2004.

17. Piller, F. and Kumar, A., For each, their own: The strategic imperative of mass customization, *Industrial Engineer,* 38, 9, 40, 2006.

18. Schmenner, R.W., *Service Operations Management,* Pearson Custom Publishing, Boston, 1995.

19. Heskett, J.L., *Managing in the Service Economy,* Harvard Business School Press, Boston, 1986.

20. Metters, R.M., King-Metters, K., and Pullman, M., *Successful Service Operations Management,* Thomson South-Western, Mason, OH, 2003.

20a. Fitzsimmons, J.A. and Fitzsimmons, M.J., *Service Management for Competitive Advantage,* McGraw-Hill, New York, 1994.

21. Klassen, K.J. and Rohleder, T.R., Combining operations and marketing to manage capacity and demand in services, *The Service Industries Journal,* 21, 2, 1, 2001.

22. Legarreta, J.M.B. and Miguel, C.E., Collaborative relationship bundling: A new angle on services marketing, *International Journal of Service Industry Management,* 15, 3/4, 264, 2004.

23. Jain, D. and Singh, S.S., Customer lifetime value research in marketing: A review and future directions, *Journal of Interactive Marketing,* 16, 2, 34, 2002.

24. Kapanen, R., Customer relationship management and service delivery, *International Journal of Services Technology and Management,* 5, 1, 42, 2004.

25. Zablah, A.R., Bellinger, D.N., and Johnston, W.J., An evaluation of divergent perspectives on customer relationship management: Towards a common understanding of an emerging phenomenon, *Industrial Marketing Management,* 33, 6, 475, 2004.

26. Crandall, R.E., A fresh face for CRM: Looking beyond marketing, *APICS Magazine,* 16, 9, 20, 2006.

27. Payne, A. and Frow, P., Customer relationship management: From strategy to implementation, *Journal of Marketing Management,* 22, 1/2, 135, 2006.

28. Zablah, A.R., Bellinger, D.N., and Johnston, W.J., An evaluation of divergent perspectives on customer relationship management: Towards a common understanding of an emerging phenomenon, *Industrial Marketing Management,* 33, 6, 475, 2004.

29. Nairn, A., CRM: Helpful or Full of Hype? *Journal of Database Management,* 9, 4, 376, 2002.

30. Chan, J.O., Towards a unified view of customer relationship management, *Journal of American Academy of Business,* 6, 1, 32, 2005.

31. Blackstone, J.H., *APICS Dictionary* (13th ed.), 2010, Chicago, IL, APICS—The Educational Society for Resource Management.

32. Chan, J.O., Towards a unified view of customer relationship management, *Journal of American Academy of Business,* 6, 1, 32, 2005.

33. Lambert, D.M., Garcia-Dastugue, S.J., and Croxton, K.L., An evaluation of process-oriented supply chain management frameworks, *Journal of Business Logistics,* 26, 1, 25, 2005.

34. Lambert, D.M., Garcia-Dastugue, S.J., and Croxton, K.L., An evaluation of process-oriented supply chain management frameworks, *Journal of Business Logistics,* 26, 1, 25, 2005.

35. Compton, J., It's not business as usual, *Customer Relationship Management,* 8, 12, 32, 2004.

36. Bois, R., Live, From Boston! It's Microsoft CRM, http://www.amrresearch.com, July 12, 2006.

36a. CRM Café 2012. http://www.crmcafe.com/crm-software.php (accessed September 7, 2012).

37. Doyle, S., A sample road map for analytical CRM, *Journal of Database Marketing & Customer Strategy Management,* 12, 4, 362, 2005.

38. Ling, R. and Yen, D.C., Customer relationship management: An analysis framework and implementation strategies, *The Journal of Computer Information Systems,* 41, 3, 82, 2001.

39. Payne, A. and Frow, P., Customer relationship management: From strategy to implementation, *Journal of Marketing Management,* 22, 1/2, 135, 2006.

40. Winer, R.S., A framework for Customer Relationship Management, *California Management Review,* 43, 4, 89, 2001.

41. Nairn, A., CRM: Helpful or full of hype? *Journal of Database Management,* 9, 4, 376, 2002.

42. Winer, R.S., A framework for Customer Relationship Management, *California Management Review,* 43, 4, 89, 2001.

43. Lambert, D.M., Garcia-Dastugue, S.J., and Croxton, K.L., An evaluation of process-oriented supply chain management frameworks, *Journal of Business Logistics,* 26, 1, 25, 2005.

44. Kennedy, M.E. and King, A.M., Using customer relationship management to increase profits, *Strategic Finance,* 85, 9, 36, 2004.

45. Nairn, A., CRM: Helpful or full of hype? *Journal of Database Management,* 9, 4, 376, 2002.

46. Keiningham, T.L., Vavra, T.G., Aksoy, L., and Wallard, H., *Loyalty Myths: Hyped Strategies That Will Put You out of Business—and Proven Tactics That Really Work,* John Wiley & Sons, Hoboken, NJ, 2005.

47. Lambert, D.M. and Pohlen, T.L., Supply chain metrics, *International Journal of Logistics Management,* 12, 1, 1, 2001.

48. Giltner, R. and Ciolli, R., Re-think customer segmentation for CRM results, *The Journal of Bank Cost & Management Accounting,* 13, 2, 3, 2000.

49. Hayes, M., Get close to your clients, *Journal of Accountancy,* 201, 6, 49. 2006.

50. Feinberg, R.A., Kadam, R., Hokama, L., and Kim, I., The state of electronic customer relationship management in retailing, *International Journal of Retail & Distribution Management,* 30, 10, 470, 2002.

51. West, J., Customer relationship management and you, *IIE Solutions,* 33, 4, 34. 2001.

52. Shah, J.R. and Murtaza, M.B., Effective customer relationship management through Web services, *The Journal of Computer Information Systems,* 46, 1, 98, 2005.

53. Ko, E. and Kincade, D.H., The impact of Quick Response technologies on retail store attributes, *International Journal of Retail & Distribution Management,* 25, 2, 90, 1997.

54. Crandall, R.E., Beating impossible deadlines, *APICS Magazine,* 16, 6, 20, 2006.

55. Hunter, N.A. and Valentino, P., Quick Response—Ten years later, *International Journal of Clothing Science and Technology,* 7, 4, 30, 1995.

56. Kurt Salmon Associates, Quick Response mandates today, *Online Viewpoint,* March 1997.

57. Lummus, R.R. and Vokurka, R.J., Defining supply chain management: A historical perspective and practical guidelines, *Industrial Management + Data Systems,* 99, 1, 11, 1999.

58. Frankel, R., Goldsby, T.J., and Whipple, J.M., Grocery industry collaboration in the wake of ECR, *International Journal of Logistics Management*, 13, 1, 57, 2002.
59. Angulo, A., Nachtmann, H., and Waller, M.A., Supply chain information sharing in a vendor managed inventory partnership, *Journal of Business Logistics*, 25, 1, 101, 2004.
60. Smaros, J., Lehtonen, J.-M., Appelqvist, P., and Holstrom, J., The impact of increasing demand visibility on production and inventory control efficiency, *Organizational Dynamics, International Journal of Physical Distribution & Logistics Management*, 33, 4, 336, 2003.
61. Poirier, C.C. and Quinn, F.J., How are we doing? A survey of supply chain progress, *Supply Chain Management Review*, 8, 8, 24–31, 2004.
62. http://www.cpfr.org, VICS 2000.
63. Barratt, M. and Oliverira, A., Exploring the experiences of collaborative planning initiatives, *Organizational Dynamics, International Journal of Physical Distribution & Logistics Management*, 31, 4, 266, 2001.
64. Crum, C. and Palmatier, G.E., Demand collaboration: What's holding us back? *Supply Chain Management Review*, 8, 1, 54, 2004.
65. Harrington, L.H., 9 Steps to success with CPFR, *Transportation & Distribution*, 44, 4, 50, 2003.
66. Poirier, C.C. and Quinn, F.J., How are we doing? A survey of supply chain progress, *Supply Chain Management Review*, 8, 8, 24–31, 2004.
67. Sullivan, L. and Bacheldor, B., Slow to sync, *Information Week*, 992, 55, 2004.
68. Corsten, D. and Kumar, N., Do suppliers benefit from collaborative relationships with large retailers? An empirical investigation of Efficient Consumer Response adoption, *Journal of Marketing*, 69, 3, 80, 2005.
69. Rosenblum, P., The Business Benefits of Advanced Planning and Replenishment, Aberdeen Group, 2005, white paper.
70. Sparks, L. and Wagner, B.A., Retail exchanges: A research agenda, *Supply Chain Management*, 8, 3/4, 201, 2003.
71. Pine, B.J., II, *Mass Customization, The New Frontier in Business Competition*, Harvard Business School Press, Boston, 1993.
72. Quinn, J.B., *Intelligent Enterprise: A Knowledge and Service Based Paradigm for Industry*, The Free Press, New York, 1992.
73. Quinn, J.B., Anderson, P., and Finkelstein, S., Leveraging intellect, *The Academy of Management Executive*, 10, 3, 7, 1996.
74. Quinn, J.B., *Intelligent Enterprise: A Knowledge and Service Based Paradigm for Industry*, The Free Press, New York, 1992.
75. Lovelock, C.H. and Young, R.F., Look to consumers to increase productivity, *Harvard Business Review*, 57, 3, 168, 1979.
76. Klassen, K.J. and Rohleder, T.R., Combining operations and marketing to manage capacity and demand in services, *The Service Industries Journal*, 21, 2, 1, 2001.
77. Mentzer, J.T., A telling fortune, *Industrial Engineer*, 38, 4, 42, 2006.
78. Morash, E.A., Droge, C., and Vickery, S., Boundary-spanning interfaces between logistics, production, marketing, and new product development, *Organizational Dynamics, International Journal of Physical Distribution & Logistics Management*, 27, 5/6, 350, 1997.
79. Rhyne, D.M., The impact of demand management on service system performance, *The Service Industries Journal*, 8, 4, 446, 1988.

80. Boudreau, J., Hopp, W., McClain, J.O., and Thomas, L.J., On the interface between operations and human resource management, *Manufacturing & Service Operations Management,* 5, 3, 179, 2003.

81. Lankford, W.M. and Johnson, J.E., EDI via the Internet, *Information Management & Computer Security,* 8, 1, 27, 2000.

82. Kalakota, R. and Whinston, A.B., *Frontiers of Electronic Commerce,* Addison-Wesley, Reading, MA, 1996.

83. Sawabini, S., EDI and the Internet, *The Journal of Business Strategy,* 22, 1, 41, 2001.

84. Angeles, R., Revisiting the role of Internet-EDI in the current electronic commerce scene, *Information Management,* 13, 1, 45, 2000.

85. Senn, J.A., Expanding the reach of electronic commerce, *Information Systems Management,* 15, 3, 7, 1998.

86. Kalakota, R. and Whinston, A.B., *Frontiers of Electronic Commerce,* Addison-Wesley, Reading, MA, 1996.

87. Jessup, L.M. and Valacich, J.S., *Information Systems Today* (2nd ed.), Pearson Education, Inc., Prentice Hall, Upper Saddle River, NJ, 2006.

88. http://www.ediuniversity.com.

89. Bury, S., Piggly Wiggly's doing it, *Manufacturing Business Technology,* 23, 2, 42, 2005.

90. Bednarz, A., Internet EDI: Blending old and new, *Network World,* 21, 8, 29, 2004.

91. Anonymous, AS2 is A-OK at Wal-Mart, *Chain Store Age,* 80, 2, 38, 2004.

92. Jessup, L.M. and Valacich, J.S., *Information Systems Today* (2nd ed.), Pearson Education, Inc., Prentice Hall, Upper Saddle River, NJ, 2006.

93. Jessup, L.M. and Valacich, J.S., *Information Systems Today* (2nd ed.), Pearson Education, Inc., Prentice Hall, Upper Saddle River, NJ, 2006.

94. Grossman, M., The role of trust and collaboration in the Internet-enabled supply chain, *Journal of American Academy of Business,* 5, 1/2, 391, 2004.

95. Ruppel, C., An information systems perspective of supply chain tool compatibility: The roles of technology fit and relationships, *Business Process Management Journal,* 10, 3, 311, 2004.

96. Jessup, L.M. and Valacich, J.S., *Information Systems Today* (2nd ed.), Pearson Education, Inc., Prentice Hall, Upper Saddle River, NJ, 2006.

97. Juran, J.M. and Gryna, F.M., *Quality Planning and Analysis, From Product Development through Use,* McGraw-Hill, New York, 1993.

98. Fleming, J.H., Coffman, C., and Harter, J.K., Manage your Human Sigma, *Harvard Business Review,* 83, 7, 106, 2005.

99. Evans, J.R., An exploratory study of performance measurement systems and relationships with performance results, *Journal of Operations Management,* 22, 3, 219, 2004.

100. Belkin, G., Business Intelligence in Retail: Bringing Cohesion to a Fragmented Enterprise: A Benchmark Report, Aberdeen Group, June 2006.

101. Crandall, R.E., Looking to service industries, new resources for learning a thing or two, *APICS Magazine,* 16, 10, 21, 2006.

102. Drucker, P.F., *Management: Tasks, Responsibilities, Practices,* Harper & Row, New York, 1974.

103. Bund, B., *The Outside-In Corporation: How to Build a Customer-Centric Organization for Breakthrough Results,* McGraw Hill, New York, 2006.

104. Bowen, J. and Ford, R.C., Managing service organizations: Does having a "thing" make a difference? *Journal of Management,* 28, 3, 447, 2002.

105. Chase, R.B. and Garvin, D.A., The service factory, *Harvard Business Review,* 67, 4, 61, 1989.
106. Bund, B., *The Outside-In Corporation: How to Build a Customer-Centric Organization for Breakthrough Results,* McGraw Hill, New York, 2006.
107. Juran, J.M. and Gryna, F.M., *Quality Planning and Analysis, From Product Development through Use,* McGraw-Hill, New York, 1993.
108. Parasuranam, A., Zeithaml, V.A., and Berry, L.L., *Delivering Quality Service and Balancing Customer Expectations,* Free Press, New York, 1990.
109. Bowen, J. and Ford, R.C., Managing service organizations: Does having a "thing" make a difference? *Journal of Management,* 28, 3, 447, 2002.
110. Quinn, J.B., *Intelligent Enterprise: A Knowledge and Service Based Paradigm for Industry,* The Free Press, New York, 1992.
111. Rayport, J.F. and Jaworski, B.J., *Best Face Forward: Why Companies Must Improve Their Service Interfaces with Customers,* Harvard Business School Press, Boston, 2005.
112. Piller, F. and Kumar, A., For each, their own: The strategic imperative of mass customization, *Industrial Engineer,* 38, 9, 40, 2006.

Appendix 6A: GE—An Example of How to Blend Services into a Manufacturing Company

Background

General Electric (GE) is often cited as a company that has grown from being known primarily as a manufacturing company to one that blends services with its manufacturing base.[1,2] Why did GE decide to add services? Once having decided, how did they achieve it successfully? These are straightforward questions; however, as we will describe, the answer is not as straightforward.

GE started as the Edison Electric Light Company in 1878, named after its founder, Thomas Edison. By 1879, they introduced the carbon filament incandescent lamp to replace gas-fired lamps. Edison pushed for the use of direct current (DC) while George Westinghouse, a prime competitor, promoted the use of alternating current (AC). Westinghouse won that battle but one of the early investors in Edison's company, Henry Villard, assumed management of the company and acquired the Thomson-Houston company, which had capabilities in both DC and AC, thereby ensuring continuation in the lighting business. From that somewhat uncertain beginning, GE has grown into the major business it is today. It is No. 6 in total sales volume in the Fortune 500 list, trailing only Walmart, three energy companies (Exxon Mobil, Chevron, and ConocoPhillips), and Fannie Mae, with sales of over $150 billion. It is larger than General Motors, Ford, IBM, and even Apple, to list a few of the more prominent names.[3] How did a lightbulb company achieve this level of prominence?

The Early Years

Although incandescent lamps were the starting point, Edison, and the company, soon moved into other products for the lighting industry—dynamos in 1879 and power stations in 1882. In 1892, the company was renamed General Electric and continued to build products that used electricity—generating it and using it in thousands of applications. For example, in 1895, they introduced an electric locomotive and transformers; in 1896, an x-ray machine; in 1902, electric fans; and in 1903, large steam turbines. Part of their success in developing new products came from the establishment of the GE Research and Development Laboratory in 1900.[4]

With all of the emphasis on products that use electricity, GE recognized there would be a need for utilities to generate electricity. Consequently, they began to develop products they could sell to electric utilities. Even at the beginning, they recognized that it would be beneficial to develop relationships with their customers that would last. They favored private companies for this, as opposed to government-owned businesses, and supported small companies by providing financing. This financing of customers has carried through to modern times and was the genesis of today's GE financial services sectors. So services are not new at GE! Management also used another strategy to build a tightly linked supply chain to their ultimate consumers; they built products for homeowners that used electricity, such as electric ranges (1910) to be followed with a wide array of other home appliances. If homeowners bought and used the product, they would use more electricity, causing the electric utilities to need more capacity. In turn, the utilities would buy more products from GE. GE built relationships with their customers but never tried to vertically integrate forward into their customers' businesses. From the beginning, GE recognized the need to make products their customers needed, a critical success factor in service businesses.

Why Are Services Important to a Manufacturing Company?

In Chapter 6, we have described how manufacturing companies are becoming more customer focused and recognizing the need to supplement their products with a service package. In many cases, adding services generates more revenue and income for companies. As a minimum, they enable the company to remain competitive. While adding services can be good, they can also become a negative if not done correctly. GE has had a great deal of success with their service additions; however, they have also experienced some less than successful service ventures.

Ways for Manufacturers to Add Services

Manufacturing companies can move into service businesses, in rational and systematic ways. These strategies include the following:

- To help develop a market for the product
- To extend known technology into a new product area
- To provide post-sales support
- To help sell products by financing sales
- To provide infrastructure in emerging countries
- To enter a more profitable business venture.

GE has used these approaches to expand their service offerings.

Develop a Market for the Product

As described above, GE looked downstream in their supply chain to focus on the ultimate consumer, the homeowner. The following examples described by Rothschild[5] illustrate this approach, which was used by GE during the first half of the twentieth century.

Swope and Young enhanced and perfected the process of stimulating electrical demand. Their strategies included the following:

- **Benign cycle strategy.** This was a strategy to develop products for the home that would enable people to "Live Better Electrically," the slogan GE used to convey this idea to the public. Some of the products developed included electric stoves, clothes washers and dryers, monitor top refrigerators, vacuum cleaners, and electric fans.
- **Retailer franchising.** GE did not own the retail stores but set up carefully selected merchants to represent GE products in exclusive territories. The dealers were given business, accounting, and sales training and were supported with strong advertising and sales promotion programs.
- **Consignment selling.** To reduce the independent business capital and cash needs, GE initiated consignment selling and trained the dealers to provide after-sales repair and maintenance with GE factory-authorized parts. GE also imposed "fair-trade" pricing, later declared illegal.
- **GE Credit** was formed in 1932 to provide financing to the franchised dealers to buy inventory and devise attractive floor displays for customers to view. Franchised dealers were successful until the mid-1950s when discounters and outlawing of fair-trade practices made them obsolete.
- **Aggressive advertising and promotion.** Using the GE brand as the attractor, advertising and sales promotion programs were designed to entice the prospective customer into the store while dealers were trained to get the order.
- **General Electric Supply Company.** This wholesale function was added in 1929 to handle a complete line of GE consumer, lighting, industrial, and commercial products, focusing primarily where GE distribution was weak or nonexistent.
- **Relationship selling (supporting the retailers).** "It is interesting to note that GE has never been a true 'consumer company' in the sense that it has always relied on intermediaries to deal with the consumer. In fact, GE has treated the consumer business more like an industrial business, and it has prospered by establishing a network of selected intermediaries rather than by trying to sell directly to consumers" (p. 63).[5]

Extend Known Technology into a New Product Area

GE continued to develop products that use, or could use, electricity. From their early concentration in lighting, this approach carried them from incandescent lamps into fluorescent lamps (1938), then into halide lamps, and on into light-emitting diodes (LED) lights in more recent years.

This same pursuit to develop products that use electricity produced electric locomotives in the early 1900s that led them on into transformers and other products for locomotives, x-ray machines that extended into a variety of products for health care, vacuum tubes that led to radio and television sets, Calrod heating rods for electric ranges, magnetrons as a forerunner to microwave ovens, superchargers for airplane engines, generators for Niagara Falls, alternators for transoceanic radio, turbine generators to power ships, invisible glass for cameras and optical devices, and a host of other products that stemmed from their search for products that use electricity.[6]

In the materials sector, GE started studying materials to make better insulators for transformers. This led them into resins and polymers that later resulted in the development of Lexan in 1953 and Noryl in 1956. Two Nobel Prize winners came out of this research—Dr. Irving Langmiur for surface chemistry (1932) and Dr. Ivar Giaever for superconductive tunneling (1973). Lexan has been used in coatings for electronic display screens, paint replacement on cars, and in the development of superhydrophobic nano-coatings.[4]

By adding new products, GE has opened up the opportunity to add services to support those new products, especially those products that are complex, such as jet engines, locomotives, and nuclear power plants.

Provide Post-Sales Support

As GE developed new products, they found the need to provide post-sale services in the form of maintenance and repair, as well as increasing the efficiency of the product's usage. This need was especially true in high-technology, high-complexity product areas, such as nuclear plants, jet engines, and hospital imaging equipment. This was a natural extension of services along the supply chain downstream toward the ultimate consumer.

Help Sell Products by Financing Sales

From its beginning, GE has been in the business financing business. They started by helping small electric utilities to grow by providing capital and, in some cases, taking an equity position in the utility company.[5] They extended the financing of small retailers through which GE started selling home appliances. Because they learned how to extend financing to other businesses, it was a natural evolution to extend this form of service, largely through what became GE Capital. GE Capital was led by aggressive management and extended their financing services into other areas, such as buying airplanes and leasing them back to airlines. They also entered the real estate business, with mixed results, especially in the past few years when the real estate market has been dismal.

Provide Infrastructure in Emerging Countries

GE has an extensive network of businesses outside the United States. Many are extensions of existing products; however, some are new product areas developed for specific needs in other countries, such as mobile water purification systems.

Enter a More Profitable Business

Some service areas have historically carried higher profit margins than established manufactured products. IBM demonstrated this transition by moving from their dominant, but declining, position in large computers to becoming predominantly a software and consulting provider of information technology services. GE has not changed as dramatically although the increasing presence of GE Capital has almost overshadowed the presence of manufactured products in GE's product portfolio. This mix appears to be changing under the current CEO, Jeffrey Immelt, as he moves to reestablish the emphasis on products, and away from service sectors not directly related to products, such as GE Capital's venture into real estate and entertainment.

GE Progress over Time

Another view of GE's evolution in products and services is to look at snapshots of their product lines over time. Rothschild[5] has provided insight into the influence of different CEOs on not only product selection and development but also the company's approaches to relationships with external entities, such as government, society, and unions.

One of GE's strengths has been their top management. They have had stability in the CEO and president positions and have always promoted from within. The company has long been known as a staunch advocate of management training. One of their secrets has been to require the existing CEO to develop a succession plan, usually with the objective of selecting someone young enough to have an extended tenure as CEO. The objective is to enable a CEO to establish and develop "his program" (to date they have had only male CEOs). Often, the CEO and president have performed as a team. Table A6a.1 provides a view of how products and services have progressed under different leaders.

Growth during World War II

During two periods in GE's history they made deliberate attempts to broaden their product lines, not only in new and innovative products but also in services. The first was during World War II when the U.S. government urged companies to produce equipment and supplies for the war effort. GE responded with a major emphasis in the following:

- **Propulsion systems**—For ships, aircraft and locomotives. Led to the jet engine for airplanes (the aerospace and defense businesses were sold by Welch in 1980).
- **Electronics**—Industrial controls, industrial computers, numerous telecommunication businesses, substituting electronics for many electromechanical devices.
- **Nuclear**—"GE led to the development of the boiling water reactor (BWR) for electric utility applications, and it was a major contributor to the development of the pressure water reactors for Admiral Hyman G. Rickover's nuclear submarine fleets" (p. 78).[5]

Table A6a.1 Evolution of Products and Services over Stages of GE History

Stages of GE History	Organization and Culture	Products	Services
Stage One, "Living Better Electrically" — Edison through Coffin and Rice, and Swope and Young (1879–1945)	Tolerant and cordial relationship with unions and employee	Lighting and products for electric utilities. Also the beginning of materials research as they looked for better transformer insulating materials.	Financing for electric utilities grew into financing for retailers who sold appliances.
Stage Two, Diversification and Decentralization, "Progress is our Most Important Product," Wilson, Reed, Cordiner, Phillippe, and Swope (1946–1970)	Took a much harder line with unions; introduced "Boulwarism" to establish rigid union relationships. Instituted hierarchy with groups, divisions and department structure. Battled "discounters" who opposed Fair Trade laws that enabled manufacturers to control pricing of their products.	Wilson and Reed were caught up in the need to develop products for the U.S. government during WW II. Swope later moved aggressively into jet engines and polymer materials. By 1956, GE manufactured 200,000 products in 100 departments: 35% consumer products, 25% for business and industry, 20% highly engineered defense, electronics and atomic products, and 20% components and materials for other businesses.	As GE used computers, they developed a capability in their applications. Swope also put emphasis on personal and business financing, and hospital services. Created a new distribution center to sell directly to home and apartment builders. Offered a variety of services to electricity generators, such as demand forecasting, lobbying for rates based on ROI, and training.

Table A6a.1 (continued) Evolution of Products and Services over Stages of GE History

Stages of GE History	Organization and Culture	Products	Services
Stage Three, Portfolio Leadership, "We Bring Good Things to Life," Swope, Jones, and Welch (1971–2001)	Jones emphasized managerial training and formalized strategic planning. Welch shifted strategic planning from strategies to plans and from corporate to SBU as the focal point.	See Table A6a.2. Jones focused on managing better the existing product lines. Welch weeded out lower performing sectors, such as small appliances and electronic components.	Jones acquired Utah International for stable cash flow and added emphasis on energy (It didn't work!). Welch also relied heavily on services, especially GE Capital for growth and earnings.
Stage Four, Back-to-the-Future, "Imagination at Work," and "Ecoimagination," Immelt (2001–present)	Reduced power of GE Capital and increased the emphasis on research, especially in health care, sustainability, and management of complex systems. Heavy emphasis on global markets, including emerging markets.	Heavy emphasis on hospital imaging equipment and disease detection equipment. Introduction of wind power farms.	Sold reinsurance business, spun off insurance as Genworth. Provide facility management services, especially in health care and nuclear.

Source: Adapted from Rothschild 2007.

- **Industrial automation and productivity systems**—Used to improve its own operations and to sell to other industrial companies.
- **Materials**—Used its chemical and materials knowledge to improve silicones and to make new materials that could be used in military applications. Led to GE's innovations in plastics and artificial diamonds.[5]

Although these efforts concentrated on new products, they also created opportunities in service areas such as maintenance and repair, with the need for replacement parts. This became evident in the nuclear sector later during the period when GE's nuclear business was profitable solely from after-market sales despite no new nuclear plants being built.

Growth Council Product Proposals (1970s)

The second major product development emphasis was during Borch's early tenure during the 1970s. Borch established the Growth Council to "identify market opportunities that were growing faster than the GNP and in which GE had unique abilities and might gain a competitive advantage." Rothschild[5] identifies the following recommendations by the Growth Council in the product area:

- **Nuclear energy**—Provide enrichment services and help utilities dispose of highly toxic nuclear waste safely. This venture was never profitable, and the 1979 Three Mile Island accident cooled the movement toward nuclear power plants. However, GE remained in the nuclear business by selling replacement parts plus maintenance and repair.
- **Business computers**—GE was the largest IBM computer user, and the Growth Council recommended that GE enter the commercial and business computer market and compete with IBM. GE became one of the "seven dwarfs," a term used to designate competitors of IBM that held a small percent of the market in computers. In this venture, GE learned that even good managers can't manage everything.
- **Commercial jet engines**—Based on its knowledge acquired during World War II, GE expanded its jet engine business. Initially, GE (25 percent of the market) competed with Pratt & Whitney (75 percent of the market). They set up two facilities: Cincinnati, Ohio, for large commercial jet engines and Lynn, Massachusetts, for smaller military engines. Success came when GE provided an attractive financing package: GE would buy the airplanes and lease them to the airlines. This reversed the share of market, and GE controlled 75 percent with P&W left with 25 percent.
- **Polymer chemicals**—GE's initial entry into materials was in attempting to develop a new and better way of insulating generators or transformers. The growth strategy, led by a chemical engineer, Charlie Reed, was to solve high-value problems faced by key industries but to avoid competing in the

high-volume community materials sector dominated by DuPont and others. "Reed's team developed a number of new materials, including Lexan and Noryl, and the business has been very profitable for many years" (p. 135).[5]

In Rothschild's opinion, one of the reasons the chemical business was successful was that it felt it was "different" from the rest of GE. "In fact, I don't believe that there has been one 'GE culture' since the company became highly diversified. Each business is so different that it develops its own culture and style. Unlike GM, Ford, or the pre-Gerstner IBM, there was no one GE style or culture. In fact, one reason the company has been able to be both diversified and successful is that more than one culture is tolerated" (p. 135).[5]

> One of the common characteristics of these ventures was that they were going after huge potential markets, which required perseverance and financial resources to make them successful. All of them required long-term relationship building, and they forced GE to provide the customer not only with great products but also with the skills, financing, and management needed to be successful. You can see that these are very similar to the characteristics required in the electric utility industry, which was so well known and understood by the GE management. These are the types of businesses in which GE has been able to win (p. 136).[5]

Two of these ventures succeeded—jet engines and polymer chemicals, most likely because they were extensions of earlier technology. Building nuclear plants was not a profitable venture because GE offered a turnkey package and underestimated the costs. However, the company has reopened their efforts in this area as alternative energy options have become more actively discussed. Business computers also failed to become a profitable business, perhaps because GE trailed IBM and others and never developed a viable strategy for this business.

Growth Council Service Proposals (1970s)

The Growth Council suggested five major ventures in the services sector, two of which became major business segments, one of which has had modest success, and two of which never became viable businesses for GE.

■ **Entertainment**—An opportunity to create special programming for the television networks and independent stations and to provide films for distribution through traditional movie theater channels. This venture failed because show business was too different from what GE knew. However, in recent years, GE has continued to pursue this high-risk, high-reward business by acquiring Universal Studies (2003). This involved acquisitions, not internal growth.

- **Community development and housing**—The idea was to build complete communities and supply the houses with GE products, not only for the building but also to the consumers. The venture was unsuccessful because GE didn't recognize the need to acquire the land, the difficulty in getting government approvals, and the complexity of dealing with trade unions.
- **Personal and financial services**—The concept was to use the insurance, pension, investment, and lending skills of the company and create a package to be offered by the GE Credit Company (GECC). "This financial venture was the most successful of all of the recommended service ventures, and it became a major part of the company's portfolio when Welch became CEO. Even though the Growth Council started the company moving into this arena, it took more than two decades to make it a very significant contributor to, and factor in, GE's strategies" (p. 140).[5]
- **Medicine**—Expand on the x-ray tube beginning by offering financial and management services to hospitals and other medical systems that would tie in with its diagnostic position. In addition, GE could sell health care services to other companies to manage their health care programs. This venture stalled until Jack Welch became its champion and invested in new technologies, including magnetic resonance imaging (MRI). Jeff Immelt continues to be a strong advocate in this area.
- **Education**—GE sought to use its computer technology and programmed learning systems to enhance the educational process. This venture failed because of the power of the educational unions and the resistance of state and federal educational agencies to support new ways of teaching (p. 141).[5]

Two of these service areas have become strong business sectors—personal and business financing, and health care. Business financing grew out of the early efforts in financing electric utilities and has been extended to other industries. Health care services have complemented product developments, particularly in imaging and resonance equipment. The total system approach to community development was a grand vision but proved to be too far removed from the company's core competencies. The education initiative was also a conceptual home run but was never realized because of lack of readiness in the marketplace. The entertainment option has been an area that appears to be far removed from GE's core strengths; however, top management has steadily reduced the company's holdings in the entertainment sector.

Conclusions about Growth Council Ventures

- GE was too early in some markets. The community housing concept has come into fruition in the form of gated communities and self-contained buildings where people work, play, and live within the same complex.
- GE attempted to pursue too many ventures. Even a company the size of GE can reach the limit of their resources in adding major new product areas.

■ GE did not have the knowledge in some ventures. Although GE had one of the strongest and most effective managerial training programs, they found that even well-trained managers cannot manage everything, especially those requiring knowledge and experience unique to the product.[5]

■ It was difficult for GE to seriously consider small business areas, even those which could have great potential. This is the "innovator's dilemma" described by Christiansen.[7]

More Recent Product Profiles

GE management has actively pursued new products and services, and, at the same time, they have not been reluctant to dispose of, or terminate, product areas considered no longer attractive. This was evident in their termination of their computer business and the idea of harvesting and discontinuing businesses continues today. Table A6a.2 compares the strategic analyses of Jones and Welch as to the desirability of product lines.

In Table A6a.2, the horizontal axis shows Jones's analysis of the product lines. He classified them as Priority I: Invest and Grow; Priority 2: Selectively Grow and Defend; Priority 3: Harvest (phase out); and Priority 4: Exit/Divest. Welch's analysis is shown on the vertical axis and consists of three categories: Services, High Technology, and Core. While the two CEOs used different approaches, there was much agreement in their final decisions. The area in the upper left section of the table shows the agreement between the two analyses. While Welch initially included Aerospace in his Technology group, he later sold this business. Jones would have kept small appliances, but Welch did not include this in his segments to be retained and subsequently sold them. Both Jones and Welch agreed that those products shown in the lower right corner of the table were candidates for disposition. While there was much agreement, the differences also indicate the fact that CEOs at GE are not selected to be caretakers, but innovators.

The Welch Years

Table A6a.3 provides financial results for the period from 1982 through 2001. This covers the twenty years when Welch was CEO. It is interesting to look at this total of twenty years in five-year increments to see the shifts in the mix of products and services.

The results for 1986 show that products represented 77 percent of total sales, services 19 percent, GE Capital 1 percent, and other 3 percent. The gross margin for goods was 26 percent and 23 percent for services. Net earnings were 6.8 percent of sales.

During Welch's second five years as CEO (1987–1991), the mix changed rapidly. Sales increased from $36.7 billion to $60.2 billion, a 64 percent increase, with most of the increase coming from GE Capital, which represented 27 percent

Table A6a.2   Comparison of Jones and Welch Product Strategies

Corporate Strategy by Jones → Corporate Strategy by Welch ↓	Invest/Grow	Select Growth/ Defend	Harvest	Exit/Divest
Services	Financial Informational Construction & Engineering Nuclear			
High Technology	Materials Aircraft engine	Medical systems Industrial Electronics	Aerospace[a]	
Core	Lighting Turbines	Major appliances Transportation Motors Contract equipment		
Outside the Circles (Question Marks)		Small appliances[a]	Central air-conditioning Large transformer	TV/audio Switchgear Wire/cable

Note: The shaded areas were strategies supported by both Welch and Jones.

[a] Strategies on which Welch and Jones differed.

of total sales, with products contributing 55 percent and services 16 percent, with other only 1 percent. The gross margin for goods was 26 percent and 20 percent for services, indicating that services were not necessarily a more profitable business. Net earnings, as a percent of sales, were around 7 percent, except for a low of 4.4 percent in 1991.

From 1992 through 1996 (Welch's third five-year stint), total sales increased from $60.2 billion to $79.2 billion, an increase of $19.0 billion, or 31 percent. Of the total increases, $16.2 billion came from GE Capital, or 85 percent of the increase. Products represented 43 percent of total sales, with services and GE Capital

Table A6a.3a   GE Sales and Income for the Years 1982–1991

Welch's First Ten Years from 1982 until 1991
In Millions of Dollars

	Dec 31 1982	Dec 31 1983	Dec 31 1984	Dec 31 1985	Dec 31 1986	Dec 31 1987	Dec 31 1988	Dec 31 1989	Dec 31 1990	Dec 31 1991
Sales	26,500	26,797	27,947	28,285						
Goods					28,139	29,937	28,953	31,314	33,321	33,127
Services					7,072	9,378	9,840	9,673	9,630	9,911
Other					1,010	648	675	690	752	857
GE Capital					504	552	10,621	12,897	14,711	16,341
Total Revenues	26,500	26,797	27,947	28,285	36,725	40,515	50,089	54,574	58,414	60,236
Cost of goods sold	18,605	18,701	19,460	19,775	20,757	22,359	21,155	23,059	24,815	24,635
Cost of services sold					5,430	7,298	7,676	7,314	7,515	7,937
Interest & Finance Charges					625	645	4,817	6,591	7,392	7,404
Insurance Losses							1,501	1,614	1,599	1,623
Provision Financing Receivables							434	527	688	1,102
Other Overhead	5,142	5,063	5,131	4,970	6,224	7,006	9,724	9,682	10,176	11,027
Miscellaneous							61	84	82	72

continued

Table A6a.3a (continued)   GE Sales and Income for the Years 1982–1991

	Welch's First Ten Years from 1982 until 1991 In Millions of Dollars									
	Dec 31 1982	Dec 31 1983	Dec 31 1984	Dec 31 1985	Dec 31 1986	Dec 31 1987	Dec 31 1988	Dec 31 1989	Dec 31 1990	Dec 31 1991
Total Costs	23,747	23,764	24,591	24,745	33,036	37,308	45,368	48,871	52,267	53,800
Earnings from Operations	2,753	3,033	3,356	3,540	3,689	3,207	4,721	5,703	6,147	6,436
Total Taxes and Adjustments	936	1,009	1,076	1,204	1,197	292	1,335	1,764	1,844	3,800
Net Earnings	1,817	2,024	2,280	2,336	2,492	2,915	3,386	3,939	4,303	2,636
% Gross Margin										
Goods					26%	25%	27%	26%	26%	26%
Services					23%	22%	22%	24%	22%	20%
% of Sales										
Goods					77%	74%	58%	57%	57%	55%
Services					19%	23%	20%	18%	16%	16%
Other					3%	2%	1%	1%	1%	1%
GE Capital					1%	1%	21%	24%	25%	27%
Total Percent of Revenues					100%	100%	100%	100%	100%	100%
% Net Earnings to Revenues	6.9%	7.6%	8.2%	8.3%	6.8%	7.2%	6.8%	7.2%	7.4%	4.4%

Source: Adapted from Mergent Online.

Table A6a.3b GE Sales and Income for the Years 1992–2001

Welch's Second Ten Years from 1992 until 2001
In Millions of Dollars

	Dec 31 1992	Dec 31 1993	Dec 31 1994	Dec 31 1995	Dec 31 1996	Dec 31 1997	Dec 31 1998	Dec 31 1999	Dec 31 2000	Dec 31 2001
Sales										
Goods	29,575	29,509	30,740	33,157	34,180	40,675	43,749	47,785	54,828	52,677
Services	8,331	8,268	8,803	9,733	11,791	12,729	14,938	16,283	18,126	18,722
Other	799	735	793	752	638	2,300	649	798	436	234
GE Capital	18,368	22,050	19,773	26,386	32,570	35,136	41,133	46,764	56,463	54,280
Total Revenues	57,073	60,562	60,109	70,028	79,179	90,840	100,469	111,630	129,853	125,913
Cost of goods sold	22,107	22,606	22,748	24,288	24,578	30,889	31,772	34,554	39,312	35,678
Cost of services sold	6,273	6,308	6,214	6,682	8,293	9,199	10,508	11,404	12,511	13,419
Interest & Finance Charges	6,860	6,989	4,949	7,286	7,904	8,384	9,753	10,013	11,720	11,062
Insurance Losses	1,957	3,172	3,507	5,285	6,678	8,278	9,608	11,028	14,399	15,062
Provision Financing Receivables	1,056	987	873	1,117	1,033	1,421	1,609	1,678	2,045	2,481
Other Overhead	12,494	13,774	12,987	15,429	19,618	21,250	23,477	27,011	30,993	28,162
Miscellaneous	53	151	170	204	269	240	265	365	427	348
Total Costs	50,800	53,987	51,448	60,291	68,373	79,661	86,992	96,053	111,407	106,212

continued

Table A6a.3b (continued)   GE Sales and Income for the Years 1992–2001

Welch's Second Ten Years from 1992 until 2001
In Millions of Dollars

	Dec 31 1992	Dec 31 1993	Dec 31 1994	Dec 31 1995	Dec 31 1996	Dec 31 1997	Dec 31 1998	Dec 31 1999	Dec 31 2000	Dec 31 2001
Earnings from Operations	6,273	6,575	8,661	9,737	10,806	11,179	13,477	15,577	18,446	19,701
Total Taxes and Adjustments	1,548	2,260	3,935	3,164	3,526	2,976	4,181	4,860	5,711	6,017
Net Earnings	4,725	4,315	4,726	6,573	7,280	8,203	9,296	10,717	12,735	13,684
% Gross Margin										
Goods	25%	23%	26%	27%	28%	24%	27%	28%	28%	32%
Services	25%	24%	29%	31%	30%	28%	30%	30%	31%	28%
% of Sales										
Goods	52%	49%	51%	47%	43%	45%	44%	43%	42%	42%
Services	15%	14%	15%	14%	15%	14%	15%	15%	14%	15%
Other	1%	1%	1%	1%	1%	3%	1%	1%	0%	0%
GE Capital	32%	36%	33%	38%	41%	39%	41%	42%	43%	43%
Total Percent of Revenues	100%	100%	100%	100%	100%	100%	100%	100%	100%	100%
% Net Earnings to Revenues	8.3%	7.1%	7.9%	9.4%	9.2%	9.0%	9.3%	9.6%	9.8%	10.9%

Source: Adapted from Mergent Online.

providing the remaining 57 percent. The remaining increase came from products ($1.0 billion) and services ($2.8 billion). Clearly, GE's growth was tied to the growth in services and GE Capital. The gross margin of services also improved to 30 percent, slightly better than the 28 percent for products. Net income, as a percent of sales, ranged from a low of 7.1 percent in 1993 to a high of 9.4 percent in 1995.

During Welch's last five-year period (1997–2001), sales increased from $79.2 billion in 1996 to $125.9 billion in 2001, an increase of $46.7 billion, or 59 percent. This growth was spread among all three sectors with product sales increasing by $18.5 billion, services by $7.0 billion, and GE Capital by $21.7 billion, while other sales declined by $0.4 billion. Product sales declined as a percent of sales to 42 percent, while services and GE Capital represented 58 percent. Gross margin for products increased to 32 percent in 2001, while the gross margin for services remained fairly steady in the 28–30 percent range for the period. Net earnings, as a percent of sales, rose throughout the five-year period to a high of 10.9 percent in 2001.

Figures A6a.1 and A6a.2 show the mix of revenues during the period from 1987 through 2011. By 1987, CEO Welch had trimmed some of the products, such as aerospace, that he felt did not fit his strategic objectives for the company and was counting heavily on the growth of services to build total volume and earnings. As Figure A6a.1 shows, the total revenues under Welch tripled from 1987 through most of 2001, from $40 billion to over $120 billion. Product sales increased from

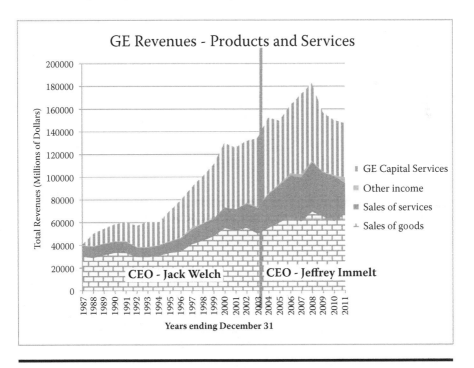

Figure A6a.1 Total revenues by products and services (Mergent Online, 2011).

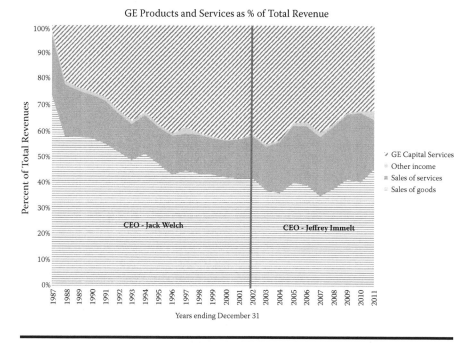

Figure A6a.2   GE revenues as percent of total (Mergent Online 2011).

about $30 billion to $52 billion, while sales from services and GE Capital increased from about $10 billion to $73 billion. Of the $73 billion, GE Capital accounted for $54 billion. A significant portion of GE Capital's expansion was in aggressive new areas of financing, such as real estate.

It is obvious that this twenty-year period was an extremely successful era for GE and Welch gained the reputation as a masterful CEO. It is also obvious that he transformed GE from a primarily goods-producing company to one almost equally divided between products and services, with particular emphasis on the financing element of services—GE Capital.

The Immelt Years

Table A6a.4 shows the financial results for the time Jeffrey Immelt has been CEO, from 2002 through 2011. It shows that the first five years (2002–2006) continued the period of rapid growth, while during the second five years growth peaked in 2008 and declined during 2009–2011.

From 2001 through 2006, total sales increased by $37.5 billion, or 30 percent. Product sales increased by 22 percent, while service sales increased by 94 percent, and gross margin of service sales increased from 28 percent in 2001 to 35 percent in 2006. Services were providing attractive returns during this period. GE Capital sales increased by $12.3 billion from 2001 to 2004, then declined by 2006 so that

Table A6a.4   GE Sales and Income during the Years 2002–2011

| | Immelt's First Ten Years from 2002 until 2011 | | | | | | | | | |
| | Immelt | | | | | | | | | |
	Dec 31 2002	Dec 31 2003	Dec 31 2004	Dec 31 2005	Dec 31 2006	Dec 31 2007	Dec 31 2008	Dec 31 2009	Dec 31 2010	Dec 31 2011
Sales										
Goods	55,096	49,963	55,005	59,837	64,297	60,670	69,100	65,068	60,812	66,875
Services	21,138	22,391	29,700	32,752	36,403	38,856	43,669	38,709	39,625	27,648
Other	1,013	1,297	1,064	1,683	2,734	3,019	1,586	1,006	1,151	5,063
GE Capital	54,451	60,536	66,594	55,430	59,957	70,193	68,160	52,000	48,623	47,714
Total Revenues	131,698	134,187	152,363	149,702	163,391	172,738	182,515	156,783	150,211	147,300
Cost of goods sold	38,833	37,189	42,645	46,169	50,588	47,309	54,602	50,580	46,005	51,455
Cost of services sold	14,023	14,017	19,114	20,645	23,522	25,816	29,170	25,341	25,708	16,823
Interest & Finance Charges	10,216	10,432	11,907	15,187	19,286	23,787	26,209	18,769	15,983	14,545
Insurance Losses	17,608	16,369	15,627	5,474	3,214	3,469	3,213	3,017	3,012	2,912
Provision Financing Receivables	3,087	3,752	3,888	3,841	3,839	4,546	7,518	10,928	7,191	4,083
Other Overhead	28,714	31,727	38,148	35,271	37,414	40,297	42,021	37,804	38,104	37,384
Micellaneous	326	797	928	986	908	916	641			
Total Costs	112,807	114,283	132,257	127,573	138,771	146,140	163,374	146,439	136,003	127,202

continued

Table A6a-4 (continued)   GE Sales and Income during the Years 2002–2011

| | Immelt's First Ten Years from 2002 until 2011 | | | | | | | | | |
| | Immelt | | | | | | | | | |
	Dec 31 2002	Dec 31 2003	Dec 31 2004	Dec 31 2005	Dec 31 2006	Dec 31 2007	Dec 31 2008	Dec 31 2009	Dec 31 2010	Dec 31 2011
Earnings from Operations	18,891	19,904	20,106	22,129	24,620	26,598	19,141	10,344	14,208	20,098
Total Taxes and Adjustments	4,773	4,902	3,513	5,776	3,791	4,390	1,731	–897	2,029	5,655
Net Earnings	14,118	15,002	16,593	16,353	20,829	22,208	17,410	11,241	12,179	14,443
% Gross Margin										
Goods	30%	26%	22%	23%	21%	22%	21%	22%	24%	23%
Services	34%	37%	36%	37%	35%	34%	33%	35%	35%	39%
% of Sales										
Goods	42%	37%	36%	40%	39%	35%	38%	42%	40%	45%
Services	16%	17%	19%	22%	22%	22%	24%	25%	26%	19%
Other	1%	1%	1%	1%	2%	2%	1%	1%	1%	3%
GE Capital	41%	45%	44%	37%	37%	41%	37%	33%	32%	32%
Total Percent of Revenues	100%	100%	100%	100%	100%	100%	100%	100%	100%	100%
% Net Earnings to Revenues	10.7%	11.2%	10.9%	10.9%	12.7%	12.9%	9.5%	7.2%	8.1%	9.8%

Source: Adapted from Mergent Online.

the increase for the five-year period was only 10 percent. In total, product sales declined to 39 percent of total sales by 2006, with services, GE Capital, and other representing the remaining 61 percent. At this point, GE appeared to be transitioning into a predominantly services company. Table A6a.5 shows the composition of the product lines in 2006. It is obvious that the service element is strong, both in support of manufactured products and as an independent financing entity.

The five-year period from 2007 through 2011 was tumultuous for GE, especially for the GE Capital portion of their business. Sales for GE Capital peaked at $70.2 billion in 2007 and then declined significantly during the remaining four years to $47.7 billion, evidence that the real estate market had collapsed. Sales from services also declined although the gross margin continued to be higher than for product sales. Product sales increased by about 10 percent during the five-year period, indicating support for the core businesses. CEO Immelt has taken steps to reduce the risk of GE Capital by focusing it more on financing that relates to the core product businesses of the company.

Table A6a.6 shows the latest reported product segments for GE. Although still heavy in services, it is apparent that Immelt is shifting the emphasis away from GE Capital by selling parts of the business and transferring other parts to the product-oriented segments. Recent television ads also indicate that GE is refocusing their efforts on presenting themselves as a manufacturer of established product lines such as jet engines and refrigerators. They are also active in products for the future, such as alternative energy generation (wind, solar, and nuclear), environmentally sensitive products such as water and composite materials, and socially oriented products such as health care equipment and services.

As of the end of 2011, GE had sales of approximately $150 billion. Of this total, they generated approximately 55 percent of its revenue from services, including GE Capital, with 45 percent from sale of products.[8]

During Immelt's tenure as CEO, he has placed a renewed interest in developing revenues in high-technology products, especially in transportation and health care. He has also reduced the dependence on GE Capital by selling the reinsurance segments, spinning off the insurance into a separate company—Genworth—and transferring health care and finance to the Technology Infrastructure (Rothschild 2007, p. 242). As a result, at the end of 2011, GE Capital represented only 32 percent of total revenues versus a high of 43 percent during Welch's last year. Product sales, as a percentage of total revenues, declined during Immelt's early years, even though total dollars of revenue increased; however, they increased to 45 percent in 2011. The remainder of the revenues comes from services, amounting to 23 percent.

Table A6a.7 is taken from Immelt's letter to shareholders in the 2011 Annual Report.[6] It shows his commitment to moving the company strongly toward high-technology products and a simpler, more stable GE Capital.

Table A6a.8 summarizes the position of GE's product and service segments as of the end of 2011. It shows that the remnants of GE Lighting are slowly fading as a percentage of total revenues while the growth areas are in energy, transportation, and health care.

Table A6a.5 Product Line in 2006

	Primarily Products			Primarily Services		
	Health Care	*Infrastructure*	*Industrial*	*Commercial Finance*	*Consumer Finance*	*NBCU*
	$15.1 billion	$41.7 billion	$32.6 billion	$20.6 billion	$19.6 billion	$14.7 billion
	$2.8 billion	$7.8 billion	$2.6 billion	$4.3 billion	$3.1 billion	$3.1 billion
	Diag. imaging	Aircraft engine	Consumer/indus.	Leasing	Europe	Network
	Biosciences	Energy	Plastics	Real estate	Asia	Film
	Clinical systems	Oil and gas	Silicones/quartz	Corp. fin. serv.	Americas	Stations
	Info technology	Water	Security	Health care fin. serv.	Australia	Entertainment
	Services	Energy finance	Sensing	Insurance		Cable
		Aviation finance	Fanuc			TVPD
			Inspect tech			Sports/Olympics
			Equip. services			Theme parks

Source: Adapted from Rothschild (2007), p. 246.

Table A6a.6 Segment Revenues and Profits 2011

	Primarily Products				Primarily Services		
	Energy Infrastructure	Aviation	Health Care	Transportation	Home & Business Solutions	GE Capital (GECC)	Other (NBCU plus)
% of Total Revenues	30%	13%	12%	3%	6%	31%	5%
% Segment Profit	33%	17%	14%	4%	1%	32%	–2%

Source: Adapted from GE 2011 Annual Report.

Table A6a.7 Portfolio Strategy for GE (2011)

	Primarily Products			Primarily Services		
	Energy Infrastructure	*Technology Infrastructure*	*Home & Business Solutions*	*GE Capital (GECC)*	*NBC Universal*	
Revenues	25%	25%	6%	34%	10%	
Segment Profit	33%	33%	2%	20%	12%	
	Power generation	Aviation	Lighting	Leasing	Network	
	Sensing & insp.	Health care	Appliances	Real estate	Film	
	Digital energy	Transportation		Corp fin. serv.	Stations	
	Wind, oil, gas	Water		Health care fin. serv.	Entertainment	
	Water	Energy finance		Insurance	Cable	
	Services	Aviation finance			TVPD	
					Sports/Olympics	
					Theme parks	

Source: From GE 2010 Annual Report (pp. 32–33).

Table A6a.8 Examples of Segment Products and Services (2011)

	Product Examples	*Service Examples*
Energy Infrastructure (29.7% of Revenues)		
Energy production, distribution, and management	Wind turbines, steam and gas turbines; generators, nuclear reactors, IGCC systems to convert coal and other hydrocarbons into synthetic gas, motors, and control systems	Aftermarket services, maintenance service agreements, repairs, equipment installation, monitoring and diagnostics, asset management, performance optimization tools, and remote performance testing
Water treatment solutions for industrial and municipal water systems	Specialty chemicals, water purification systems, pumps, valves, filters, and fluid handling equipment	Supply and related services to the products
Integrated electrical equipment and systems to distribute, protect, and control energy and equipment	Electrical distribution and control products, lighting and power panels, switchgear and circuit breakers, and commercial lighting systems	Customer-focused solutions centered on the delivery and control of electric power; services that increase the reliability of electrical power networks
Helps oil and gas companies make more efficient and sustainable use of the world's energy resources	Surface and subsea drilling and production systems, equipment for floating production platforms, compressors, turbines, turboexpanders, high-pressure reactors, and industrial power generation	Equipment from drilling and completion through production, liquefied natural gas (LNG) and pipeline compression, pipeline inspection, and downstream processing in refineries and petrochemical plants
Aviation (12.8% of Revenues)		
Aircraft and airports	Jet engines, turboprop and turbo shaft engines; replacement parts; global aerospace systems, airborne computing systems, mechanical actuation products and landing gear, and engine components	Provide maintenance, component repair, and overhaul services (MRO), including sales of replacement parts for many models of engines and repair and overhaul of engines manufactured by competitors
Health Care (12.3% of Revenues)		
Medical equipment and service	Medical imaging and information technologies, medical diagnostics, patient monitoring systems, disease research, drug discovery, and biopharmaceutical manufacturing technologies	Remote diagnostic and repair services for medical equipment manufactured by GE and by others, as well as computerized data management, information technologies, and customer productivity services

continued

Table A6a.8 (continued) Examples of Segment Products and Services (2011)

	Product Examples	*Service Examples*
Transportation (3.3% of Revenues)		
Railroads	High-horsepower diesel-electric locomotives; drive technology solutions to the mining, transit, marine and stationary, and drilling industries; motors and engines	Services to improve fleet efficiency and reduce operating expenses, repair services, locomotive enhancements, modernizations, and information-based services like remote monitoring and diagnostics
Home and Business Solutions (5.7% of Revenues)		
Appliances and lighting	Refrigerators, freezers, electric and gas ranges, dishwashers, clothes washers and dryers, microwave ovens, room air conditioners, residential water systems, softening and heating, and hybrid water heaters	In-home repair and aftermarket parts
Lighting products	Variety of lamp products for commercial, industrial, and consumer markets, including full line of incandescent, halogen, fluorescent, high-intensity discharge, light-emitting diode, automotive, and miniature products	
Intelligent platforms	Plant automation, hardware, software, and embedded computing systems, including advanced software, controllers, embedded systems, motion control, and operator interfaces	
GE Capital (31.0% of Revenues)		
Commercial lending and leasing (CLL)		Collateralized loans, leases, and other financial services in many industries, including construction, manufacturing, transportation, media, communications, entertainment, and health care

Table A6a.8 (continued)   Examples of Segment Products and Services (2011)

	Product Examples	*Service Examples*
Consumer		Private-label credit cards; personal loans; bank cards; auto loans and leases; mortgages; debt consolidation; home equity loans; deposit and other savings products; and small and medium enterprise lending
Real estate		Equity capital for acquisition or development for acquisitions or re-capitalizations of commercial real estate—office and apartment buildings, retail facilities, hotels, parking facilities, and industrial properties
GE Capital Aviation Services (GECAS)		Leases and secured loans on commercial passenger aircraft, freighters, and regional jets; engine leasing and financing services; aircraft parts solutions; and airport equity and debt financing

Source: Adapted from GE 2011 10-K Report.

Recent Changes

One of the newest service areas for GE is in what is being called "big data." This is a "fast-growing market for information technology systems that can sift through massive amounts of data to help companies make better decisions."[9] GE customers include Norfolk Southern, who use the customized software to monitor rail traffic, which enables trains to move faster and reduces congestion. Another area in which GE hopes to capitalize on their strong market position is to help airlines that buy GE jet engines monitor their performance and anticipate maintenance needs, reducing costly flight cancellations.[9] GE trails IBM in this area but hopes their expertise in making industrial equipment, such as gas-fired electrical turbines and locomotives, will help them. As Bill Ruh, vice president of GE, explains: "If you don't have deep expertise in how energy is distributed or generated, if you don't understand how a power plant runs, you're not really going to be able to build an analytical model and do much with it. We have deep insight into several very specific areas. And that's where we're staying focused."[9]

GE announced in July 2012 that they intended to split the GE Energy Infrastructure into three separate operating units reporting directly to the CEO, Jeffrey Immelt. This change is designed to reduce organizational complexity and enable these segments to grow more rapidly. The three business segments will be as follows:

- GE Power and Water—Supplies full lifecycle solutions for power generation customers, including renewable energy and water process technologies (41,000 employees and projected 2012 revenues of $28 billion).
- GE Oil and Gas—Provides equipment and services for offshore and onshore oil and gas industries, including turbomachinery and drilling and surface, subsea, and pipeline equipment and services (33,000 employees and projected 2012 revenues of $15 billion).
- GE Energy Management—Develops technologies for the delivery, management, conversion, and optimization of electrical power for customers across multiple energy-intensive industries (27,000 employees and projected 2012 revenues of $7 billion).[10]

GE has also initiated programs in wind and solar renewable energy programs.[6]

A Model to Illustrate GE's Strategy

From the foregoing examples, it is obvious that GE's product lines have grown in scope and size over the decades. Is there a logical explanation for the growth? Many of the strategic choices were influenced by CEO preference and by existing circumstances at the time of the decision. Not all choices have been complete successes; however, GE has a very good batting average, as evidenced by their high ranking among global enterprises.

It is presumptuous to attempt to reduce GE's product line experiences to a simple model; however, it is also interesting to see that their success does have a pattern—a pattern shown in Figures A6a.3 and A6a.4. Figure A6a.3 shows the progression of manufactured goods, from the lightbulb through connecting links that have led to today's product lines. The original products were lightbulbs and small kitchen appliances. Later major product lines included today's major product segments: Aviation, Health Care, Transportation, and Energy. Figure A6a.3 shows that, in most cases, new products at GE have been extensions of products or services that were developed early in the company's history. The early recognition by Edison of the future of electricity is evident in today's products.

The bottom portion of Figure A6a.3 shows the evolution of service businesses at GE. Almost from the beginning, GE was providing financing to customers, primarily privately owned electric utilities. This early financing spawned GE Capital, which grew into a major portion of the company's revenues, reaching a peak of approximately 43 percent of revenues around 2005. In the past few years, GE has

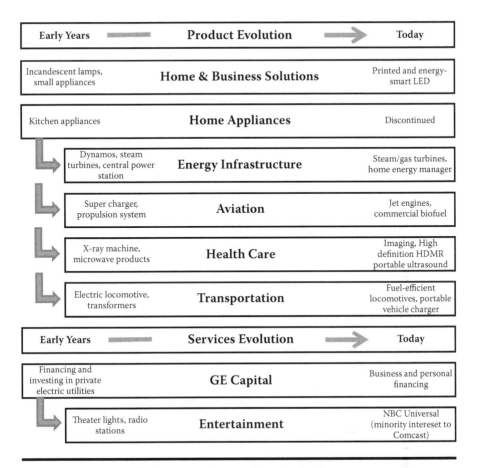

Early Years	Product Evolution	Today
Incandescent lamps, small appliances	**Home & Business Solutions**	Printed and energy-smart LED
Kitchen appliances	**Home Appliances**	Discontinued
Dynamos, steam turbines, central power station	**Energy Infrastructure**	Steam/gas turbines, home energy manager
Super charger, propulsion system	**Aviation**	Jet engines, commercial biofuel
X-ray machine, microwave products	**Health Care**	Imaging, High definition HDMR portable ultrasound
Electric locomotive, transformers	**Transportation**	Fuel-efficient locomotives, portable vehicle charger
Early Years	Services Evolution	Today
Financing and investing in private electric utilities	**GE Capital**	Business and personal financing
Theater lights, radio stations	**Entertainment**	NBC Universal (minority intereset to Comcast)

Figure A6a.3   Growth of GE's product and service business segments.

reduced its emphasis on GE Capital, especially in the real estate area. GE has also moved to reduce its dependence on the entertainment industry by selling a majority interest in NBC to Comcast, where GE retains a 49 percent interest but with Comcast the majority owner and manager.[11]

Figure A6a.4 shows that products go through three life cycles: manufacture and sale; service and repair; and recovery, or recycling. Many manufacturing companies have focused only on the first life cycle, that of making and selling their product. Astute companies have recognized the second life cycle and, as manufacturers, they have extensive knowledge about how best to service those products to ensure effective and efficient performance. They can provide preventive maintenance, supply spare parts, and prescribe optimum operating conditions, with the result being increased output, lower maintenance costs, and longer effective product life.

Figure A6a.4 also shows that a third life cycle exists for most manufactured products—the recovery or recycling part of the product's existence. With the

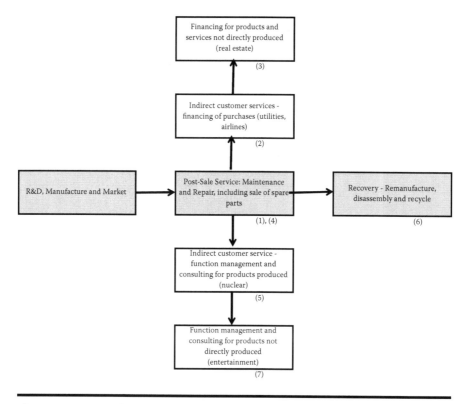

Figure A6a.4   Product life cycles.[12]

increased awareness of the need to use the earth's resources more prudently, companies are increasingly mindful of the opportunities to use their knowledge in ensuring that their products are carefully managed during the recovery life cycle.

Figure A6a.5 shows the predominant strategies of GE in developing their services businesses.

1. Because they made assembled products, it was natural to provide spare parts and repair services for those products.
2. Almost from the beginning, they recognized the opportunity to increase the sale of their existing products—lighting to consumers and transformers to electric utilities—by helping the utilities to grow by providing financing. If utilities grew, they would provide power and encourage consumers to buy products from GE. This also motivated GE to produce more products using electricity, such as kitchen appliances, for the consumer. This was the "Benign Cycle" strategy described earlier.
3. Once they were in the financing business, they were able to extend financing to other types of businesses, the origin of GE Capital.

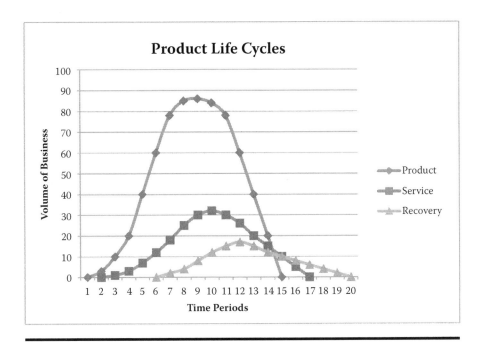

Figure A6a.5   GE's product-to-service progression.

4. As GE's product lines became more technically complex, such as in nuclear power plants and imaging technologies, they recognized the opportunity to provide after-sales services to customers—repair, maintenance, and spare parts. They also offered process consulting services in this area.

5. This, in turn, led to the complete facilities management of complex systems. Eventually, these services expanded to providing consulting services in the procurement and management of complex technical systems.

6. Finally, with the emergence of sustainability issues in the latter part of the twentieth century, GE moved to do more in the reverse logistics phase of a product's life by introducing more efficient products and by providing services in facilities management.

7. One venture that has not proved completely successful is the entertainment business, a service distinctly different from manufacturing. Acquisitions in these areas are being quietly moved to minority positions of ownership, or divested completely.

The strategy displayed in Figure A6a.5 shows that services exist throughout the supply chains of products. It is a natural extension to move downstream in the supply chain and provide some of the producer services that other industrial firms require.

Figure A6a.6 shows that GE has concentrated its service efforts primarily in Tier 1 and Tier 2 type of services. The numbers in the diagram correspond with

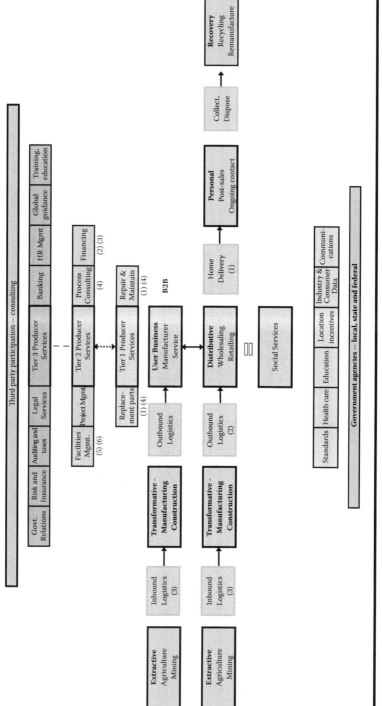

Figure A6a.6 Evolution of services.

the numbers of the product strategies used in Figure A6a.4. By closely aligning their services with their physical products, they have been successful. They have experienced less success when they have offered services not directly aligned with their products.

Conclusions

GE is not the only company to blend the manufacture of products with the providing of services. However, they have been at it for most of their more than 130-year history. Based on their position as a major global company, they have learned the value of supplementing products with services well. Rothschild[5] has captured what he believes is the essence of GE with the following summary of their critical success factors.

> GE's ability to grow and prosper over this long period can be summarized in five themes. As in earlier chapters, each still ties to the acronym LATIN: leadership, adaptability, talent, influence, and networks.
>
> 1. *Leadership*: No cookie cutters. GE has had many diverse types of leaders, who were clearly the right leaders for the right time—in part because they recognized that the world changes continually and therefore requires different types of leadership.
> 2. *Adaptability*: Nothing is sacred or indispensable. The second factor was the ability to challenge every business and policy—even when they were working. This mindset has enabled the company to make changes before it was too late.
> 3. *Talent*: Cultural evolutions. GE's culture has evolved, which has permitted it to continue to attract, retain, and motivate a strong and deep talent pool.
> 4. *Influence*: Being politically incorrect when necessary. GE has long recognized that there are multiple stakeholders that must be considered, some of whom are friends and others of whom are adversaries, at least at the moment.
> 5. *Networks*: It's all about expectations. Setting viable expectations that it has met consistently has been a hallmark of the company, and it has helped minimize the number of major surprises (pp. 253–345).[5]

GE is an example of a company that has learned to blend services with its products and manufacturing capabilities. It also shows a history of evolution and innovation, traits that enable a company to maintain a leadership role in a dynamic business environment. While they have sometimes strayed from their core businesses in the attempt to add revenue, such as in real estate financing and entertainment media, they have been successful in blending services with their product manufacturing core business.

They have also had stability at the CEO position as a result of their practice of carefully planning for CEO succession. As part of this strategy, they try to provide the CEO with a long enough tenure to effect needed changes and then to see those changes through to successful conclusions.

References

1. Wise, R. and Baumgartner, P., Go downstream: The new profit imperative in manufacturing, *Harvard Business Review*, 77(5), 133–141, 1999.
2. Oliva, R. and Kallenberg, R., Managing the transition from products to services. *Journal of Service Management*, 14(2), 160–172, 2003.
3. Largest U.S. corporations, *Fortune*, May 21, 2012, F1–F25.
4. General Electric, *GE Timeline of History*, http://www.ge.com/company/history/index.html, 2012.
5. Rothschild, W. E., *The Secret to GE's Success*, McGraw-Hill, New York, 2007.
6. GE 2011 Annual Report, http://www.ge.com/ar2011/index.html#!section=ge-2011-annual-report.
7. Christiansen, C. M., *The Innovator's Dilemma*, Harper Business, New York, 2000.
8. Mergent Online, http://0-www.mergentonline.com.wncln.wncln.org/compsearch.asp, 2012.
9. Catts, T., GE heads west with $1 billion to spend, *Business Week*, http://www.businessweek.com/articles/2012-04-26/ges-billion-dollar-bet-on-big-data, April 26, 2012 (accessed August 6, 2012).
10. Brooks, R., GE to split energy into three businesses, *American Machinist*, http://www.americanmachinist.com/Classes/Article/ArticleDraw.aspx?CID=88953&Refresh=1, July 23, 2012 (accessed July 23, 2012).
11. Hamill, K., Comcast and GE complete NBC deal, *CNNMoney*, http://money.cnn.com/2011/01/29/news/companies/comcast_ge_nbc/index.htm, 2011 (accessed August 6, 2012).
12. Crandall, R.E., An Expanded Perspective on Product Life Cycles, *APICS Magazine*, 22(3), 20–23, 2012.

Appendix 6B: Hewlett-Packard—From Scientific Instrumentation to Business and Consumer Products and Services

Introduction

Hewlett-Packard, or HP as it is known today, was started by William Hewlett and David Packard in 1939 in a one-car garage in Palo Alto, California. Their first product was a resistance-capacity audio oscillator, an electronic instrument used to test sound equipment. The HP200A used an incandescent bulb as part of its wiring to provide variable resistance, a breakthrough in stability of oscillator design.[1] From that beginning, HP has grown into one of the major technology companies in the world, ranking No. 10 on the 2011 Fortune 500 list of U.S. companies.[2]

Table A6b.1 was adapted from the comprehensive book about HP by Charles H. House and Raymond L. Price,[3] both employees at HP during its formative years. It shows the transition of HP from a high-technology instrument maker to today's emphasis on computing, peripherals (printers), and services. The authors divide the evolution of HP as carrying over six stages:

- First transformation (1949–1959). From audio-video test to microwave test
- Second transformation (1959–1968). From frequency domain test to scientific test
- Third transformation (1968–1976). From electrical engineering test to scientific systems
- Fourth transformation (1976–1986). From scientific systems to business computing
- Fifth transformation (1986–1996). From computing to printing and imaging
- Sixth transformation (2006–2008). From enterprise to professional services.

The years 1997–2005 are not included in their summary, perhaps because this was a period in which HP was going through an unstable period, especially at the CEO level, and its business direction was also uncertain.

HP put its emphasis on developing products, primarily testing and measurement equipment in the beginning. It was a company founded by engineers who developed and manufactured equipment for engineers to use in laboratory and manufacturing applications. It only gradually moved away from producing equipment to meet individual customer specifications to making larger-volume products for mass markets.

Product innovations came quickly from the engineering-oriented employees, sometimes as a result of planned product development strategies and sometimes in spite of top management support. "HP cannibalized its roots repeatedly, replacing established products with bold new ideas. Focusing on contribution rather than

Table A6b.1 Timeline of Product Line Evolution at HP

Hewlett-Packard Product Line Distribution as Percentage of Total Revenues

Stage	Year	Audio Video Microwave	Time Domain	Scientific Test	Scientific Computing	Business Computing	Personal Computing	Peripheral (Printers)	Services	Other	Total
Early	1949	82%								6%	100%
1	1949–1959	28%	24%							4%	100%
2	1959–1968	14%	20%	28%						8%	100%
3	1968–1976	6%	10%	26%	34%	8%				2%	100%
4	1976–1986	30%			14%	32%		24%		0%	100%
5	1986–1996	–4%				20%	34%	30%	20%	0%	100%
	1996–2006	To Agilent				Not covered in book					
6	2006–2008					14%	28%	24%	34%	0%	100%

Source: Adapted from Appendix A, *The HP Phenomenon*, by Charles H. House and Raymond L. Price, pp. 517–518.[3]

endeavor, HP morphed six times into something else, changing its leading products each decade. Such transformation is unparalleled in modern business. Even more stunning is the discovery that Hewlett resisted three transformations, and Packard is on record at some point opposed to each of the six. So much for today's conventional leadership wisdom!" (p. 5).[3]

One of the more interesting examples of this occurred in the 1960s when HP was working on its first computer displays. Based on an unproven production process and dismal market forecasts, David Packard told the division manager and staff, "When I come back next year, I don't want to see that project in the lab," referring to the CRT display box. Chuck House, division manager, opted to accelerate the development process and put the product into production. He also spearheaded its sale to key accounts and the product became a successful launch. When Packard heard of it, he was naturally furious at House—until he found that the product was being sold in profitable quantities. Years later, in 1982, Packard awarded House a "Medal of Defiance to Charles H. House: Awarded in recognition of extraordinary contempt and defiance beyond the normal call of engineering duty."[3] Neither Hewlett nor Packard was afraid to adapt and take the company in new directions.

This example illustrates the environment, and the resulting culture, that Hewlett and Packard created and nurtured. Engineers were encouraged to be aware of existing and potential customer needs, and to propose, and develop, products that meet those needs. This also included providing services where needed, in the form of training in the use of and maintenance of the equipment sold. Services were a complement to the product.

HP was eminently successful in developing innovative products. However, both Bill Hewlett and David Packard were also innovative in developing management practices that were part of the fabric of the organization's success in attracting, motivating, and retaining some of the greatest talent existent in a single company. "A leader in product sector after product sector, the company became better known for its cultural practices than its products. Profit sharing, flexible work hours, extended medical coverage, and 'Management by Wandering Around'—and especially belief in the dignity of the individual employee—were all part of the HP Way long before these concepts were embraced by other companies" (p. 7).[3]

Transformation One (1939–1959)

HP's first product was an audio oscillator, the HP200A, developed by Bill Hewlett. This product proved to be such an advance that it was adopted by Disney Productions in their filming of the movie Fantasia.[4]

For years, HP developed and sold electronic equipment to businesses that needed test and measurement equipment. The equipment was low unit-volume, highly customized to fit the needs of individual companies. HP rapidly built a reputation for high-quality products that were leading edge technically; the company was also valued because they completed their projects on time. During World War II, they added new products and moved into their first manufacturing facility. David Packard bore most of the management load during the early 1940s because Bill Hewlett was serving in the United States Army. Packard also demonstrated his engineering capabilities by developing an innovative voltmeter.[4]

In 1943, the company entered the microwave field by supplying signal generators to the Naval Research Laboratory along with a radar-jamming device. This led to additional microwave products to go along with their products for use in the audio and video fields. Table A6b.1 shows that microwave products became the major product sector by 1959.

In 1959, the company also extended its reach outside of California by establishing the European Marketing Organization in Geneva, Switzerland, and built their first manufacturing plant outside the United States in Böblingen, West Germany.

Transformation Two (1959–1968)

While Hewlett and Packard continued their steady growth in the test and measurement fields, they also branched out into related fields, such as medical electronics and analytical instrumentation. They also developed other innovative products, including the Cesium Bean "Atomic" Clock. The 8551 Spectrum Analyzer became the first product to provide over $1 million of revenue per month.[1]

They put more emphasis on the medical field by purchasing the Sanborn Company in 1961 and introduced a noninvasive fetal heart monitor in their West German plant. Other new products included their first computer, an all-solid-state component oscillator, and light-emitting diodes.

Table A6b.1 shows the dominant products to be in the microwave area, although closely followed by scientific test equipment and time domain products. The company was becoming more diversified in product lines, in industries served, and in geographic areas in which they marketed. They also took a tentative step in the computer field by introducing the HP 1000 minicomputer.

Transformation Three (1968–1976)

Table A6b.1 shows an ever-expanding product line with an increasing emphasis on scientific testing equipment. In addition, during this period, the company made a significant increase in their commitment to scientific computing products. The movement into computers was not necessarily a planned strategy as much as the invisible hand of the marketplace. David Packard made this observation: "It would be nice to claim that we foresaw the profound effect of computers on our business and that we prepared ourselves to take early advantage of the computer age. Unfortunately, the record does not justify such pride. It would be more accurate to say that we were pushed into computers by the revolution that was changing electronics."[3]

In the mid-1960s, HP viewed itself as an electronic instrument company building tools for engineers and scientists. They were reluctant to enter a world served by back-office machinery because (1) IBM and AT&T were HP's largest customers, and (2) IBM was already known for ruthlessly competing against the minor participants in the computer industry—the Seven Dwarfs.[3]

> Yes, Hewlett-Packard took a long time to target computing in the classical senses. But HP created more pioneering and trend-setting computers than generally recognized, products that identified the elements of and the issues around personal computing long before such words were used. HP's foray into computing, a classic application of the next-bench syndrome, changed HP irrevocably.[3]

HP continued to extend their work in minicomputers with the Model 2000 and HP 3000, although the latter had a rocky start, possibly because it coincided with the transition of leadership from Hewlett and Packard to Young, who became the first CEO after the retirement of Hewlett and Packard.[4] William Hewlett was fascinated by this product but challenged the designers to come up with a design with "one-tenth the size but ten times the capability." Although initially stymied, the HP research team developed the HP-35, a shirt-pocket-size calculator that every engineer and many others wanted. They followed the HP-35 with the HP-65, a programmable calculator. These products further endeared HP to the scientific community and was the beginning of a movement toward the personal computer market.

One of the interesting side stories of this period is that Steve Wozniak, who worked for HP for about four years, made a proposal to HP management about developing a computer (that later became the first Apple computer) but was turned down.

Although perhaps not obvious at the beginning of computer development, becoming a computer company spawned the emergence of services as a separate segment of the business.

Table A6b.2 shows the product transformations achieved during the three transformation periods from 1949 through 1976. The impact of computing, especially for scientific applications, was rapidly becoming the dominant product line for the company.

As shown in Table A6b.3, this was a period of rapid growth, as revenues increased from approximately $250 million in 1967 to over $1.1 billion by 1976. Earnings increased proportionately, further establishing HP as one of the leading electronic manufacturing companies in the world. It is also of interest to note that research and development (R&D) expenditures were $107 million in 1976, or 9.7 percent of revenues, indicating HP's commitment to continued new product development.

Transformation Four (1976–1986)

This period solidified HP's position in the commercial computer market and was the introductory phase for peripherals, especially printers. HP developed the hardware and software needed to convert data into dots for the mechanism to print.

Table A6b.2   Transformation in Product Groups (1949–1976)

Product Groups (as Percent of Total Revenue)				
	1949	1959	1968	1976
Audio-video	84	29	14	5
Microwave	12	42	28	14
Time-domain		23	24	10
Scientific instrumentation			30	28
Scientific computing				32
Other	4	6	4	11
Total	100	100	100	100

Source: Adapted from Figure 5.1, p. 151 (House and Price 2007).[3]

The test and measurement line of instrumentation continued to be a major portion of the product portfolio; however, it was undergoing changes with the need to link the test and measurement equipment with computers for data processing and with printers for graphing and printing outputs.

The 1986 HP Annual Report listed the product groups, shown in Table A6b.4, and their revenues in millions of dollars and as a percentage of the total. Services are spelled out as a separate product segment and also listed as a complement to other product groups.

In the ten-year period since 1976, revenues increased from $1.1 billion to $7.1 billion. Although the product categories are different, it is evident that computers and peripherals—printers, plotters, and mass-storage devices—contributed a large share to the growth. It is also relevant that Hewlett-Packard Finance Company (HPFC), a wholly owned, unconsolidated U.S. subsidiary, "has helped spur demand for the company's products by providing competitively priced lending options to growing numbers of HP customers" (p. 4).[5]

Transformation Five (1986–1999)

The leadership at HP was stable during this period as the CEOs were promoted from within the organization. John A. Young was CEO from 1978 through October 1992, and he was succeeded by Lewis Platt, who remained CEO until July 18, 2009, when Carly Fiorina was brought in as CEO.

Table A6b.5 shows that revenues increased sevenfold during this period from $6.5 billion in 1985 to over $47 billion in 1998. The decline in revenues for 1999 reflects the divestiture of Agilent Technologies. The percentage of revenues derived from services increased from 20 percent in 1985 to a high of 25 percent in 1992, and then declined to 15 percent by 1999, indicating that product revenues tripled

Table A6b.3   Product Segments Revenues and Income, 1972–1976

Years Ending October 31	Revenues (Dollars in Millions)					Operating Income (Dollars in Millions)				
Product Segments	*1972*	*1973*	*1974*	*1975*	*1976*	*1972*	*1973*	*1974*	*1975*	*1976*
Test measuring and related items	259	320	394	442	487	48	49	63	47	84
Electronic data products	155	255	372	386	448	21	41	71	55	53
Medical electronic equipment	43	57	78	99	119	4	4	7	11	18
Analytical instrumentation	22	29	40	54	58	1	1	3	6	6
Total	479	661	884	981	1112	74	95	144	119	161
Product Segments	*1972*	*1973*	*1974*	*1975*	*1976*	*1972*	*1973*	*1974*	*1975*	*1976*
Test measuring and related items	54%	48%	45%	45%	44%	65%	52%	44%	39%	52%
Electronic data products	32%	39%	42%	39%	40%	28%	43%	49%	46%	33%
Medical electronic equipment	9%	9%	9%	10%	11%	5%	4%	5%	9%	11%
Analytical instrumentation	5%	4%	5%	6%	5%	1%	1%	2%	5%	4%
Total	100%	100%	100%	100%	100%	100%	100%	100%	100%	100%

Source: Adapted from I-P 1976 Annual Report.[21]

Table A6b.4   Product Groups Revenues, 1985–1986

	Revenues (Millions of $)	
Product Segments	*1985*	*1986*
Measurement, design, information, and manufacturing equipment and systems	2,929	2,995
Peripherals and network products	1,560	1,761
Service for equipment, systems, and peripherals	1,125	1,323
Medical electronic equipment and service	448	513
Analytical instrumentation and service	248	303
Electronic components	195	207
Total	$6,505	$7,102
Years Ending 10/31	Percentage of Revenues	
Product Segments	*1985*	*1986*
Measurement, design, information, and manufacturing equipment and systems	45%	42%
Peripherals and network products	24%	25%
Service for equipment, systems, and peripherals	17%	19%
Medical electronic equipment and service	7%	7%
Analytical instrumentation and service	17%	19%
Electronic components	3%	3%
Total	100%	100%

Source: Adapted from 1986 HP Annual Report, p. 37.[5]

while service revenues increased by only about 50 percent. Net earnings, as a percentage of revenues, declined from the 7–8 percent range at the beginning of the period to a low of 3.3 percent in 1992 and then improved back into the 7–8 percent range by the end of the period.

The spin-off of Agilent Technologies in 1999 marked the end of the scientific instrument era of HP. Agilent took with it the core products that had been carryovers from HP's origin, leaving HP as primarily a computer products and imaging business. Table A6b.6 shows the product mix for HP, including Agilent. Table A6b.7 shows the product mix, without Agilent.

Table A6b.5   Revenues and Income, 1985–1999

Report Date—October 31	Young								Platt						
	1985	1986	1987	1988	1989	1990	1991	1992	1993	1994	1995	1996	1997	1998	1999[a]
Products	5,204	5,622	6,315	7,709	9,404	10,214	11,019	12,354	15,533	19,307	27,125	33,114	36,672	40,105	36,178
Services	1,301	1,480	1,775	2,122	2,495	3,019	3,475	4,056	4,784	5,684	4,394	5,306	6,223	6,956	6,192
Total net revenue	6,505	7,102	8,090	9,831	11,899	13,233	14,494	16,410	20,317	24,991	31,519	38,420	42,895	47,061	42,370
Operating income	758	780	962	1,142	1,212	1,162	1,210	1,404	1,879	2,549	3,568	3,726	4,339	3,841	3,688
Other income (expenses)	269	264	318	326	383	423	455	855	702	950	1,135	1,140	1,220	896	197
Net earnings (loss)	489	516	644	816	829	739	755	549	1,177	1,599	2,433	2,586	3,119	2,945	3,491
Earnings (as % of revenues)															
Operating Income	11.7%	11.0%	11.9%	11.6%	10.2%	8.8%	8.3%	8.6%	9.2%	10.2%	11.3%	9.7%	10.1%	8.2%	8.7%
Other income (expenses)	4.1%	3.7%	3.9%	3.3%	3.2%	3.2%	3.1%	5.2%	3.5%	3.8%	3.6%	3.0%	2.8%	1.9%	0.5%
Net earnings (as % of revenues)	7.5%	7.3%	8.0%	8.3%	7.0%	5.6%	5.2%	3.3%	5.8%	6.4%	7.7%	6.7%	7.3%	6.3%	8.2%
Revenues (as % of total)															
Products	80%	79%	78%	78%	79%	77%	76%	75%	76%	77%	86%	86%	85%	85%	85%
Services	20%	21%	22%	22%	21%	23%	24%	25%	24%	23%	14%	14%	15%	15%	15%
Total net revenue	100%	100%	100%	100%	100%	100%	100%	100%	100%	100%	100%	100%	100%	100%	100%

Source: Adapted from Mergent Online (2012).

Note: Dollars in millions.

[a] 1999— Reflects divestiture of Agilent Technologies with revenues of approximately $7.6 billion.

Table A6b.6 Segment Results before Agilent Divestiture

	Revenues (in Millions of $)		
Report Date (as of October 31)	1996	1997	1998
Computer products, support, and service	31,447	35,407	39,466
Test and measurement products and service	3,910	4,339	4,169
Medical electronic equipment and service	1,287	1,265	1,408
Electronic components	918	975	1,052
Chemical analysis and service	858	909	966
Total Revenues	38,420	42,895	47,061
	As Percentage of Revenues		
Report Date (as of October 31)	1996	1997	1998
Computer products, support, and service	81.9%	82.5%	83.9%
Test and measurement products and service	10.2%	10.1%	8.9%
Medical electronic equipment and service	3.3%	2.9%	3.0%
Electronic components	2.4%	2.3%	2.2%
Chemical analysis and service	2.2%	2.1%	2.1%
Total	100.0%	100.0%	100.0%

Source: From Mergent Online (accessed July 12, 2012).

Although the financial results were impressive, there was increasing evidence that the company was coasting along, feeding off of products that had been developed earlier. HP had failed to recognize that the computer industry was changing. The industry was moving rapidly into a services-dominated environment where products were becoming commodities and higher-level services were destined to be the driver of growth.[4]

Transformation Six (1999–Present)

This was a period that was filled with turmoil, especially at the CEO level. During this period, HP had seven different persons at the CEO level (five permanent and two interim). The sequence was as follows:

- CEO: Carly Fiorina (July 19, 1999–February 9, 2005)
- Interim CEO: Robert Wayman (February 9, 2005–March 28, 2005)
- President and CEO: Mark Hurd (April 1, 2005–August 6, 2010)

Table A6b.7   Segment Results after Agilent Divestiture

	Revenues (in Millions of $)		
Report Date (as of October 31)	1997	1998	1999
Imaging and printing systems	15,822	17,006	18,832
Computing systems	14,958	17,313	17,877
IT services	4,747	5,164	5,880
All other	120	157	79
Total Revenues	35,647	39,640	42,668
	As Percentage of Revenues		
Report Date (as of October 31)	1997	1998	1999
Imaging and printing systems	44.4%	42.9%	44.1%
Computing systems	42.0%	43.7%	41.9%
IT services	13.3%	13.0%	13.8%
All other	0.3%	0.4%	0.2%
Total	100.0%	100.0%	100.0%

Source: From HP 1999 Annual Report.

- Interim CEO: Cathie Lesjak (August 6, 2010–September 30, 2010)
- President and CEO: Leo Apotheker (September 30, 2010–September 22, 2011)
- President and CEO: Meg Whitman (September 22, 2011–present).

Table A6b.8 shows the financial results for the thirteen-year period from 1999 through 2011. The results are shown for the CEOs who were active at the end of the fiscal years ending October 31, although there were interim CEOs at the end of both Fiorina's and Hurd's tenures. Needless to say, this kind of turnover at the highest position was not conducive to stability of direction or employee effectiveness. Despite this, HP increased revenues from $47 billion in 1998 to $127 billion in 2011. There were three major events that influenced the revenue levels:

- The spin-off of Agilent Technologies as a separate company in 1999. This was an $8 billion company with 47,000 employees.[6] Consequently, there was a decline in HP's revenues in the fiscal year of 1999 with a decrease from $40.1 billion in 1998 to $36.2 billion. Part of the lost revenues from Agilent's departure was offset by increase in revenues from HP's remaining product lines.

Table A6b.8 Revenues and Income, 1999–2011

Report Date—October 31	Fiorina						Hurd						Apotheker
	1999a	2000	2001	2002	2003	2004	2005	2006	2007	2008	2009	2010	2011
Products	36,178	41,446	37,498	45,955	58,939	64,127	68,945	73,557	84,229	91,697	74,051	84,799	84,757
Services	6,192	7,336	7,325	10,178	13,657	15,389	17,380	17,773	19,699	26,297	40,124	40,818	42,039
Financing income	—	—	403	455	465	389	371	328	358	370	377	418	449
Total net revenue	42,370	48,782	45,226	56,588	73,061	79,905	86,696	91,658	104,286	118,364	114,552	126,033	127,245
Earnings (loss) from continuing operations	3,688	3,889	1,439	–219	2,897	4,264	3,276	6,612	8,392	10,518	10,143	11,479	9,677
Other income (expenses)	197	192	1,031	684	358	767	878	414	1,128	2,189	2,483	2,718	2,603
Net earnings (loss)	3,491	3,697	408	(903)	2,539	3,497	2,398	6,198	7,264	8,329	7,660	8,761	7,074
Earnings (as % of revenues)													
Operating Income	8.7%	8.0%	3.2%	–0.4%	4.0%	5.3%	3.8%	7.2%	8.0%	8.9%	8.9%	9.1%	7.6%
Other income (expenses)	0.5%	0.4%	2.3%	1.2%	0.5%	1.0%	1.0%	0.5%	1.1%	1.8%	2.2%	2.2%	2.0%
Net earnings (as % of revenues)	8.2%	7.6%	0.9%	–1.6%	3.5%	4.4%	2.8%	6.8%	7.0%	7.0%	6.7%	7.0%	5.6%
Revenues (as % of total)													
Products	85%	85%	83%	81%	81%	80%	80%	80%	81%	77%	65%	67%	67%
Services	15%	15%	16%	18%	19%	19%	20%	19%	19%	22%	35%	32%	33%
Total net revenue	100%	100%	99%	99%	99%	100%	100%	100%	100%	100%	100%	100%	100%

Source: Adapted from Mergent Online (2012).

Note: Dollars in millions.

a 1999—Reflects divestiture of Agilent Technologies with revenues of approximately $7.6 billion.

- The acquisition of Compaq Computers in 2002. Compaq included sales of both products and services. The combined revenues were projected to be approximately $87 billion, $47 billion from HP and $40 billion from Compaq. However, sales for both companies dropped significantly in 2002, and the resultant combined revenues for 2002 were $46 billion and $59 billion in 2003, which fully reflects the Compaq contribution. Revenues continued to increase through the 2004–2007 fiscal years by close to 10 percent per year.
- The acquisition of EDS in 2008 brought a service business with revenues of approximately $20 billion. This addition was expected to double HP's services revenues, as it did, when comparing the service revenues of $19.7 billion in 2007 with the $40.1 billion in 2009, the first full year of the combined operations. It also increased services business percentage of total revenues from 20 percent to about 33 percent.
- During the period from 2000 through 2011, HP spent approximately $20 billion on acquisition, restructuring, and amortization expenses as a result of these and other acquisitions.

Fiorina as CEO (July 19, 1999, to February 9, 2005)

During the first decade of the twenty-first century, HP moved out of its high-technology test and measurement instrument products (their base for over one-half century) with the spin-off of Agilent Technologies. This essentially removed all the products associated with the early days of the company and set the stage for moving aggressively into the computer and computer-related businesses. The newly appointed CEO, Carly Fiorina, "unveiled her vision for transforming the company from a product-driven colossus to a pioneer of Internet services and ideas."[7]

With the acquisition of Compaq, HP moved into a leading position in the business and personal computing industry. Table A6b.8 shows that the percentage of revenues from services began to increase during Fiorina's reign as CEO. In 2004, she announced the Adaptive Enterprise strategy, which was directed at increasing the bonds with channel partners to increase HP's competitiveness with IBM.[8] It was also likely intended to move into higher-margin services such as consulting.[9] Following Fiorina's replacement as CEO in 2005, some felt that HP had been overly concerned with the "direct sales" model and had neglected their channel partners.[10]

Others blamed HP's problem as being the lack of a coherent strategy, causing them to compete in too many areas with less than desired results. In personal computers, HP fought a price war with Dell, a competitor with a more efficient direct sales model and supply chain. In corporate computers, they were squeezed between Dell at the low end and IBM at the high end, partially because they found it difficult to explain their vague slogan of adaptive enterprise, the program announced by Fiorina in 2004. In information technology services, they lagged far behind IBM, Accenture, and EDS.[11]

During Fiorina's tenure, the strategy seemed to vacillate between competing with Dell and its direct sales strategy and IBM and its channel strategy. The Clancy article[8] outlines Fiorina's push of the channel strategy; however, Zarley's article suggests that Fiorina focused primarily on the direct sales strategy. Regardless, HP seemed to have a lack of positive direction during the Fiorina reign and a definite lack of concentration on operations management (they didn't have a COO from 2002 through 2005).[10]

Hurd as CEO (April 1, 2005, to August 6, 2010)

One author advised Mark Hurd soon after his appointment as CEO that HP should not try to compete with IBM in the services area as IBM led with services and backed it up with their products. Instead, "HP needs to do the reverse: build on its legacy of innovation by investing more in R&D and prove it's worth buying HP and having HP on hand to help make the most of those purchases."[12] This was a suggestion that HP should revert to its legacy of developing valued products and helping their customers gain greater value by using HP to provide support services.

One year after Fiorina's departure as CEO, Alan Murray of the *Wall Street Journal* suggested that the merger with Compaq was not being rejected by either Hurd or the HP Board of Directors. Rather, he suggested that the Board was disenchanted with Fiorina, not the merger, which was working out quite well. Murray summed it up by writing, "H-P's directors went through hell together. In the end, they got the best of both worlds—a charismatic CEO who brought about a hotly contested but transformational merger, and a no-nonsense, operations-oriented CEO determined to make the combined company work."[13]

In Hurd's first year, he focused his efforts on improving the profitability of all product lines, as opposed to making major divestitures, although he did make some smaller acquisitions to boost its software, services, and printing divisions. He was portrayed as an "understated operations geek" who could excel at HP.[14]

Other writers viewed HP's pursuit of Dell's direct sales approach as inferior to a strategy that viewed PCs as a commodity and emphasized the aftermarket service model as one worthy of HP's broad product lines. "Instead of trying to emulate Dell's online sales genius, HP should be creating and selling a comprehensive hardware and software strategy."[15]

Revenues increased rapidly from 2005 through 2008, and HP became the largest IT company in the world, largely "on the strength of high-profile acquisitions like Mercury Interactive in 2006 and EDS in 2008, while doubling down on its reseller channel engagements with its increasingly muscular Solution Partners Organization (SPO)."[16] HP was attempting to expand its IT services to be a formidable competitor to IBM, Dell, Oracle, and Cisco.[16]

By 2010, Hurd as well as HP were viewed in a favorable light: "Under Hurd's watch, HP has worked to transform itself from an underperforming printer manufacturer into a profitable IT-services supermarket."[17] Hurd had taken the acquisition of Compaq, and through focused execution, "took Carly's strategy and finally got it done."[17]

There was some indication the favorable results under Hurd may have been achieved by sacrificing the future by reducing the expenditures on R&D, thereby reducing the flow of new products for which HP had been noted during its first fifty years of existence. Part of the increased revenues came from acquisitions instead of internal product developments. HP bought eighty-six companies under Fiorina and Hurd.[18]

Hurd's dismissal came not from poor company performance but from his inappropriate behavior as CEO "for false explanations on expense accounts relating to interactions with contractor/actress Jodie Fisher."[18] However, there was also increasing evidence of dissatisfaction with Hurd's difficult management style by senior managers, and some members of the Board were concerned by what they perceived as deteriorating performance of the company.[18]

Apotheker as CEO (September 30, 2010, to September 22, 2011)

Léo Apotheker had a tumultuous year as CEO of HP. He moved from SAP, where he was CEO of a company primarily involved in software, most notably enterprise resource planning (ERP) systems, where SAP was the world leader. With this background, he moved to transform HP from a hardware company to a software company and added his major acquisition influence to HP by acquiring UK-based Autonomy for $10 billion in 2010. This acquisition, coupled with the acquisition of Compaq (a PC manufacturer) for $25 billion and EDS for $13.9 billion (IT systems consulting), further fragmented the strategic direction of a company that had been largely product-oriented up until the arrival of Carly Fiorina in 1999. To add to the confusion, Apotheker also proposed selling off the PC division, where HP was the world leader.[19]

During 2010, HP was plagued with a lawsuit by Oracle stemming from Apotheker's tenure as CEO at SAP. In addition, a major new product—a tablet computer using webOS operating system developed by Palm Inc. (a July 2010 acquisition by Hurd)—was a complete disaster, further weakening HP's position in the industry.

Apotheker was not viewed as a good manager of people or operations. This, coupled with his proposed change in strategic direction for HP from a product-oriented company to a software company, à la IBM, led to his dismissal just eight days before his one-year anniversary with the company.[18]

Whitman as CEO (September 22, 2011, to present)

In her first Letter to the Stockholders, Whitman conceded that 2011 had been a difficult year for HP because of an uncertain macroeconomic environment, especially in the United States and Europe, a series of natural disasters, including the tsunami in Japan and flooding in Thailand, and discontinuity within HP. In her words, "However, amid a number of strategic announcements and changes, we did not clearly articulate our direction. Our execution during the year proved inconsistent, and we had to manage through a CEO transition."[20] Whitman went on to report that HP would retain the PC business, reversing Apotheker's announcement earlier in 2011; finalize the acquisition of Autonomy, an acquisition initiated by Apotheker; and rejuvenate the webOS operating system, which had been discontinued earlier in 2011 when the tablet computer introduction failed. Whitman describes HP as follows:

> HP is the world's largest provider of information technology infrastructure, software, services, and solutions to individuals and organizations of all sizes. That's what we are and we are proud of it. It's straightforward and clear.
>
> At the core of HP are our infrastructure products. Representing about 70 percent of our revenues, this is an area where HP excels. We are the market leader in PCs, industry-standard servers, and imaging and printing, and we are well positioned across key enterprise infrastructure segments. Everything else we do either amplifies or builds on this massive differentiating strength.
>
> On top of our infrastructure products, we layer software. Our software allows us to manage and secure technology environments consisting of our infrastructure as well as others'—something customers really need. Our Autonomy acquisition also enables us to solve a host of new customer problems, which will provide new growth opportunities for HP and our customers. Software at HP is about solving customer problems.
>
> Services wrap our core infrastructure and software businesses, and help customers get the most value from HP. Many of our services customers have relationships with HP that have lasted for decades. Services position HP to do what we do best: act as a strategic partner to customers by leveraging our full portfolio and making sure that HP technology is meeting their needs.
>
> Finally, our solutions make it all work by combining our technologies to advance customers' organizational objectives in a holistic and compelling way.[20]

An article in *Barron's* in early 2012 portrayed Whitman's role as follows: "Whitman must coach the company back to health after more than a decade of multibillion-dollar blunders, or face the slow demise of one of America's great stories."[19] The article goes on to say that "over the past dozen years, HP's engineers have been whipsawed by ever-changing strategies, not to mention scandals and the ousters of three successive CEOs."[19]

Table A6b.9 shows that HP had the following major business segments in early 2012:

- PCs: Although the leader, this is becoming a commodity business subject to intense competition, especially in pricing pressures. Competitors are Dell and Lenovo.
- Enterprise (servers, storage, and networking products): Threatened by the rapid rise in cloud computing with the need to package integrated systems. Cisco is a major competitor.
- Imaging and Printing: Rapidly moving to a rental service market instead of a sale and post-sale service. Major competitors for products are Canon and Samsung while Xerox is the leading rental company.
- IT Services: Here, HP has also focused on selling hardware and then adding services, where maintenance contracts can be lucrative. The major competitor is IBM, which concentrates on high-end services, such as consulting and systems management.[19]

Financial Results during 1999–2011

What is so amazing about this recent history is that HP was able to continue to generate positive net earnings, except for 2002, which absorbed most of the added overhead expenses from the acquisition of Compaq and before the merger was integrated into the HP operation. Table A6b.10 shows that they averaged about 7 percent net earnings during 1998–2000, before the Compaq acquisition, and about 7 percent during 2006–2011, including the EDS acquisition.

Figure A6b.1 shows the relative contribution of products and services to total revenue over the period from 1982 through 2011. While there is some income from financing activities, most of the services revenue appears to come from consulting and system management activities.

Figure A6b.2 shows the decline in R&D expenditures during this period. Instead of developing products internally, the company was attempting to buy those products and services through acquisitions.

Services since 2000

Under Bill Hewlett and David Packard, HP appeared to encourage services as a means of enhancing the value of the manufactured product to customers. As new products were developed, services were added to complement those products.

Under Fiorina, the strategy seemed to shift to adding revenues in an attempt to overcome Dell as the leading computer manufacturer. The acquisition of Compaq made this a reality, although the bitter infighting among the Directors and key

Table A6b.9 Summary of Revenues and Earnings by Major Product Segment, 2011

	Revenue	Earnings	%	Revenue	Earnings
				% of Total	
	Revenue	*Earnings*	*%*	*Revenue*	*Earnings*
Notebooks	21,319			16.3%	
Desktops	15,260			11.7%	
Workstations	2,216			1.7%	
Handhelds and Other	779			0.6%	
Personal Systems Group	$39,574	$2,350	5.9%	30.3%	16.9%
Infrastructure technology outsourcing	15,189			11.6%	
Technology services	10,879			8.3%	
Application services	6,852			5.2%	
Business process outsourcing	2,672			2.0%	
Other	362			0.3%	
Services	$35,954	$5,149	14.3%	27.5%	37.0%
Supplies	17,154			13.1%	
Commercial hardware	5,790			4.4%	
Consumer hardware	2,839			2.2%	
Imaging and Printing Group	$25,783	$3,973	15.4%	19.7%	28.5%

continued

Table A6b.9 (continued) Summary of Revenues and Earnings by Major Product Segment, 2011

	Revenue	Earnings	%	% of Total Revenue	% of Total Earnings
Industry standard servers	13,521			10.3%	
Storage	4,056			3.1%	
Business critical systems	2,095			1.6%	
HP Networking	2,569			2.0%	
Enterprise Storage and Servers	$22,241	$3,026	13.6%	17.0%	21.7%
HP Software	$3,217	$698	21.7%	2.5%	5.0%
HP Financial Services	$3,596	$348	9.7%	2.8%	2.5%
Corporate Investments	$322	$(1,616)	−501.9%	0.2%	−11.6%
Total Products	$87,598	$9,349	10.7%	67%	67.1%
Total Services	$43,089	$4,579	10.6%	33%	32.9%
Total	$130,687	$13,928	10.7%	100%	100%
Eliminations	$(3,442)				
Net Revenue	$127,245				

Source: Adapted from HP 2011 10-K Report.

Table A6b.10   Financial Results for the Years Ending October 31, 1998–2011

Major Acquisitions					Compaq						EDS			Autonomy	
CEOs		Fiorina						Hurd						Apotheker	Whitman
	Oct 31 1998	Oct 31 1999	Oct 31 2000	Oct 31 2001	Oct 31 2002	Oct 31 2003	Oct 31 2004	Oct 31 2005	Oct 31 2006	Oct 31 2007	Oct 31 2008	Oct 31 2009	Oct 31 2010	Oct 31 2011	Oct 31 2012
Products															
Sales	40,105	36,178	41,446	37,498	45,955	58,939	64,127	68,945	73,557	84,229	91,697	74,051	84,799	84,757	
Cost of sales	27,477	25,498	29,727	28,370	34,573	43,689	48,359	52,550	55,248	63,435	69,342	56,503	65,064	65,167	
Gross margin	12,628	10,680	11,719	9,128	11,382	15,250	15,768	16,395	18,309	20,794	22,355	17,548	19,735	19,590	
% Gross margin	31%	30%	28%	24%	25%	26%	25%	24%	25%	25%	24%	24%	23%	23%	
Services															
Sales	6,956	6,192	7,336	7,325	10,178	13,657	15,389	17,380	17,773	19,699	26,297	40,124	40,816	42,039	
Cost of sales	4,595	4,222	5,137	4,870	6,817	9,959	11,791	13,674	13,930	15,183	20,250	30,695	30,723	32,056	
Gross margin	2,361	1,970	2,199	2,455	3,361	3,698	3,598	3,706	3,843	4,516	6,047	9,429	10,093	9,983	
% Gross margin	34%	32%	30%	34%	33%	27%	23%	21%	22%	23%	23%	23%	25%	24%	
Other	0	0	0	403	455	465	389	371	328	358	370	377	418	449	

continued

Table A6b.10 (continued Financial Results for the Years Ending October 31, 1998–2011

Major Acquisitions					Compaq							EDS			Autonomy	
CEOs	Fiorina								Hurd						Apotheker	Whitman
	Oct 31 1998	Oct 31 1999	Oct 31 2000	Oct 31 2001	Oct 31 2002	Oct 31 2003	Oct 31 2004	Oct 31 2005	Oct 31 2006	Oct 31 2007	Oct 31 2008	Oct 31 2009	Oct 31 2010	Oct 31 2011	Oct 31 2012
Total															
Sales	47,061	42,370	48,782	45,226	56,588	73,061	79,905	86,696	91,658	104,286	118,364	114,552	126,033	127,245	
Cost of sales	32,072	29,720	34,864	33,240	41,390	53,648	60,150	66,224	69,178	78,618	89,592	87,198	95,787	97,223	
Gross margin	14,989	12,650	13,918	11,986	15,198	19,413	19,755	20,472	22,480	25,668	28,772	27,354	30,246	30,022	
% Gross margin	32%	30%	29%	27%	27%	27%	25%	24%	25%	25%	24%	24%	24%	24%	
Other overhead	11,148	8,962	10,029	10,547	16,210	16,517	15,528	16,999	15,920	16,949	18,299	17,218	18,767	20,345	
Other costs	−250	−506	−736	737	40	8	31	−176	−631	−458	0	721	505	695	
Operating earnings	4,091	4,194	4,625	702	−1,052	2,888	4,196	3,649	7,191	9,177	10,473	9,415	10,974	8,982	
% Operating earnings	9%	10%	9%	2%	−2%	4%	5%	4%	8%	9%	9%	8%	9%	7%	
Other															
Taxes and adj.	1,146	703	928	294	−149	349	699	1,251	993	1,913	2,144	1,755	2,213	1,908	
Net Earnings	2,945	3,491	3,697	408	−903	2,539	3,497	2,398	6,198	7,264	8,329	7,660	8,761	7,074	
% Net Earnings	6%	8%	8%	1%	−2%	3%	4%	3%	7%	7%	7%	7%	7%	6%	

Source: Adapted from Mergent Online (2011).

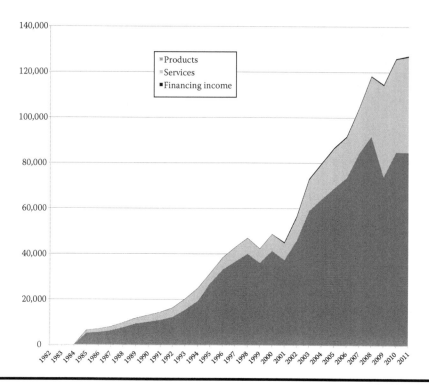

Figure A6b.1   Distribution of revenues between products and services.

	1999	2000	2001	2002	2003	2004	2005	2006	2007	2008	2009	2010	2011
Revenues	42,370	48,782	45,226	56,588	73,061	79,905	86,696	91,658	104,286	118,364	114,552	126,033	127,245
R&D Expense	2,440	2,646	2,670	3,312	3,652	3,506	3,490	3,591	3,611	3,543	2,819	2,959	3,254
R&D Expense (%of Rev)	5.8%	5.4%	5.9%	5.9%	5.0%	4.4%	4.0%	3.9%	3.5%	3.0%	2.5%	2.3%	2.6%

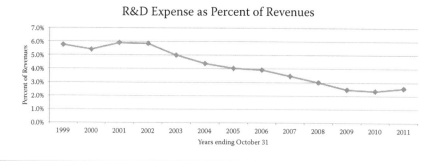

Figure A6b.2   R&D expenses (as percent of revenues). (Adapted from Mergent Online, 2012.)

shareholders extinguished any hopes that Fiorina would realize her dream. R&D expenses began to decline as a percentage of revenues.

Under Hurd, the strategy again shifted to one of milking the product line for maximum income. R&D spending was cut as part of the cost-cutting moves; therefore, new products were not developed. See Figure A6b.2 for a diagram of R&D expenses, as a percentage of total revenues. In the interest of pursuing higher-margin business because the PC business had moved steadily to commodity status, Hurd moved to acquire EDS. This acquisition put HP into services that did not directly support their products. This appears to have been a move to compete with IBM in the IT consulting business.

Apotheker also moved further away from the "services to support products" thinking by putting increased emphasis on software. This put HP on a collision course with other heavy software companies.

The last decade at HP seems to have been more of an attempt to be like some other company rather than in building on the strengths of an already well-established company.

Summary

During Bill Hewlett's and Dave Packard's reigns as CEOs, HP was primarily a product-oriented company. Its services component was designed to support their products by providing customers with the services that would make the products more useful and effective. By keeping the emphasis on new product development, it just naturally followed that the service business would also grow. It was clearly a "services to support the products" period.

Another product evolution was HP's approach to the computer. Although HP developed semiconductors in the early 1960s for use in their scientific instruments, developing computers was not a primary emphasis. In fact, David Packard was less than enthusiastic about entering the computer business, perhaps because of IBM's dominant position. However, HP did develop a minicomputer—the HP 2100/HP 1000 series—in 1966 and continued to develop subsequent versions. Packard was inclined to call them calculators, and HP became a leading developer of programmable handheld calculators such as the HP-35 (the first scientific electronic calculator in 1972), the HP-65 (the first programmable calculator in 1974), the HP-41C (the first alphanumeric, programmable, expandable calculator in 1979), and the HP-28C (the first symbolic and graphing calculator). In the 1990s, HP made a major shift in their target market for computers. They expanded their computer product line, originally targeted to university, research, and business users, to reach consumers. This shift from business-to-business (B2B) to business-to-consumer (B2C) would mean a significant transition in the thinking of top management and their strategies to capitalize on this new market. This set the stage for HP to move from its original product focus on instrumentation to computers.

With the shift to computers and printers, which had been developed during the 1980s to supplement desktop computers, the decision was made to spin off all the businesses not related to computers, storage, and imaging to form Agilent, which at that time was the largest initial public offering in the history of Silicon Valley. It also set the stage for the most tumultuous period in HP's history.

The commitment to computers and printers required HP to take a new look at the services side of the business. In order to be competitive, services could no longer be viewed as an add-on, or post-sale service of products. Computer hardware required software and added services that were of a higher level of sophistication, such as consulting and systems management, areas in which IBM was already dominant.

As HP neared the end of the twentieth century, they faced a dilemma: They had market share in computers, computer infrastructure (servers and storage equipment), printers, and IT services. In each of these areas, they had different competitors. How could they best compete? Fiorina chose to expend resources to strengthen HP's position in PCs to battle Dell; Hurd chose to acquire EDS to strengthen HP's position in IT services, primarily consulting, in order to better compete with IBM; Apotheker was heading toward the elimination of the PC products line, and with the acquisition of Autonomy, was positioning HP to become a software titan. These dramatic shifts in strategy had a demoralizing effect on employees and diluted the internal efforts to develop new products, which was the strength of the company for most of its first fifty years.

Figure A6b.3 suggests that HP operated primarily in Tier 1 services when they viewed services to be complementary to their products. They are finding it a more difficult transition to move into Tier 2 services, such as process consulting, and even more difficult to move to Strategic Consulting in the Tier 3 level of services.

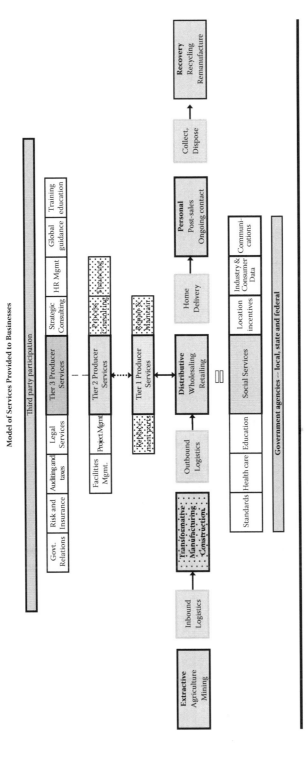

Figure A6b.3   Hierarchy of services provided to businesses.

References

1. Agilent Technologies, http://www.agilent.com/about/companyinfo/history/timeline_1930s.html, 2012.
2. Largest U.S. corporations, *Fortune*, F1–F25, May 21, 2011.
3. House, C.H. and Price, R.L., *The HP Phenomenon: Innovation and Business Transformation*, Stanford University Press, Stanford, CA, 2009.
4. Malone, 2009.
5. HP 1986 Annual Report, http://www.hpmuseum.net/pdf/HPAnnualReport_1986_41pages_OCR.pdf.
6. Wikipedia 2012.
7. HP reverts to its inventive roots, *Journal Record*, 1, Nov. 16, 1999.
8. Clancy, H., Faletra, R., and Demarzo, R.C., HP adapting to a new. *CRN*, 1089, 18–22, 2004.
9. McDougall, P., Fixing HP with strong operations strategy, *InformationWeek*, 1026, 24–26, 2005.
10. Zarley, C. and Campbell, S., Channel execs offer HP some advice, *CRN*, 1134, 47, 2005.
11. Exit Carly; Hewlett-Packard, *The Economist*, 374, 63–64, Feb. 12, 2005.
12. Schick, S., A voice from the past, *Computer Dealer News*, 21, 18, Apr. 15, 2005.
13. Murray, A., H-P lost faith in Fiorina, but not in merger, *Wall Street Journal*, A.2, May 24, 2006.
14. The cash-register guy; face value, *The Economist*, 378, 72–72, Mar. 18, 2006.
15. Schwartz, E., HP's time warp, *InfoWorld*, 28(40), 8, 2006.
16. Poeter, D., Never enough, *CRN*, 1296, 24–30, 2010.
17. Kingsbury, K., HP vs. everybody, *Time*, 175, 1, Apr. 26, 2010.
18. Bandler 2012.
19. Ray, T., Game plan for HP, *Barron's*, 92(8), 23-24, 26, 2012.
20. HP 2011 Annual Report, President's Letter to Stockholders, Mergent Online (accessed July 15, 2012).
21. HP 1976 Annual Report, http://www.hpmuseum.net/pdf/HPAnnualReport_1976_41pages_OCR.pdf.

Chapter 7

The Role of Technology in Continuous Improvement

In this chapter, we try to describe the role of technology in implementing continuous improvement programs. We say "try" because the task is daunting, if not overwhelming. To cover the entire topic is impossible; to select what is relevant to our discussion is challenging. Consequently, we are selective in what we present and supplement it with a number of references that you can refer to if you want additional information. Technology is important; however, it is only one of the three essential elements in making continuous improvement programs successful. In Chapter 8 we describe the need for building the infrastructure necessary to support the improvement programs, and in Chapter 9 we discuss the importance of organizational culture in the change management process.

Paul Kidd points out the importance of recognizing the limitations of technology. Without technology, there would be no manufacturing, but manufacturing is more than technology. It is also about people and their relationship with the technical resources of an organization. Manufacturing is about organization, people, technology, management accounting, business strategy, and the list goes on. It is also about the connections between all these dimensions. In the past, we have tended to ignore not only the connections but also some of the dimensions. We have placed too much faith in technology, using technology to compensate for inadequacies elsewhere, and trying to solve all problems as though they were technical problems.[2]

What do you think of when you hear the word *technology*? It probably depends on your background and interests, but chances are that some of your thoughts will include machines, information systems, computers, vehicles, electronic communications, magnetic resonance imaging (MRI) devices, virtual reality, and mobile communication devices (cell phones). All of these examples and more are items of technology. So are such things as enterprise resources planning (ERP) software, self-directed work teams, and supply chains, to name just a few that may not at first come to mind. If it is a new way of thinking about or doing something, it probably falls under the umbrella of technology. Technology is also about ideas, people, and how to blend them together.

In many improvement efforts, the use of technology outpaces the redesign of the company's infrastructure, which we discuss more thoroughly in Chapter 8. For example, the new technology may require a change in organizational structure to facilitate horizontal communications with customers and suppliers. The company may also need the effective participation of its employees to accept and implement the technology changes, a subject we discuss in Chapter 9. The new technology may require the involvement of self-directed teams. If the company is still using extreme job specialization, the employees will not be intellectually or emotionally prepared for the change.

In a typical improvement program, technology often precedes changes in infrastructure and organizational culture. Technology innovation is a proactive exercise; the changing of infrastructures and cultures are reactive exercises. Consequently, they lag the development of technology. For example, new technologies include EDI and Six Sigma. However, successful implementation requires changes in infrastructure and culture. The technical part of the improvement program is often implemented before the changes in infrastructure and culture. Because they lag, the changes in infrastructure and culture are often implemented under significant time pressures, which may mean they are done piecemeal or inappropriately.

One area of growing importance is inter-company communication, leading to collaboration. This goal is now possible because of new developments in technology. Although information and communication technology are important to collaborative networks, technology can never overcome a lack of trust and ineffective goal setting between key cross-company partners.[3]

Technology is often linked with innovation.[4] Innovation implies change, so the introduction of new technologies in continuous improvement programs is accepted as a natural and worthwhile progression. However, infrastructure and culture convey the impression of permanence, and as a result can be impervious to change.

Although there is often a resistance to new technologies, the evolution of technology has a persistent forward movement that makes it almost inevitable that there will be a continuous wave of new technologies. Resistance may delay the adoption of new technologies but seldom prevents a worthwhile technology from finding its niche in business practice.

Organizations develop new technologies, such as the introduction of new clothing styles, because they want to and can. On the other hand, some new technologies, such as a new vaccine or a new management program, are developed as a result of identifying a need.

Technology is anything that enables a person or an entity to do something differently, hopefully better than before. In this sense, it is a driver of change. Perhaps that explains some of the resistance to the introduction of new technology: the introduction of change. New technology may cause a change in how we do things. We may use PCs, or mobile devices, instead of typewriters. For example, setting a goal of zero defects, instead of a range of allowable defects, may change the way we think about the scope of a problem. Using self-directed teams instead of job specialization can even change the way we manage.

In Chapter 3 we describe the Input–Transformation–Output (ITO) Model and its role in developing the supply chain. In the ITO Model, technology relates primarily to the transformation process, so here we describe the role of technology in the evolution of both manufacturing and service transformation processes. We show how technology enhances the effectiveness of the resources used in the transformation processes—employees, equipment, facilities, systems, and information; then we describe the implications of technology in strategic planning, and describe some of the issues concerned with new technology implementation.

What is technology? People may conjure up visions of spaceships and automated production lines. Those are examples of high technology, but the fact is, a handsaw or a hammer is also considered technology. So is the thought process you use to solve a math problem. Krajewski and Ritzman define technology as "any manual, automated, or mental process used to transform inputs into products or services."[5] Operations managers are concerned with the level or degree of technology used and the corresponding capital intensity of the operations process.

The fast pace of technology change and the many technologies available can tempt managers to upgrade process technology without thoroughly analyzing their actual needs. Changes in technology affect all aspects of an organization—human, technical, and financial—and are usually long lasting. Because of the high stakes involved in the technology game, it is important that decisions about upgrading technology are based on the needs of the organization and customers, and that these decisions are supported by an appropriate analysis of needs. Changing the technology of a service delivery system for reasons other than productivity, service quality, or delivery time improvements can be a serious management mistake.

Technology can be related to the ITO Model described in Chapter 3. Technology can be an input (data), a transformation agent (database), or an output (information). An extension of this is to view the input as information, the transformation process as learning, and the output as knowledge. It can be a driver of innovation (the computer used in system design). Technology has many faces. We concentrate primarily on its role in improvement programs in this chapter.

Definitions

The *APICS Dictionary* defines technologies as the "terms, concepts, philosophies, hardware, software, and other attributes used in a field, industrial sector, or business function."[6] This definition is so broad that we will have to look further.

The *American Heritage Dictionary*'s definition of technology is almost as broad. It says that technology is:

    a. The application of science, esp. to industrial or commercial objectives, or

    b. The entire body of methods and materials used to achieve such objectives.[7]

Science is the investigation of natural phenomena to develop theoretical explanations that engineering can use to design and build tools and systems for practical use. Often, technology is thought of as the combination of science and engineering.

Technologies can be described as "hard"—cellular telephony, railway signaling, or electricity generation or "soft"—management competencies or governmental policy development processes. Technology may be defined as the construction and use of machines, systems, or engineering. "Socio-technologists," take a broader view and consider technology meaningful only when it becomes a social fact.[8]

We therefore conclude that technology has a wide variety of meanings. In the following sections, we examine subsets of technology to show how technology relates to continuous improvement programs. To focus our discussion, we consider technology as consisting of two major categories—tangible products, or "hard" technologies and concepts, or "soft" technologies.

The Role of Technology in Continuous Improvement

Advances in technology are commonplace today, so we seldom stop to think about what impact technology has had and will have on our lives. Every economy in the industrialized world is experiencing the effect of this phenomenal explosion of technological advances. New technology offers new opportunities and new threats to both individuals and businesses. Those who seize the opportunities will move forward; those who fail will probably not be around to compete in the next decade.

The primary use of technology in business and industry is to improve competitiveness. At the customer-facing side, it means recognizing the need for new or redesigned products and services. Internally, it includes increasing productivity, improving quality, and enhancing service delivery. We stated previously that technology alone is not the key to productivity improvement. The key to redefining the tasks includes clarifying what needs to be done and eliminating what does not need to be done. This is a good beginning because many organizations apply advanced technology to tasks that do nothing to add value for the customer. When considering the application of technology in operations, Rule 1 is to streamline the tasks before applying higher technology. Do not automate tasks that do not need to be done. A corresponding fact is that technology will not correct poor management. Our discussions here assume that Rule 1 is in effect and that you understand this fact of technological life.

Technology can be separated into three categories: information technology, materials sciences, and process technology. These are key resources for any organization, and computer technology plays a significant role in all three. In this chapter we discuss the functions of equipment and process technology in organizations, and the major decision areas involving technology. We explore the criteria used in decision making, the steps in adding technology to the process, and key issues in the use of technology.

Technology and the Infrastructure

For purposes of discussion, we refer to the structure of an organization and the systems used in managing the firm as the infrastructure. Does the introduction of higher technology in a firm influence the infrastructure, and if so, how? Certainly, with more and higher technology, there is a corresponding increase in the rate of capital intensity, and this creates a new awareness of the cost of capital. Applying technology for productivity improvements implies changes, perhaps unwelcome ones, in work tasks. Skinner and Chakraborty discuss the impact on an organization's structure and business strategies as "changes in the structure of jobs, changes in the content of jobs, and the need to redefine the environment at work."[9]

Even though it is difficult to generalize about technology's impact on the infrastructure throughout the various service industries, there are some common threads among industries. Initially, the introduction of new equipment and process technology increased the responsibilities of managers and tended to increase middle management ranks. Staff specialists were hired to support the needs of this growing cadre of technology specialists and managers. Ironically, the more recent introduction of powerful information and communications technology has reversed that trend. If information can be made more meaningful and widely distributed quickly, fewer middle managers are required. Upper-level managers now have access to management information that was previously the territory of middle managers. This trend is creating flatter organizations and facilitating easier information flow along the supply chain.

Technology has implications in other ways in an organization. Many employees have to learn new skills and absorb the duties of those who are re-engineered out of the firm. There is a potential for new technology to affect the quality of working life both positively and negatively. On the positive side, word processors and spreadsheet technology have expanded the abilities of and made work easier for their users. On the negative side, new technology, especially in the form of computer hardware and software upgrades, may be stressful to employees who are trying to stay current with the latest developments. These new skill sets primarily affect job design and work assignments.

Another infrastructure area that is affected is the system used in managing the organization. Again, introducing new technology has the potential to affect the operating system either positively or negatively. When management recognizes that individual technologies cannot operate independently, new technology that can integrate functions will have a higher positive impact.

There is an approach in the manufacturing sector referred to as "a focused factory" approach, whereby a factory focuses on producing a single product or family of products. The advantages of this approach include narrowing the number of demands, which will result in higher productivity, fewer levels of management to coordinate, employee-led teams that are more effective in solving problems, communications that are easier and less likely to be misunderstood, and a closer-knit culture. This approach is also evident in some service firms. Fast-food restaurants, business schools with specialized majors, niche dominant retailers, and specialty medical practices are a few examples. This trend necessitates organization changes, often requiring a decentralization of management responsibilities.

There is also an approach that attempts to link specialized functions together to smooth the flow of goods and services from the point of origination to the ultimate consumer. This is the concept of a supply chain, a topic more fully explored in Chapter 10.

The trend to outsource more of the tasks performed by organizations will disrupt the existing organizational structure and systems and require that companies develop new organizational forms to accommodate the new arrangements. This practice of outsourcing may prove to have more impact on the infrastructure than technology changes.

Technology and Organizational Culture

The successful implementation of continuous improvement programs involves knowledge transfer and subsequent knowledge management. Companies must adapt their organizational structure and systems to accommodate the new improvement program; in addition, they must adapt their corporate culture to fit the new program, or adapt the program to fit their culture.

What is corporate culture? It is the belief system of the organization. It results in a feeling among employees about how they should act in carrying out the company's work. It is not the same as the vision, mission, strategies, and plans of a business. Companies could have similar strategies but differing cultures. Culture is the cumulative result of blending strong leadership, environmental influences, personal characteristics of the region in which the company operates, and the manner in which the company competes in its marketplace. Cultures form over long periods of time and are highly influenced by the founder(s) of the business.

Corporate culture should be carefully assessed when implementing new technology, such as a continuous improvement program. Although technology may be the driver of the program, it will not be successful unless it is properly assimilated into the culture of the business. Some cultures actively accept and endorse change; other cultures resist and even oppose change.

That corporate cultures exist is a reality. To be able to implement new technology, companies must successfully manage their corporate cultures. There are two diametrically opposite approaches—change the culture to accommodate the proposed changes or modify the program to fit the culture. The following examples illustrate this dilemma.

One approach is to accept the culture as it is and adapt the technology to fit the culture. Consider a business that wants to implement a quality improvement program. The corporate culture is heavily oriented toward low cost as the competitive advantage. If the quality improvement program were presented as a way to reduce costs of scrap and rework, it would probably be more acceptable in the existing culture.

Another approach is to attempt to change the corporate culture. Consider the same quality improvement program. If presented as a way to encourage greater employee involvement and job enrichment, employees would have to reorient their approach to their work as well as changing their relationships with their supervisors. Some employees would readily change, others would change more slowly, and some would resist the change. Working through this transition takes careful planning and recognition of individual differences.

284 ■ Vanishing Boundaries

A third alternative is to replace completely the existing culture with a new one. This is a radical approach that usually causes major disruptions. The quality program is intended not just to improve the quality of existing products; it is a movement to replace the entire product line with a new emphasis on high quality instead of low cost. Such a change within the culture of existing employees, at both the production level and the manager level, requires extensive preparation. Some employees, especially at the middle management level, may not be able to adjust and will have to be replaced or moved to other positions. In the most extreme cases, it may be more advantageous to move the operation to a new site, hire new employees, and build a new corporate culture at that location. IBM did something like this when deciding to start its PC operation in a remote location in Florida rather than risk its rejection in the existing mainframe culture at established facilities.

John Kotter and James Heskett were among the first to study the impact of corporate cultures on a firm's financial performance. They found that firms that included all the key managerial constituencies and leadership from managers at all levels outperformed firms that had less well-developed cultures. They also concluded that cultures would inhibit firms from adopting strategic or tactical changes, a serious negative effect in a rapidly changing world. Corporate cultures formed over a long period are the most difficult to change. Although difficult to change, corporate cultures can enhance performance with good leadership.[10]

Corporate cultures also have an effect on knowledge management (KM) practices. Standard KM tools may be used differently by diverse organizational cultures and, as a result, achieve different outcomes. Management's choices are to change how the different subcultures use the tools or encourage uses that fit the cultures. One study concluded that it was best to "avoid KM initiatives to achieve uniform targeted outcomes across the firm. Rather, facilitate KM initiatives that appeal to a wider range of more favorable outcomes."[11] Another study found that one of four enablers in developing a KM system is the forming of a culture of sharing, a major shift from the present thinking of many individuals that knowledge is power and not to be shared with others.[12]

Enterprise resource planning (ERP) systems have been one of the major technological developments in the past decade. Studies indicate that many companies are less than completely satisfied with the results of their ERP implementations despite the potential benefits. Although some of the less than favorable results may be a result of the technology, researchers are finding that "managers as well as scientists in many cases attribute the problems associated with the development and implementation of ERP to aspects of organizational culture. However, in general, culture has not yet been a topic of explicit concern in studies of ERP."[13] Because ERP systems are designed to standardize processes and centralize control, they tear at the framework of existing cultures, especially among multi-division companies.

They also attack the job content of the individual employee at all levels of an organization, affecting the worker's skills and attitudes.

Many companies are moving toward globalization of their businesses. When a business moves into another geographic area or country, it must confront the issue of national cultures. In some cases, the cultures are markedly different and the literature is full of missteps that companies have made in violating the cultural norms of the country in which they operate. The culture of an individual country may be well entrenched; however, the globalization of business is having an effect. Although businesses may adapt to different cultures, the cultures are also adapting to global management practices. "Thus, as much as the managerial values are changing, so much so the organizations are cross-verging with different cultures, and a new cultural community is emerging in global business."[14]

As part of the dynamic changes in businesses, there is a revolution in entrepreneurship as large corporations are finding it difficult to keep up with the innovations required, in part because of the deeply entrenched cultures and their resistance to change. Consequently, many of the innovations are being developed by start-up companies, or as separate entities under the umbrella of the parent company. The R&D Limited Partnership model also offers a route for large companies to evolve sequentially into these new ways of operating, without threatening their entrenched bureaucracies. Major cultural shifts are required, but these can be managed in incremental steps.[15]

Traditional software development relied on the fundamental assumption that systems can be fully defined ahead of time. It used a mechanistic command-and-control form of organization that was formal and included a project manager and specialized team members. It involved formal communication and developed extensive documentation, or implicit knowledge. A newer approach to software development involves the use of agile methodologies where the basic assumption is that high-quality, adaptive small teams using an iterative design and test approach that relies on rapid feedback and change can develop software. It is people-centric and builds tacit knowledge using self-organizing teams with informal communication and heavy reliance on customer input. An organic organizational form encourages flexible and collaborative interaction in the development process. Bureaucratic and formal organizations with entrenched cultures are not as likely to be successful in using agile methodologies as organizations that support change and innovation.[16]

The growth of individually developed "apps" for mobile communication devices illustrate the creativeness and uniqueness of individuals that is being displayed in social media outlets.

Corporate culture is a key ingredient in the successful implementation of technology innovations. This is discussed more fully in Chapter 9.

Technology Transfer

Technology is a powerful aid to businesses; however, before they can take advantage of it, businesses must know that the technology exists and they must learn how to use it. In general, knowledge first resides in individuals, as tacit knowledge. Individuals can learn new technology in several ways:

- Read about the technology in a publication
- Attend a workshop or some other educational program
- Work as an aide to someone who has deeper knowledge about the technology.

As individuals learn, they can pass that knowledge along to others in the organization either informally or through training classes. As a greater number of individuals in an organization learn about the technology, they prepare memos, manuals, and other documents that help to establish a store of knowledge within the organization, called explicit knowledge.

It may be possible to learn enough about a simple technology by reading about it. However, as the complexity of the technology increases, most businesses find that one of the following scenarios is advantageous:

- Have selected employees attend training sessions
- Hire a person with the technology experience in another company
- Hire a third party to help introduce the technology into the organization.

Continuous improvement programs are generally of sufficient complexity that businesses find it necessary to use more formal methods to acquire knowledge about the new technology.

Thomas Davenport and Laurence Prusak prepared one of the more popular books about knowledge management. It included chapters on knowledge generation, knowledge codification and coordination, knowledge transfer, knowledge roles and skills, and technologies for knowledge management, along with some examples of knowledge management projects.[17]

Technology for Process Improvement

An almost universal expectation of technology is that it will be used to improve the transformation processes used by businesses. Technology enables companies to be more effective and efficient. Although the introduction of new technology is often disruptive, the availability of new technology eventually prevails if it adds basic value to the operation of a business.

Agriculture, Mining, Construction, and Manufacturing: Goods Producers

The goods-producing industries have been revolutionized by hard technology. Reapers and baling machines in agriculture, automated cutting machines and conveyors in mining, factory-built trusses and high-rise cranes in construction, and flexible manufacturing systems (FMS) and robots in manufacturing. These are just a few examples of how hard technologies have enabled these industries to reduce costs, improve quality, and provide other benefits.

Table 7.1 is a representative list of technologies that have enhanced production and productivity in goods-producing industries. These industries have also improved their processes with management concepts, or soft technologies. In Chapter 5, we described a number of management programs that originated in manufacturing. Here we briefly describe some of the major concepts that made it possible to apply technology to the manufacturing sector. Table 7.2 provides a sampling of these concepts.

There is an abundance of literature about technologies that apply to the goods-producing area. We include just a sampling of the sources here.

Lean production was pioneered by James Womack and Daniel Jones.[18] Since then, there has been a steady stream of books and articles about lean production. It has replaced JIT as the program of choice when companies want to streamline their operations to reduce costs and customer response times.

Mohsen Attaran and Sharmin Attaran studied the re-engineering movement. They found that re-engineering initiatives have produced a range of results. They believe that information technology, with a major emphasis on the use of the Internet, will play an increasingly important role in re-energizing business process improvement efforts.[19]

Manufacturing technology innovations continue to be developed. One study found that early adopters gained a financial advantage because they were able to combine the technology, such as CAD and cellular manufacturing, with knowledge transfer capabilities and culture adjustments to assimilate the technology into their operations successfully.[20]

RFID is a new technology still in the early stages. Although it has great promise, high tag costs, evolving reading reliability, and the need for industry standards are making the move to RFID tags a more deliberate process. As with most technology innovations, there is a need to develop the infrastructure to facilitate the change and the culture to accept it.[21] Ariel Markelevich and Ronald Bell studied the effect that RFID will have on accounting practices. They expect that accounting will have to make significant changes to capitalize on the added information provided by RFID.[22]

Table 7.1 Process Technologies in Manufacturing

Type of Technology	Description
Automated storage and retrieval system (ASRS)	A high-density rack inventory storage system with vehicles automatically loading and unloading the racks
Automated guided vehicle system (AGVS)	A transportation network that automatically positions materials handling devices at predetermined destinations without operator assistance
Automatic identification system (AIS)	A system for transforming data into electronic form with the use of bar codes or RFID tags
Computer-aided inspection and test (CAIT)	The use of computer technology in the inspection and testing of manufactured products
Computer numerical control (CNC)	A technique in which a machine tool controller uses a computer or microprocessor to store and execute numerical instructions
Electronic data interchange (EDI)	The paperless (electronic) exchange of trading documents, such as purchase orders, advanced shipping notices, and invoices
Flexible manufacturing system (FMS)	A group of numerically controlled machine tools connected by a central computer system and an automated transport system
Process control	The use of information technology to control a physical process such as temperatures, pressures, and quantities
Rapid prototyping	The transformation of product designs into physical prototypes through cross-functional teams, data sharing, and computer technology
Robots	A flexible machine with the ability to hold, move, or grab items; electronic impulses activate motors and switches that operate a robot
Servo system	A control mechanism linking a system's input and output, designed to feed back data on system output to regulate the operation of the system
Vision systems	Using video cameras and computer technology in inspection roles to assure that cans of soup are full

Source: Adapted from Blackstone, J.H., *APICS Dictionary* (13th ed., revised), APICS, Chicago, IL, 2010.[48]

Table 7.2   Soft or Conceptual Technologies

Concept	Description
Advanced planning and scheduling (APS)	Techniques that deal with analysis and planning of logistics and manufacturing over the short, intermediate, and long-term periods
Business-to-business commerce (B2B)	Business being conducted over the Internet between businesses through electronic communications networks
Cellular manufacturing	A concept that produces families of parts within a single line or cell of machines controlled by operators who work within the line or cell
Collaborative planning, forecasting, and replenishment (CPFR)	A collaboration process whereby supply chain trading partners can jointly plan key activities required for effective supply chain operations
Design for manufacturability	Simplification of parts, products, and processes to improve quality and reduce manufacturing costs
Design for manufacture and assembly (DFMA)	A product development approach that involves the manufacturing function in product design to ensure ease of manufacturing and assembly
Design–measure–analyze–improve–control (DMAIC)	A Six Sigma improvement process with a structured approach to the problem-solving process
Concurrent engineering or participative design	Involves the participation of all the functional areas of the firm in the product design activity, often including customers and suppliers
Flowchart	The output of a flowcharting process, a chart that shows the operations, transportation, storages, delays, and inspections related to a process
Value engineering and/or analysis	A disciplined approach to the elimination of waste from products or processes by deciding if such functions add value to the good or service
Computer-aided design (CAD)	The use of computers in interactive engineering drawing and storage of designs
Computer-aided manufacturing (CAM)	The use of computers to program, direct, and control production equipment in the fabrication of manufactured items

Source: Adapted from Blackstone, J.H., *APICS Dictionary* (13th ed., revised), APICS, Chicago, IL, 2010.[49]

Supply chains will benefit from new technologies. In addition to new forms of materials handling, there will be new inter-organizational communications systems that will more closely link companies with one another. We describe this more fully in Chapter 10. One study found that firms have to incorporate new product and process technologies and, at the same time, develop capabilities that allow them to be flexible and agile. To accomplish this, they have to increase their effectiveness, exploit the synergies, and learn throughout the organization.[23]

Research and development (R&D) centers develop much of the new technology for manufacturing and service operations. Consequently, R&D centers have a heavy responsibility to manage an effective technology transfer system. Jeffrey Lind describes a process developed by The Boeing Company that "combines insights from systems engineering, software process improvement, organizational psychology, and anthropology to provide a coherent approach to innovation in a large enterprise."[24]

Simulation is a technology that has been around for several decades but has probably never reached the level of use that it merits. A study by Sameer Kumar and Promma Phrommathed looked at how process mapping and simulation could be combined with some popular software tools—Visio, Excel, and Arena—to develop a way to optimize the scheduling of a manufacturing process.[25]

Technology has been a major contributor to the success of manufacturing operations, as we saw in Chapter 5. In the next section, we describe some of the technologies used in service operations.

Services

One of the significant differences between services and manufacturing is that the customer often participates in the service process. This means that the success of technology changes depends on customers and their acceptance or rejection of the change. For example, many people will not give up the personal attention of a bank teller for the convenience of an ATM or an even more advanced development—online banking. Some people refuse to use self-checkout lines at the grocery store because they do not want to learn how. Some callers do not have the patience to go through the 1, 2, 3, or # sequence to leave a voice mail. The role of the customer in the service process must be considered when making changes in the service delivery system. Customers often rebel and find new sources for services when forced to make a change they resist.

Service operations delivery systems usually consist of two parts: the front end, where there is high customer contact, and the back room, which supports the front-end customer encounter. Bank tellers work in the front end; check clearing is a back-room operation. There appears to be a trend of pushing more tasks into the back-room operations to facilitate productivity improvements. Moving tasks away from the customer contact part of the process offers more opportunities for technology upgrading and automation projects. However, this move does take away some of the long-term value of "personalized service," and you need understanding customers who are willing to move to less contact with people and more participation with electronic systems. A real, or perceived, improvement in delivery, timing, quality or a price reduction may be necessary to facilitate this move.

Over the past few years, many service industries have taken a different approach to productivity improvement by involving customers more in the service delivery system. ATMs, salad bars, fast-food clean-your-own-table restaurants, and self-service gas stations are a few examples of involving customers more to gain productivity improvements. The furniture retailer IKEA provides many of its products in flat boxes so the consumer can take it home in his own vehicle.

Service operations may not be thought of as candidates for technology upgrades and automation, but there is a wide range of opportunities for applying automation in the varied services sector. Table 7.3 gives some examples of automation in service industries.

The Japan Industrial Robot Association and the American Robot Industry have classified equipment available for automation. Table 7.4 offers a summary of these classifications with examples of the application in a service business. In addition to the automated systems listed in Table 7.4, a wide variety of manual

Table 7.3   Examples of Automation in Services

Industry	Technology
Airlines	Air traffic control systems, autopilot systems, reservation systems, self-ticketing, security checks
Health care	MRI and CAT scans, RFID tags, patient monitoring systems, remote diagnostics, online information systems
Financial services	Automatic teller machines (ATMs), credit cards, debit cards, electronic funds transfers, online transactions
Hospitality industry	Online reservation systems, automatic checkout, electronic key/lock systems
Wholesale and retail	Point-of-sale terminals (POS), computerized stock replenishment, self-checkout terminals, e-commerce
Transportation	RFID and GPS tracking, automated toll booths, containerized shipping, security checking
Restaurants	Terminals to place dining room orders with the kitchen, POS terminals, computerized employee scheduling
Government services	Online inquiry capability, one-person trash removal trucks, automated telephone response during power failure

Source: Adapted from Collier, D., *Service Management: The Automation of Services,* Reston Publishing, Reston, VA, 1985, p. 20.[50]

Table 7.4 Categories of Robots

Type of Robot	Description	Application
Fixed sequence robot	A machine that repetitively performs successive steps of a given operation according to a predetermined sequence, condition, and position, and whose set information cannot be easily changed	Automated one-person garbage trucks
Variable sequence robot	A machine that repetitively performs successive steps of a given operation according to a predetermined sequence, condition, and position, and whose set information can be easily changed	Automated teller machine
Playback robot	A machine that can produce from memory operations originally executed under human control	Telephone answering machine
Intelligent robot	A machine with sensory perception, such as visual or tactile, that can detect changes in the work environment or task itself, and has its own decision-making abilities	Medical information systems
Totally automated systems	A system of machines that performs all the physical and intellectual tasks required to produce a product or service	Space shuttle

Source: Adapted from Collier, D., *Service Management: The Automation of Services*, Reston Publishing, Reston, VA, 1985, p. 20.[50]

machines or manipulators for technology improvements are available to the operations manager.

Before ending this discussion of equipment and process technology, it is appropriate to examine the human aspect of technology changes. There are advantages and disadvantages to using people and machines for work. In general, people can think creatively and adapt to unexpected situations much better than machines can. On the other hand, machines are better suited for complex or repetitive tasks that require speed and precision. The challenge is to provide the optimum mix of people and technology.

Self-Service

Self-service is a growing part of the service economy. At one time, most retail stores made it difficult for customers to touch the merchandise because grocery items were stored on shelves behind the clerk at the counter, and tools, watches, and other valuables were locked in cases. This practice was the standard applied to reduce theft. Over time, merchandisers decided that customers might be more likely to buy the goods if they could examine them more closely, so they displayed the merchandise in open shelves and counters. This was the beginning of self-service.[26]

Over time, customers have increased the level of their participation in the service process. They collect groceries in a cart and check themselves out through a self-checkout station. They pump their own gasoline and use a credit or debit card to pay for it without ever entering the store. They check their own blood pressure or administer their insulin shots without assistance. Homeowners do more "do-it-yourself" projects, thanks to improved instructions and training courses provided by the retailer. Self-service reduces the cost of the business and is usually more convenient to the consumer. However, it does require an increasing knowledge on the part of the consumer and increased blending of human and machine skills.

Over time, a number of self-service technologies have caught on as customers conclude there is a benefit to them and they can learn to use the technologies without extreme difficulty. One of the newest and most demanding areas in which the self-service concept is being introduced is in e-business.

E-Business

The world is moving rapidly, albeit erratically, toward electronic communications systems. As new technologies evolve, participants increase. This is a case where the technology is definitely the driver, and infrastructures and cultures are scurrying to keep pace.

E-business in the broad sense is conducting business through electronic-aided means, with emphasis on the use of the Internet. Managerial Comment 7.2 includes definitions from the U.S. Census Bureau. Although e-business is generally considered a broader term that includes e-commerce, the two terms are often used interchangeably.

Managerial Comment 7.2
E-Business: Definitions and Concepts

The three primary components of our electronic economy, and the feature shared by two of them, are defined below. Each definition includes examples of its scope and content. The definitions are intentionally broad to provide an inclusive framework for planning statistical measures, and to allow flexibility to incorporate continuing changes in the electronic economy.

E-business infrastructure is the share of total economic infrastructure used to support electronic business processes and conduct electronic commerce transactions. It includes hardware, software, telecommunications networks, support services, and human capital used in electronic business and commerce. The following are examples of e-business infrastructure:

- Computers, routers, and other hardware
- Satellite, wire, and optical communications and network channels
- System and applications software
- Support services, such as website development and hosting, consulting, electronic payment, and certification services
- Human capital, such as programmers.

Electronic business (e-business) is any process that a business organization conducts over a computer-mediated network. Business organizations include any for-profit, governmental, or nonprofit entity. Their processes include production-, customer-, and management-focused or internal business processes. The following are examples of electronic business processes:

- Production-focused processes including procurement, ordering, automated stock replenishment, payment processing, and other electronic links with suppliers as well as production control and processes more directly related to the production process
- Customer-focused processes including marketing, electronic selling, processing of customers' orders and payments, and customer management and support
- Internal or management-focused processes including automated employee services, training, information sharing, videoconferencing, and recruiting.

Electronic commerce (e-commerce) is any transaction completed over a computer-mediated network that involves the transfer of ownership or rights to use goods or services. Transactions occur within selected e-business processes (e.g., selling processes) and are "completed" when agreement is reached between the buyer and seller to transfer the ownership or rights to use goods or services. Completed transactions may have a zero price (e.g., a free software download). Examples of e-commerce transactions include the following:

- An individual purchases a book on the Internet.
- A government employee reserves a hotel room over the Internet.
- A business calls a toll-free number and orders a computer using the seller's interactive telephone system.
- A business buys office supplies online or through an electronic auction.
- A retailer orders merchandise using an EDI network or a supplier's extranet.
- A manufacturing plant orders electronic components from another plant within the company using the company's intranet.
- An individual withdraws funds from an ATM.[25]

E-business requires a great deal of technology, such as described in Managerial Comment 7.2. However, it requires more than technology. Table 7.5 contains a number of characteristics found in successful e-business programs.

Table 7.5 Characteristics of Successful E-Commerce Implementations

Leadership	Strategy
Commitment at the top	Well-positioned online brand
Thorough competitive analysis	Online-friendly offerings
Significant financial investment	Reliable customer service
Cultural transformation	Cross-channel coordination
Structure	**Systems**
Internal investment	Modernized internal processes
Integrated management teams	Incentive-laden HR practices
IT know-how from within	Aligned performance measures
Strategic partnerships	Improved customer management

Source: Adapted from Epstein, M.J., *Implementing E-Commerce Strategies, A Guide to Corporate Success after the Dot.com Bust,* Praeger, Westport, CT, 2005, p. 20.[51]

Technology for Resource Enhancement

In addition to improving processes, technology can enhance the effectiveness of the resources used in the transformation process.

Human Resources

Technology can affect human resources in a number of ways. In a study of a major aerospace company, Robert Kazanjian and Robert Drazin concluded that there was adequate technology to enable the manufacturing sector to meet the market requirements of flexibility and quality in the face of increased worker demands and resource scarcity. However, they also concluded "successful process innovation is a function of the nature of the innovation itself and the amount of new knowledge to be developed, the appropriate configuration of organizational structure, and the existence of a constellation of critical staffing roles."[28]

Advocates of technology have sometimes promoted its use as a way to overcome the weaknesses of the worker. In that case, the success of a technology depended only on training employees so they could overcome organizational deficiencies, implying that the technology was all right. A different perspective is "fix the process, not the people."[29] This approach recognizes the need to consider organizational structure and culture when implementing new technology.

In the early stages, technology applications in services aided employees in performing their work. An electric-powered lift truck in warehouses is an example. As competitive pressures increased, technology improved and replaced employees. Automated materials-handling systems is an example of replacement. Today, with demands for high customization and flexibility, there is a trend toward person/machine combinations as an integral part of the service delivery system. The fixed-sequence robot and one-person trash removal trucks are examples of interdependent person/machine combinations.

As an Aid to the Employee

Some of the early efforts in productivity improvement in services centered on using technology to make work less difficult and less time consuming. As is the case today, much of this occurred in the back-room operations and offices of service firms. One of the early technology introductions in offices was the electro-mechanical calculator. Previous methods of performing calculation functions were time consuming and difficult, and accuracy was a problem. The evolution to electronic calculators was a further enhancement to the speed and accuracy of calculation functions. Computers, spreadsheets, and other specialized application software have moved calculation functions to higher ground.

The technology of digital document systems offers desktop control over the production, management, and finishing of documents. An employee can print, fax, copy, collate, staple, and distribute documents without leaving his or her desk. This technology closed the gap between electronic and paper documents and significantly improved workflow productivity. In general, managers and workers welcomed technology that aided work performance.

Technology can help employees and the service firm in ways other than as specific work aids. Technology can change the total service delivery system and help create jobs. Consider the use of plastics and refrigeration in the food business. The ability to prepackage, transport, and store predetermined portions of food has allowed fast-food service businesses to adopt a production line approach to service delivery. The next time you opt for a Big Mac lunch, observe the process of preparing your order. This is an example of a division of labor production line system made possible by the application of technology in a service environment.

Examples of technology that helps people accomplish their work are everywhere in the workplace. Some of the more common are:

- Telephone headsets
- Word processors and spreadsheets
- Electronic meter readers
- Electronic data interchange (EDI), both Internet and traditional
- Bar codes and RFID tags
- Retail computerized checkouts
- Police cars with computers on board
- Weather reporting systems
- Magnetically leveled passenger movers
- Mobile phones with text messaging
- Notebook computers with wireless connections
- Intranet and extranet systems
- Electronic downloads—music and books.

As a Substitute for the Employee

The one overriding characteristic of higher technology and automation is the capability to produce standardized goods and services rapidly and consistently at a reduced cost. In the manufacturing sector, this prompted a drive to integrate computer-assisted machines into larger clusters to assume even greater control of the production process and use fewer people. These systems are called flexible manufacturing systems, or FMS. Some of this technology has spilled over into the services sector. Automated guided vehicles (AGV) are flexible materials-handling systems that perform pick and put-away operations as well as transport materials from receiving to storage and then from packing to shipping.

In a mistaken belief that more computerization is the ultimate answer to success, many U.S. firms have leap-frogged basic FMS in favor of computer-integrated manufacturing (CIM). This powerful technology is appealing, but for the wrong reasons. "Its appeal is the lights-out factory, no more labor troubles, and a quick edge on the competition. That's why CIM, touted as the big manufacturing breakthrough of the age, has turned into a disaster at many companies. And the reason is simple: to implement computer-integrated manufacturing requires the kind of sweeping organizational changes that many companies investing in CIM have yet to make."[30]

What are these sweeping organizational changes? Organizational structures, groups, and people need to be formed into teams that operate as an integrated unit. This requires changes in attitude, work practices, and employee skills, and few companies are willing to make such drastic changes. Chapter 8 describes more fully the organizational infrastructure changes required, and much of Chapter 9 is dedicated to people-oriented actions that support these needed changes.

Just as in manufacturing, automation technology is being used in many service companies for the wrong reasons—to replace people in service jobs. This replacement function of technology has created a controversy among some sociologists. Displacement of people frequently hurts the individual worker and creates a burden on society as a whole. Some have gone so far as to predict that the majority of the total workforce will eventually be trapped in low-skill, low-pay job positions. Others contend that there are both social and economic obstacles to automation. The economic obstacles are well understood: high tech is high cost. The social obstacles are more complex.

In the past, the rapid growth of services absorbed many of those displaced by automation in other sectors. If there are fewer and fewer jobs in the service sector, can there be continued job growth? If fewer people have jobs, who will bear the increased burden on the Social Security system and other social systems? Can companies find new markets in the global economy to provide jobs for those displaced by automation? Will the skills of those who are displaced be adequate or appropriate to satisfy the requirements of the new jobs being created? Alternatively, is all of this just a technical Chicken Little shouting "the sky is falling"? We do not know the answers to these questions, nor is there consensus among those sociologists who study them. The point is that operations managers' jobs and responsibilities are, and will continue to be, affected by these issues. The next part of this section discusses a trend that may impact using technology to replace people.

As an Integral Part of the Process

There is a growing consumer trend toward less standardized, more customized products and services. This trend will continue, indeed accelerate, as we move into the twenty-first century. This will require a different kind of systems integration, and it will have a significant impact on service delivery systems.

In CIM, what is integrated? The computers are integrated; the workers interface to support the system! In instances where systems have been designed to interface with people, the interface was reactionary to solve a problem with the system or to compensate for the inadequacies of the technology.[31] This approach is compatible with a service environment that is cost-driven with standard services. Today, you must have a system that can address more customization and flexibility in service delivery. A system that meets these needs requires a different kind of system design, a design that integrates people and machines rather than interfaces between people and machines. In other words, people characteristics add flexibility and cognitive thinking to a service delivery system. "Flexibility cannot be purchased by acquisitions or merger. Flexibility is something people do. Organizations of people may be combined or disassociated, but flexibility itself is not attained through things, financial or physical."[32]

We will examine the forces that influence introducing technology into a service organization and the characteristics of people/machine relationships.

Technology applications are either "pulled" or "pushed" into service organizations. Pulled applications occur when market forces demand services that cannot be provided by existing technology. From the advent of credit card use, the approval process was an obstacle to using a credit card. Improvements in this process to the point of electronic approval in seconds were pulled by the market. Pushed applications occur when technology developers create new technology and find or create a market for it. Bar-code systems were developed but were not widely used in services until a universal product code was developed. Bar codes were pushed into the market and the pushing pulled a universal product code.

In his book *Things That Make Us Smart*, Donald Norman contends that up to now, most technology development has been a push process. He illustrates this by using the motto of the 1933 World's Fair: *Science Finds, Industry Applies, Man Conforms*. It is Norman's contention that we are approaching the time when there must be human-centered technology: *People Propose, Science Studies, Technology Conforms*.[33] Paul Kidd essentially says the same thing. "Balance (people/machine) is an important conceptual basis for the new paradigm of agile manufacturing. It will help us to address the interconnected elements of our overall manufacturing enterprise. A balancing capability avoids polarization from any perspective."[34] He describes a participatory model for computer-based technologies that support people. Table 7.6 illustrates this concept.

The demand for customization is pulling a change in people/machine relationships, and social changes in the workplace are pushing for a change in service delivery system design. These forces are pushing toward technology in which people/machine combinations operate as an integral part of the service delivery system. Employees will be less available to compensate for inadequacies of technologies and will be more a part of a system where they apply their knowledge to operate flexible systems that provide customized services.

Table 7.6   Interdisciplinary Design Methodology

One of the primary concepts of agile manufacturing is concerned with the integration of organization, people, and technology. To achieve this integration we need to adopt an interdisciplinary design methodology. The design paradigm that should underlie agile manufacturing is characterized by the following:

- A holistic systems based approach
- Concurrent engineering concepts applied to the design of the manufacturing/service enterprise
- Application of insights from the organizational and psychological sciences to the design of the technology and the overall manufacturing enterprise
- Consideration of organization and people issues at all stages, from formulation of business strategy right through to the design and implementation of systems.

Source: Adapted from Kidd, P.T., *Agile Manufacturing*, Addison-Wesley, Reading, MA, 1994.[52]

Equipment

Technology is used in equipment to enhance the performance of the equipment or to use the equipment as a data collection point.

Enhance the Performance of Equipment

Much of the technology described in Table 7.1 related to improving the precision and quality of output from the equipment resource. The focus in recent years has been in using computer technology to enable equipment to do more without the aid of human operators. Although equipment can perform a number of automated functions, there is a growing realization that the best results come from a blend of human and machine skills, not from machine automation alone.

Provide a Source of Performance Information

Through sensing devices and connected software, equipment can now provide a great deal of information about the amount and quality of the products and services processed. For example, machines can count and report the number of units produced. They can sense the thickness of paper and provide a permanent record of the quality of the paper processed.

Facilities

The degree of technology employed in a service delivery system influences the design and layout of the service facility. However, just as customer contact and customer needs influence the degree of technology used, they equally influence facility design and layout. It is, then, the nature of the basic service and overall operations and marketing strategies that determine the kind of facility required. When selecting a design and layout, management must consider a number of factors, including the level of customer contact, volume of demand, range of the types of services offered, degree of personalization of services offered, skills of employees, and the costs involved as well as the technology employed.

The diversity of services makes it difficult to generalize about facilities and technology; however, we can establish two ends of a continuum. There are service facilities with limited or no customer contact where the technology employed can be at the high end. The design of these facilities focuses on technology and optimizing productivity. At the other end, there are service facilities where the entire facility is designed around customers. We previously used the example of the back-room operations of a bank as the former, and the front end of a bank as the latter. Other service facilities combine the characteristics of these two extremes. Again, the degree of customer contact versus technology for productivity depends on the services provided and the tactics employed in providing the services.

Design

The service package is a bundle of goods and services consisting of four features:

Supporting facility—the physical resources that must be available in order to offer a service, such as a golf course or a hospital

Facilitating goods—the materials purchased or consumed by the buyer or provided by the customer, such as skis or medical supplies

Explicit services—benefits readily observable by the senses, such as a smooth-running automobile or a fast response by the fire department

Implicit services—psychological benefits that the customer may sense only vaguely, such as security of personal information at a bank.[35]

The makeup of the service package is a function of the competitive service concept and strategies selected, and technology decisions are mostly associated with the supporting facility feature. These decisions are about location and physical size of the facility, layout to optimize the customer encounter and facilitate the flow of goods and information, and equipment selection. The operating capacity of the service process is determined by these decisions, and most of the decisions create "hard" results. That is, once they are made, it is difficult to undo the results of the decisions. Conversely, service process design is never final, because once the operation begins, it is necessary to modify it to fit ever-changing customer needs and to accommodate technological advances. It is important that the following design objectives be accomplished:

- Optimizing the customer encounter. This includes easy and safe access and egress; providing for customer waiting lines and service counters; good directional signs; attention to customer comfort; and aesthetic factors. Designers must also consider the level of customer involvement in the encounter: Is the customer a part of the process, acted upon, or a passive observer?
- Facilitating the flow of goods and information. This is a prerequisite to improving human and machine productivity. A necessary activity is to determine optimum flow paths and the relative location of the various areas. Flowcharting, load-distance analysis, and closeness ratings are tools that assist in these determinations. Correct closeness relationships also enhance good communications as well as reducing personnel travel times.
- Optimizing facility flexibility. Service organizations frequently find they must quickly adapt to changes in the nature and quantity of demand. How well they can adapt depends on the degree of flexibility designed into the facility. The design should focus not only for today's needs but also for possible future needs. Balancing flexibility with practicality and capital intensity is the operations manager's challenge.
- Improving human and machine productivity. The type of operation determines the means to accomplish this objective. A good approach is to find the most costly, most time-consuming, or most capacity-constrained tasks and concentrate on improving them. For example, in a retail furniture warehouse, stocking, picking, and transporting are the dominant costs. Automatic materials-handling from the storage racks to the loading dock can improve productivity. This is also an example of the impact of technology choices on facility design.
- A safe and morale-building environment. Certain safety features are a given in today's society, but the effective operations manager will provide beyond the minimum. Security for employees' personal belongings, employee parking, spaces for privacy, adequate (above building code requirements) restroom facilities, and decorating schemes are some of the items to consider.

Location

When you consider the trends in how consumers receive their services, technology plays a significant role in location decisions. Buying goods from catalogs or websites is at an all-time high and continuing to grow rapidly. Banks provide online financial services. You can purchase goods and services on the Internet. These consumer practices are having an impact on location decisions. Firms that utilize catalogs and communications technology in the service delivery process change their position on the customer contact continuum from high to low. Low customer contact creates the opportunity to locate for cost advantages rather than customer convenience.

(Also, business owners can live where they want to live.) Lands' End can sell a sweater from Dodgeville, Wisconsin, as effectively and at a lower operating cost than the department store in the mall. Dell and others sell computers, Amazon sells almost any product existing, and various companies sell cloud computing services. Think of the implications of banking via computer. Not only is there an opportunity to reduce personnel costs, but there is also the significant potential to reduce the number of branch offices. Internet banking offers an even less costly and flexible alternative to branch banking. Imagine a bank with 24-hour service that you can access from anywhere and from which you can receive customized services for a low fee.

Layout

There is a strong relationship between facility layout and technology. However, what influences the use of technology in a service facility? Once again, we emphasize that the service concept and strategies selected determine the service delivery process and, therefore, the technology and corresponding layout of the service facility. The objectives for layout are identical to those for design discussed earlier:

- Optimizing the customer encounter
- Facilitating the flow of goods and information (good communications)
- Optimizing facility flexibility
- Improving human and machine productivity
- Providing a safe and morale-building environment.

Information Technology

Information technology (IT) is the arrangement of computers, telecommunications, and other devices that integrate data, equipment, personnel, and problem-solving methods in planning and controlling business activities. IT provides the means for collecting, storing, encoding, processing, analyzing, transmitting, receiving, and printing test, audio, or video information.[36]

IT can perform three important functions:

1. Control and direct equipment and facilities to perform programmed functions, such as using CNC machines or adjusting temperatures in a chemical process.
2. Collect and use information to perform administrative functions without human intervention—for example, placing automatic reorders for stock inventory items or assigning late payment charges in credit card processing.
3. Collect and report information that requires further analysis and decision making by employees. This generally is an interactive process in which the IT system and the employee interface to facilitate the flow of information.

Table 7.7 describes the links between IT and the business operations. It distinguishes among process, information, services, and technology integration and identifies the alignment requirements and issues confronting each of these categories.

Integrated Systems

At some point, companies have to integrate a variety of processes, functions, and techniques into a system. An example is a supply chain. We described this in Chapter 3, and a detailed description of an integrated supply chain is provided in Chapter 10. In the following section are a few examples of technology that have helped to integrate systems.

Enterprise Resource Planning Systems

Enterprise resource planning (ERP) systems have been popular since the mid-1990s as an attempt to integrate a large portion of a company's information processing. Earlier, materials requirements planning (MRP) and manufacturing resources planning (MRP II) systems integrated the manufacturing operations and, to some extent, linked with marketing and financial functions. ERP was designed to finish the integration of all functions within a business. Although there are successes, there is also evidence that some companies have not realized the level of integration promised, despite substantial investments.[37] Companies are also finding that it is difficult to link their ERP systems with their partners along the supply chains because of incompatibility problems. ERP systems are well integrated and represent a large investment, in both time and money, to organizations.[38]

Inter-Organizational Systems

One of the recent IT technologies that help to link companies along the supply chain is the inter-organizational system (IOS). The IOS is designed to help organizations communicate with one another using such technologies as EDI, extranets, extensible markup language (XML), and Web services.[39] "An IOS consists of computer and communications infrastructure for managing interdependencies between firms."[40] The benefits of IOS include an increase in the competitiveness of a firm, a reduction in the cost of communicating, the ability to enable tight integration among firms, an atmosphere that facilitates knowledge sharing and trust building, and the opportunity for enhanced innovation. Although there are still compatibility problems, companies are beginning to make progress in connecting with their supply chain partners.[41] Although IOS increases the communications capability of organizations, it does not necessarily provide a way to improve individual business applications.

Table 7.7   Issues in Matching IT to Business

Process-Driven Architecture (PDA) Layer	Description	Examples
Process	A structured and measured sequence of activities designed to produce a specific output based on defined input	Process design and modeling Continuous process change Process improvement programs Process management Workflow management
Information	Managing the conditions of effective and efficient information production and delivery	Need for a single view Data warehousing Real-time information needs Consistent meaning of data Multichannel data delivery Availability of metadata Semantic Web initiatives
Services	Well-defined, self-contained functions fulfilling a particular business need provided by an application or module on request of another application	Standardization of Web services Service-oriented architecture (SOA) Linking software development with application deployment and management mobility Access for different user devices Bridging the difference between business needs and technical capability
Technology integration	Provide a means for gradually migrating to a service-oriented (SOA) and process-driven architecture (PDA)	Using the still useful logic and functions of legacy systems Functional integration: formatting, transformation, routing Technical enterprise application integration (EAI); message-oriented middleware (MOM), message queuing

Source: Adapted from Strnadl, C.F., Aligning business and IT: The process-driven architecture model, *Information Systems Management*, 23, 4, 67, 2006.[53]

Service-Oriented Architecture

One area that holds promise as an integration enabler of business applications is service-oriented architecture (SOA). SOA is not a single software package or even a neatly packaged collection of software applications. It is an approach—an architecture—to moving a company from a stand-alone entity to a member of a loosely coupled network of companies, such as the partners along the supply chain. Instead of being a tightly integrated system of applications, such as in ERP, SOA offers a way for sharing companies to "mix and match" IT hardware, software, protocols, and standards to fit their mutual needs most directly. There are numerous definitions for SOA; however, there is no standard definition as yet.[42] We describe SOA further in Chapter 12.

Structure, Trust, and Collaboration

Technology is necessary, but not sufficient in itself, for the success of integrated systems. As we describe in Chapter 8, adapting the structure of a company to fit the new technology is also a necessary ingredient. Before a company is comfortable sharing information with its customers and suppliers, it must build a trusting relationship with them. Once there is trust, cooperation and collaboration follow.

Criteria Used in Decision Making

In this section we discuss criteria to consider when making decisions about strategic needs versus short-term needs, behavioral versus scientific management issues, and costs versus benefits of added technology. All decisions about the level of technology to apply must link to the strategies selected in the strategic planning process described in Chapter 2.

Strategic Needs versus Short-Term Needs

When considering technology issues, managers often encounter conflict between strategic needs and short-term needs. Often, the urgency of the short-term needs overshadows the long-term necessity of the strategic needs. At times, priorities must be on short-term needs, particularly when customers are involved; however, decisions about short-term items should always take into consideration the effect on the strategic, long-term needs. When these situations arise, it is necessary to return to the vision statement and identify the organization's driving force. This central theme should guide current and future decisions about the company. Companies should answer such questions as the following: If we concentrate our technology resources on short-term needs, what impact will that have in the future on the company's capabilities; the company's products/services/markets; or desired financial results? Examining the impact on the elements listed below usually provides enough information on which to base an enlightened decision about possible trade-offs:

- Value added to the customer
- Service delivery—quality, speed, and timeliness
- The service or product itself
- Price
- The employees
- Other elements specific to the company.

There is an abundance of literature about the strategic role of technology. Two of the tools available to assist in evaluations are the decision tree technique and the preference matrix. A decision tree is a schematic diagram of the alternatives available that shows potential effects of the various alternatives. The preference matrix is a simple table that can rank alternatives by numerically rating various elements of two or more alternatives. For a complete explanation of these tools, see any operations management textbook, such as *Operations Management* by Jay Heizer and Barry Render.[43]

Behavioral versus Scientific Management Issues

Introducing technological improvements into a service delivery system changes the manner in which tasks are accomplished. These are often unwelcome changes as they affect the employees performing the tasks. As work becomes routine, many jobs move from the more satisfying behavioral-based design to a specialized job design. Changes in job structure and job content can be disturbing to many employees and may have an adverse effect on the service delivery system. Just witness cashiers learning how to use a new point-of-sale (POS) register system. Often, they get frustrated when relearning the process they believe has been made more complicated by the introduction of new technology.

Job structure refers to the manner in which the job is designed. The introduction of new equipment and process technology can have a double-edged sword impact on job structure. At the lower levels, job structure tends to become routine, akin to the job specialization design. Job structure in higher-level jobs tends to expand. The equipment is usually complex, requiring new skills and knowledge to set up and maintain. Supervisors spend less time working with training and scheduling and more time communicating. High-tech processes are managed differently from low-tech processes.

Job content refers to the number of and complexity of tasks required to accomplish a job. Job structure and job content are closely related, and the impact of changing technology is similar to that of job structure.

Table 7.6 compares various capabilities of people and machines. The level of customer contact is an important factor in job design. In high-contact systems, you are seeking the advantages of people. In low-contact systems, you are seeking the advantages of machines. The selection of technology improvements is influenced by the need for customer contact, the degree to which the characteristics in Table 7.6 must be present, and the judgment of the operations manager.

Costs versus Benefits of Added Technology

Improving technology to gain productivity benefits means increasing capital intensity. Few businesses can afford all of the technology available to them because this would require alternative uses of financial resources. Thus, more technology is not always better. The level of technology that supports both operating and marketing strategies provides the most benefits to the company. To achieve this optimized state, technological change must be managed through planning, analysis, and justification.

The place to make and discover mistakes in technological choices is in the planning stage. When considering the application of technology in operations, it is important to streamline the tasks before applying higher technology. Companies should not automate tasks that do not need to be done in the first place.

One of the first steps in planning for technological change is to define the expectations resulting from the change. Simply reducing costs does not adequately define an objective for an expensive change. An operations manager must set specific goals, defining not only results expected but also time schedules and responsibilities. Planning for technological change should include gaining knowledge about the available technology and determining at what time, if ever, this technology may be of value. Know where the technology can take you, and know if you need to go there. Plan for the long-term, yet implement incrementally. This strategy combines the big picture of what can be accomplished with the capabilities of the organization and provides a road map for the future.

Justifying higher technology projects cannot usually be accomplished using only traditional financial analysis techniques. Even though financial analysis techniques are a good way to start and are necessary, they do not take into consideration the intangible and qualitative benefits that are possible with technology projects. You must move beyond reducing labor and fixed costs to consider the value you can add for the customer: improved customer service, improved quality, increased speed of delivery, reduced inventory levels with no negative impact on service, and improved service delivery system flexibility should all be considered.

A number of tools are available to assist in the analysis and justification of technology projects. The decision tree technique and preference matrix approach that were previously mentioned are effective tools when evaluating qualitative factors. Break-even analysis is a handy "what if" tool, and the traditional analyses of rate of return, payback time, net present value, and inventory turns are good evaluative tools when combined with the qualitative methods.

We discussed the impact of technology on employees previously. However, it is important to remember that technology affects jobs at all levels in the business. A new technology can create cultural and organizational disruptions. Some employees gain and some will lose. These outcomes can create human relations issues that could sink the acceptance of a project. Pay attention to the people aspects of technological change. Early communications and involvement of as many employees as possible in the change process are critical.

Steps in Adding Technology to the Process

The manner in which technology is introduced in a production or service delivery system is as important as the technology itself. In fact, there are far too many instances where the appropriate technology was selected for productivity improvements, yet the potential was never reached because of the way the change was implemented. In this section we discuss a seven-step process that can enhance the probability of success.

Step 1: Communicate, Communicate, Communicate

This step is actually a corollary to the other six steps. As soon as you know that a technological change is going to be considered, tell your employees. This communication must be clear and it must be complete. The reasons why a change is being considered must be given to all those who may be affected. If reasonably well known, the potential impact on work life should be discussed. If the change will cost some jobs, let people know about the possibility in the beginning. However, do not speculate. If you do not know, say so. Any facts about estimated timing should be communicated. Make a commitment to providing education to those affected. Displaced employees should receive outplacement assistance. Make a commitment to continue communicating at the appropriate times as the project progresses.

Every new technology project needs a champion, someone who can do much of the communicating with others in the company and serve as a liaison with the implementation team. The champion does not need to be the CEO or the operations manager. The champion does need to be politically sensitive and have a keen sense of timing to know when to involve non-team employees in the project. Above all, the champion must exhibit an air of confidence, be persistent when dealing with top management, and practice patience throughout the life of the project.

Step 2: Identify Needs and Opportunities

Often the need for improvement will be painfully obvious. Market share is declining, profits are drying up, service delivery is lagging, and other such red flags may appear. In other cases, there may be symptoms that are disguising problems and the operation may have to be analyzed to find those problems or identify potential areas for improvement. (For example, high sales volume for short periods can hide a multitude of problems.) In either case, a concurrent engineering approach is useful in this and all subsequent steps. Concurrent engineering is a team-based approach in which the team members come from diverse backgrounds and usually have some interest in the outcomes of the change. In a traditional organization, such a team would typically consist of people from operations, marketing, sales, human resources, accounting, and finance. In a team-oriented organization, members would come from various work teams. It is at this point that top management should express its support of the team and the project, and their support must be evident throughout the life of the project.

Once a need is known, the team must identify potential areas for improvement. Whatever the process, measurement criteria must begin to flow into the process. Identifying the changes to be made and the technology to be applied is an integral part of the process.

Step 3: Evaluate Alternatives and Select the Optimum Alternative

In this step, examine the cost–benefit relationships among various alternatives. We previously stated that justifying higher technology projects cannot usually be accomplished using traditional financial analysis techniques alone. Even though financial analysis techniques are a good way to start and are necessary, they do not take into consideration the intangible and qualitative benefits that are possible with technology projects. You must move beyond reducing labor and fixed costs and consider the value you can add for the customer: improved customer service, improved quality, increased speed of delivery, reduced inventory levels with no negative impact on service, and improved service delivery system flexibility. Evaluation should include testing the technology either through simulation in your setting or at another company's location. This is also necessary for proper education and orientation.

Step 4: Educate and Orient

Team members and top management should become familiar with the technology and understand its future applications. Talk with others who have experience with the technology. Observe it in action under operating conditions. Visit other production and service facilities that use the technology. As soon as practical, orient other people who might be affected by the change. Renew the commitment to provide education to those affected by the change.

Step 5: Develop the Implementation Plan

Technological change projects are usually large and complex and require an implementation plan. An implementation plan should not be just a schedule of activities; it must be a comprehensive document that can be used to manage the technology change. The following items should be addressed:

- The requirements of the new technology (design specifications), with inputs from users
- The expectations from the results of the technological change, with inputs from users
- How and when the results will be measured
- Training and education required
- All the known activities required to accomplish the expectations of the project, from planning through evaluation
- Start and complete times (in days or weeks, dates can be added later)
- Responsibility for each activity, including outside participants
- The critical path of activities
- Other items specific to the project.

Two of the more useful tools that have been developed to assist in project planning and management are the project evaluation and review technique (PERT) and the critical path method (CPM). PERT was developed to handle uncertainties in activity times. CPM was developed primarily for scheduling and managing projects where activity times were known. There are computerized versions of both, and many of the distinctions between the two have disappeared. A Gantt chart can display the relationships of project activities for day-to-day project management.

Step 6: Implement the Technological Changes

If you have sufficiently practiced Step 1, implementation will go more smoothly; however, *more smoothly* is a relative term. Even diligent planning and communicating does not guarantee problem-free implementation. An accepted rule of thumb is that the higher the technology involved, the higher the probability of problems. However, the real value of having a comprehensive plan becomes evident during implementation. With a thorough plan, it is less difficult to determine the impact that changes have on other elements of the implementation process. Unexpected occurrences or something not performing properly will require revisions to the implementation plan. Knowing the details of all planned activities and the relationships to the critical path will greatly facilitate rescheduling. A word of caution: Not every problem requires a plan change, and the team should make certain that all plan changes are necessary, rather than convenient.

Implementing technological change in an organization tends to disrupt day-to-day operations. This is another reason to consider implementing the components of the technology in increments, or modules. There is often an opportunity to implement and test modules in an area of the business outside the mainstream of daily activity. This provides a venue for making changes and adjustments while minimizing disruptions to more critical operations. Once tested, the technology can be transferred to the mainstream after it is performing effectively.

The project team should track all activities against the planned schedule and other measures developed as a part of the plan. Variances should be documented as they occur and accompanied by a complete description of how they were addressed. Technology performance measures should start as early in the implementation process as practical. Corrective actions and results of each action should be documented.

Step 7: Evaluate Results, Redefine Needs, and Redefine Additional Increments

Evaluate refers to both the implementation process and the technology performance. At this point, the base for comparison is always the original expectations used for justification. Technology results may exceed expectations, or they may fall short. Seldom do results exactly match expectations. Actual results compared to the original objectives should be documented and studied to determine what caused the differences. This review can be a learning experience and an important step because it provides an opportunity to improve the implementation process. Expertise in managing technology change can provide a competitive advantage.

You may discover that implementing the first increment of technology changed the original needs; therefore, it is necessary to review the additional increments to determine if they will meet the redefined needs. Some of the activities scheduled for future increments may not be required, or additional activities may be substituted. Perhaps one or more of the planned future modules can be eliminated. There are many possibilities of changes resulting from the first implementation increment, and you will uncover them only by performing this important evaluation review step. Effective planning and control of time and cost performance is one more tool to use in the global competitive markets of today and of the future. What you learn from the review process can help you with planning and controlling other projects in other areas.

Future Considerations for Technology

We have lived in a world where we could see, touch, and feel tangible things. Change occurred, but it occurred at a rate that allowed us to be relatively certain about what to expect. Today, the scope of change is different. Change happens so rapidly that we cannot "touch" things anymore. The world is constantly changing, so we do not always get a chance to get as comfortable with situations and people as we would like. With this scenario as a backdrop, you can see why predicting the future of technology in services is a difficult endeavor, so we will stay away from predictions. However, technology is the driving force for the major changes already experienced and for experiences in the future.

Some current technology trends and issues appear to be influencing service delivery systems. The trends are less customer contact, more customer participation, increased customization, and more flexibility. At times, there will be conflicts among these four systems parameters; however, that is the environment in which businesses operate today.

Technology improves productivity, service quality, and delivery times, but one thing drives technology changes—competition in the marketplace. Because the customer is what a service business is all about, we will start there. As you read the following discussions, put yourself in the shoes of the consumer and think about your first reaction to services delivered in the manner described.

Customer Acceptance

Customers have been exposed to technological changes for centuries; yet, they began accepting less customer contact in the early 1990s. We previously discussed ATMs and pay-at-the-pump gasoline delivery as examples of a less-customer-contact service delivery system. These technological changes have been accepted by customers. Even those who reject technology early on eventually warm up to it. Customers begin to accept more participation in the service delivery process. A move toward more customization requires a more flexible service delivery system. Combining the contents of the previous sentences with technological advances in computers and communications technology yields the following scenario of consumer acceptance of technology in services.

The stage is set for consumers to drive the move from high contact/low involvement to low contact (person-to-person)/high involvement in the service delivery system. Communications and computer technologies make it possible for the consumer to never leave home yet purchase products and services that fit with her or his unique requirements. Surrounded by computers, fax machines, mobile phones, and online capability in a location where they choose to live, electronically empowered consumers can reorganize the business landscape. This new electronic heartland is spreading throughout the globe.[44]

The issue of customers' acceptance of technology that keeps them away from human contact is really one of conditioning. Those who have grown up during the transition to people-less transactions are conditioned to, and probably prefer the convenience of, automatic transactions. Older generations most likely still resist much of the technological changes. Businesses will use incentives to help move customers to the use of technology to improve productivity for both parties, as in electronic bill paying. As use of time increases as a competitive priority, it may be the customer pushing the business to use technology to improve customer productivity. We all know what the incentive will be in this instance—"do it or lose my business."

Workforce Acceptance

As the service delivery system integrates technology and people, the stage is set for the employees affected to accept the changes as "friendly." Some of the future technology will continue to be aids to employees; however, much of it will be technology that assists intellect rather than muscle. Employees will recognize that knowledge technology is for their benefit, and that having additional knowledge increases their value in the marketplace.

Experience shows that employees at the lower organizational levels have less of a problem accepting new technology and change. There is a natural apprehension at first, particularly if the grapevine rather than management is the means of communication. Once the technology is a reality, they usually accept their new roles easily and seriously and do a good job implementing the change. On the other hand, senior and middle managers have a more difficult time with the change. Their perception is that they have lost some degree of managerial control because of the change and, as a result, feel a dilution of their power and influence.

Economic Feasibility

We previously discussed the necessity of considering both qualitative and quantitative factors when analyzing the economic feasibility of technology projects. Another factor involved in much of today's economic feasibility analysis concerns the customer–vendor relationship. Frequently a technology decision made by a customer affects the economic picture of their vendors. There are also opportunities for vendors' actions affecting the customer's economic picture. In the early 1990s when electronic data interchange (EDI) became a standard rather than a novelty, many companies required their vendors to install EDI systems as a prerequisite for conducting business. Some companies could use EDI for only one customer, so it became a questionable economic benefit for those companies. Large customers with much power pressured vendors to install EDI by canceling orders, or at least threatening to cancel. This shared economic factor will become more of an issue as communications technologies entwine customer–vendor relationships.

The business world is becoming more aware of the social and environmental implications of technological changes. We mentioned previously the potential social implications of displacing employees with technology. The cost of retraining or supporting a workforce trapped in low-skill, low-paying jobs could become a heavy economic burden to society as a whole. The economic impact of technology will be quite the opposite from that predicted by those who view technology as harmful.

The environmental economic impact has been the subject of numerous debates during the early 1990s, and this trend will most likely continue. Regardless of one's position on the environmental continuum, there is a cost to society for protecting our environment and there is a longer-term cost for not protecting it. Environmental problems are viewed not just as quality-of-life issues, but by many as survival issues; and there is a significant economic cost to dealing with these issues. Social and environmental issues are economic issues, and they will become a more important part of responsible businesses. The push for sustainable development, a strategy that encourages careful economic growth while simultaneously ensuring that finite and renewable resources are available to future generations, is also gaining momentum among businesses. *Green* is now becoming synonymous with *good*.

Technical Feasibility

The emerging technologies of the digital age have the potential to transform all our lives, to change forever the way we work, buy, learn, and communicate with each other.[45] These are suppositions yet unproven, but look at the technical realities of the past few years. Systems are in place that provide, instantly and simultaneously, all the information about a customer's dealings with a business. Small businesses can compete better with larger, multinational concerns because of low-cost computer hardware and software. Collaboration among software, hardware, and chip vendors has resulted in previously unimaginable off-the-shelf applications for businesses of every size. Whatever is current today is made obsolete not by competitors but by the company that developed it. The applications on the information superhighway, as powerful as they are today, represent the "ice age" of tomorrow.

With these technological advances in place, it would be difficult to argue against future technology that supports the adage "if man can imagine it, technology can make it possible." Management Comment 7.3 gives a perspective from one of the people who brought us to where we are today, Bill Gates.

Managerial Comment 7.3   Bill Gates's View of Future Technology

The past 20 years have been an incredible adventure for me. It started on a day when, as a college sophomore, I stood in Harvard square with my friend Paul Allen and pored over the description of a kit computer in *Popular Electronics* magazine. As we read excitedly about the first truly personal computer, Paul and I didn't know exactly how it would be used, but we were sure it would change us and the world of computing. We were right.

The personal-computer revolution happened and it has affected millions of lives. It has led us to places we had barely imagined.

We are all beginning another great journey. We aren't sure where this one will lead us either, but again, I am certain this revolution will touch more lives and take us farther. The major changes coming will be in the way people communicate with each other. The benefits and problems arising from this upcoming communications revolution will be much greater than those brought about by the PC revolution.

There is never a reliable map for unexplored territory, but we can learn important lessons from the creation of the $120 billion personal-computer industry. The PC—its evolving hardware, business applications, online systems, Internet connections, electronic mail, multi-media titles, authoring tools, and games—is the foundation for the next revolution.[46]

Summary

This chapter began with a look at technology today and ended with a view of where technology may take us in the future. We believe that technology should improve competitiveness, and an operation should be streamlined before introducing new technology. Technology is "any manual, automated, or mental process used to transform inputs into products or services." The use of technology affects the capital intensity of the organization, and an operations manager must be concerned with the level or degree to which technology is applied.

The relationship of customers and technology is affected by the fact that in services, customers often participate in the service delivery system. The front-end and back-room aspects of services were discussed. It is usually less difficult to apply technology to back-room operations than to the front end where customers may be affected. Involving customers more in the service delivery system is a way to gain productivity improvements. Some examples of technology in services have been given.

Technology affects facilities and organizational infrastructure. Facility decisions should be based on the basic service and overall operations and marketing strategies. The degree of customer contact has a significant influence on facility design and location. The impact of new technology on the infrastructure is changing. In past years, new technology caused an increase in management and staff specialist ranks. As information becomes more meaningful and distributed faster, fewer middle managers and staff specialists are required. This creates "flatter" organizations as opposed to "fatter" organizations. We emphasized that new technologies must be integrated with all parts of the service delivery system to gain optimum benefits.

We discussed the functions of technology as an aid to people, as a substitute for people, and integrating technology and people into the service delivery system. The latter function was identified as the current trend and the approach that can gain the most from new technology.

Decision making is important in considering technology changes. When customer contact is at the low end of the continuum, communications technology makes it possible to locate the business for cost advantage rather than customer convenience. Customer contact also plays a significant role in facility layout decisions. Regardless of a firm's position on the customer contact continuum, operations managers have five common facility design and layout objectives: optimizing the customer encounter, good communications, optimizing facility flexibility, improving human and machine productivity, and providing a safe and morale-building environment.

The criteria used in making technology decisions were presented. Strategic needs versus short-term needs, job design, and cost versus benefits were discussed in some detail. Strategic needs must be the force that guides current and future decisions about a company. Using the driving force of the strategic plan is a good place to begin the decision-making process when conflicts arise between strategic and short-term needs. Job design decisions depend on a number of factors including job structure and job content. Again, the degree of customer contact influences job design. Because improving technology means increasing capital intensity, costs and benefits must be examined. When justifying higher technology projects, operations managers must examine qualitative as well as quantitative factors.

Steps in adding technology to a service delivery system were discussed. A seven-step process was offered with the first and all other steps including timely communications with all who might be affected by the changes. Concurrent engineering was recommended as an approach to developing and implementing a project plan. The final step identified the need to review, evaluate, and adjust. This step can also help plan and control future improvement projects, yielding another competitive advantage.

The last section of this chapter presented future considerations for technology in services. The ideas are not predictions but rather extrapolations of current trends, including the relationship of customer acceptance and technology, workforce acceptance, economic feasibility, and technical feasibility. The stage is set for consumers to drive the move from high contact/low involvement in service delivery to low contact/high involvement. Employees at the lower levels of an organization have less of a problem accepting new technology and change than do senior and middle managers. Loss of control and power of managers were cited as the major reasons. The economic feasibility of future technology was expanded beyond cost–benefit considerations to customer–vendor relationships economics, environmental/technology economics, and social economic issues. These economic issues are becoming as important as cost–benefit issues in the life of operations managers. Finally, we closed the chapter with a bridge to the next chapter, some speculation regarding the feasibility of having technology take businesses to where they want to go.

We began this chapter with a quote from John Naisbitt and Patricia Aburdene about the need for technology to be combined with humanness. We would like to end the chapter with a more contemporary quote from an equally recognized writer—Thomas Friedman—with a similar appeal to blend humanness with technology.

Managerial Comment 7.4
The Technology in Our Future

There is absolutely no guarantee that everyone will use these new technologies for the benefit of themselves, their countries, or humanity. These are just technologies. Using them does not make you modern, smart, moral, wise, fair, or decent. It just makes you able to communicate, compete, and collaborate farther and faster. In the absence of a world-destabilizing war, every one of these technologies will become cheaper, lighter, smaller, and more personal, mobile, digital, and virtual. Therefore, more and more people will find more and more ways to use them. We can only hope that more people in more places will use them to create, collaborate, and grow their living standards, not the opposite. But it doesn't have to happen.[47]

References

1. Naisbitt, J. and Aburdene, P., *Megatrends 2000, Ten New Directions for the 1990's*, William Morrow, New York, 1990.
2. Kidd, P.T., *Agile Manufacturing*, Addison-Wesley, Reading, MA, 1994.
3. Chapman, R.L. and Corso, M., From continuous improvement to collaborative innovation: The next challenge in supply chain management, *Production Planning & Control*, 16, 4, 339, 2005.
4. Rogers, E.M., *Diffusion of Innovations* (5th ed.), Free Press, New York, 2003.
5. Krajewski, L. and Ritzman, L.P., *Operations Management, Strategy and Analysis* (4th ed.), Addison-Wesley, Reading, MA, 1996.
6. Blackstone, J.H., *APICS Dictionary* (13th ed.), APICS, Chicago, IL, 2010.
7. *American Heritage Dictionary*, 2nd college ed., Houghton Mifflin, Boston, 1982.
8. Bessant, J. and Francis, D., Transferring soft technologies: Exploring adaptive theory, *The International Journal of Technology Management & Sustainable Development*, 4, 2, 10, 2005.

9. Skinner, W. and Chakraborty, K., *The Impact of New Technology: People and Organizations in Service Industries,* Pergamon Press, Oxford, UK, 1982.

10. Kotter, J.P. and Heskett, J.L., *Corporate Culture and Performance,* Free Press, New York, 1992.

11. Alavi, M., Kayworth, T.R., and Leidner, D.E., An empirical examination of the influence of organizational culture on knowledge management practices, *Journal of Management Information Systems,* 22, 3, 191, 2006.

12. Yeh, Y.J., Lai, S.Q., and Ho, C.T., Knowledge management enablers: A case study, *Industrial Management + Data Systems,* 106, 6, 793, 2006.

13. Boersma, K. and Kingma, S., Developing a cultural perspective on ERP, *Business Process Management Journal,* 11, 2, 123, 2005.

14. Kanunga, R.P., Cross culture and business practice; are they coterminous or crossverging? *Cross Cultural Management,* 13, 1, 23, 2006.

15. Merrifield, D.B., Obsolescent corporate cultures, *Research Technology Management,* 48, 2, 10, 2005.

16. Nerur, S., Mahapatra, R.K., and Mangalaraj, G., Challenges of migrating to agile methodologies, *Communications of the ACM,* 48, 5, 72, 2005.

17. Davenport, T.H. and Prusak, L., *Working Knowledge: How Organizations Manage What They Know,* Harvard Business School Press, Boston, 1998.

18. Womack, J.P. and Jones, D.T., *Lean Thinking: Banish Waste and Create Wealth in Your Corporation,* Simon & Schuster, New York, 1996.

19. Attaran, M. and Attaran, S., The rebirth of re-engineering: X-engineering, *Business Process Management Journal,* 10, 4, 415, 2004.

20. Marquardt, M., Smith, K., and Brooks, J.L., Integrated performance improvement: Managing change across process, technology, and culture, *Performance Improvement,* 43, 10, 23, 2004.

21. Fontanella, J., The RFID Benchmark Report: Scaling RFID Implementations from Pilot to Production, Aberdeen Group, Boston, MA, 2005.

22. Markelevich, A., and Bell, R., RFID: The changes it will bring, *Strategic Finance,* 88, 2, 46, 2006.

23. Hyland, P.W., Soosay, C., and Sloan, R.R., Continuous improvement and learning in the supply chain, *International Journal of Physical Distribution & Logistics Management,* 33, 4, 316, 2003.

24. Lind, J., Boeing's global enterprise technology process, *Research Technology Management,* 49, 5, 36, 2006.

25. Kumar, S. and Phrommathed, P., Improving a manufacturing process by mapping and simulation of critical operations, *Journal of Manufacturing Technology Management,* 17, 1, 104, 2006.

26. Lee, J. and Allaway, A., Effects of personal control on adoption of self-service technology innovations, *The Journal of Services Marketing,* 16, 6, 553, 2002.

27. Mesenbourg, T.L., Measuring Electronic Business: Definitions, Underlying Concepts, and Measurement Plans, U.S. Bureau of the Census, 2007.

28. Kazanjian, R.K. and Drazin, R., Implementing manufacturing innovations: Critical choices of structure and staffing roles, *Human Resource Management,* 25, 3, 385, 1986.

29. Cveykus, R. and Carter, E., Fix the process, not the people, *Strategic Finance,* 88, 1, 26, 2006.

30. Davidow, W.H. and Malone, M.S., *The Virtual Corporation,* Harper Business, New York, 1992.

31. Kidd, P.T., *Agile Manufacturing*, Addison-Wesley, Reading, MA, 1994.

32. Davidow, W.H. and Malone, M.S., *The Virtual Corporation,* Harper Business, 1992.

33. Norman, D.A., *Things That Make Us Smart*, Addison-Wesley, Reading, MA, 1993.

34. Kidd, P.T., *Agile Manufacturing*, Addison-Wesley, Reading, MA, 1994.

35. Fitzsimmons, J.A. and Fitzsimmons M.J. *Service Management for Competitive Advantage*, McGraw-Hill, New York, 1994.

36. Blackstone, J.H., *APICS Dictionary* (13th ed.), APICS, Chicago, IL, 2010.

37. Manrodt, K.B., Abbott, J., and Vitasek, K., Understanding the lean supply chain: Beginning the journey, Report on lean practices in the supply chain, APICS, 2005.

38. Crandall, R.E., The epic life of ERP, Where enterprise resources planning has been and where it's going, *APICS Magazine,* 16, 2, 17, 2006.

39. Turban, E., Leidner, D., McLean, E., and Wethebe, J., *Information Technology for Management: Transforming Organizations in the Digital Economy,* John Wiley & Sons, New York, 2006.

40. Chi, L. and Holsapple, C.W., Understanding computer-mediated inter-organizational collaboration: A model and framework, *Journal of Knowledge Management,* 9, 1, 53, 2005.

41. Crandall, R.E., Let's talk, Better communications through IOS, *APICS Magazine,* 17, 6, 20, 2007.

42. Durvasula, S. et al., SOA Practitioners' Guide Part 2: SOA Reference Architecture, 15 September 2006, http://www.soablueprint.com/whitepapers/SOAPGPart2.pdf

43. Heizer, J. and Render, B., *Operations Management* (11th ed.), Prentice Hall, Upper Saddle River, NJ, 2013.

44. Naisbitt, J. and Aburdene, P., *Megatrends 2000, Ten New Directions for the 1990's,* William Morrow, New York, 1990.

45. Gates, B., *The Road Ahead,* Penguin (Non-Classics); rev. ed., 1996.

46. Gates, B., *The Road Ahead,* Penguin (Non-Classics); rev. ed., 1996.

47. Friedman, T.L., *The World Is Flat, A Brief History of the Twenty-First Century,* Farrar, Straus, and Giroux, New York, 2006.

48. Blackstone, J.H., *APICS Dictionary* (13th ed.), APICS, Chicago, IL 2010.

49. Blackstone, J.H., *APICS Dictionary* (13th ed.), APICS, Chicago, IL 2010.

50. Collier, D., *Service Management: The Automation of Services,* Reston Publishing, Reston, VA, 1985, p. 20.

51. Epstein, M.J., *Implementing E-Commerce Strategies, A Guide to Corporate Success after the Dot.com Bust,* Praeger, Westport, CT, 2005, p. 20.

52. Kidd, P.T., *Agile Manufacturing*, Addison-Wesley, 1994.

53. Strnadl, C.F., Aligning business and IT: The process-driven architecture model, *Information Systems Management,* 23, 4, 67, 2006.

Chapter 8

The Role of Infrastructure in Continuous Improvement

In Chapter 7 we described various technologies used in pursuing continuous improvement programs. In this chapter we describe why it is also necessary to address the infrastructure of an organization to be successful. Determining the infrastructure of an organization does not always receive much attention in business books for practitioners. It is, to be perfectly honest, not the most exciting aspect of running a company. And yet it is deceptively one of the most important items that need to be "right" for the business to be successful.

What Is Infrastructure?

Many managers with business degrees might remember taking a course that dealt with the subject of infrastructure. Certain terms have probably remained in their memories, such as *span of control, centralization, organizational charts, departmentalization,* and *specialization.* Managers actually use these concepts frequently in the everyday running of the firm; they are not just textbook terms.

Infrastructure is a nebulous term. We will describe it in concrete terms and show how it has roots in the bureaucratic (or administrative) management background, introduced in Chapter 4. Its lack of definite form has prevented it from being developed sufficiently by many businesses, as contrasted with technology. This chapter shows why it is important and what kinds of changes are necessary for an infrastructure to be effective.

The term *infrastructure* has various meanings to different people. Some may think of roads, bridges, and airports, or a transportation infrastructure. Others may envision the computer networks that join a firm's headquarters and field offices together, or the IT infrastructure. We use it here to describe the components of a business that tie it together into an integrated unit.

So what are these components of infrastructure? There is no set list, as this may vary from business to business. However, as a starting point, the following components could be considered part of most business infrastructures:

- Strategies
- The four classical management functions—planning, organizing, directing, and controlling
- Organization structure
- Knowledge management.

We will look at each of these elements in more detail.

Strategies

Where do you start when looking at the infrastructure of an organization? Probably the best place is to begin with its strategies. A company's strategies drive it toward its goals. When the strategies are in place, the other components of the infrastructure can be put together. Of course, when we talk about strategies, we are talking about a number of different items, so we will clarify what the different types of strategies are.

Corporate Strategy

Corporate strategy looks at the types of industries in which the business wants to compete. Some companies—for example, McDonald's—pretty much stay in one line of business (fast food in the case of McDonald's). Other companies, like the Walt Disney Company, choose to compete in several different industries, such as parks and resorts, studio and media entertainment, and consumer goods products. Every business must decide how extensive it wishes to grow in terms of breadth of industries. Sometimes a company will decide it is competing in too many industries. When that occurs, it may decide to retrench a bit and get back to basics. One of the authors once worked for what originally started out as a vending food company. Eventually it expanded into a number of different industries including trucking, contract food service, uniforms and linen, and even day care. Although it later retrenched from the trucking and day care industries, that company, ARAMARK, remains one of the premier service contract companies in the areas of facilities management, food service, and uniform/career apparel.

Corporate strategy also looks at the issue of growth for the company. In general, there are only three ways to go—stay the same size, grow bigger, or get smaller. Some industries, such as those in new technologies, offer plenty of growth opportunities so that companies competing in these industries can choose to grow larger if they wish. Staying the same size may be an option for other companies that do not wish to complicate operations any more than necessary. These companies may feel that expanding too much could sacrifice the quality of their product or service. Furthermore, if the company is in a slow growth industry, trying to expand may turn out to be costly.[1] Retrenchment, or growing smaller, is also an option if a company has encountered financial problems. Some big-name companies have cut back a bit over the years, including IBM, Proctor & Gamble, Heinz, Nokia, and Union Carbide.[2] However, sometimes the industry may dictate the corporate strategy. If a company is in a growth industry, such as mobile phones, they have little choice but to plan a growth strategy. If they are in a declining industry, such as manufacturing typewriters, they must plan to retrench and look for other businesses.

Business Strategy

Once a company knows what industry (or industries) it wants to compete in, it needs to decide how to best compete in those industries. This may seem obvious, but actually, many companies do not survive because they choose the wrong business strategies. A firm can be successful for many years but then eventually fail because it does not update its business strategies. A number of formerly large, successful companies have fallen by the wayside due to changes in their environment and the selection of poor business strategies. Some of these include Eastern Airlines, Howard Johnson's restaurants, and F.W. Woolworth's.

Michael Porter, a Harvard business professor and leader of the Institute for Strategy and Competitiveness, has led the thinking on what constitutes effective business strategies. His original book, *Competitive Strategy* (1980), has been acclaimed as the premier guide to business strategy. According to Porter, a company can follow one of three strategies: low cost, differentiation, or focus.[3] A company following a low-cost strategy (sometimes referred to as a cost leadership strategy) seeks to be efficient in the production and offering of its goods and services. Cost savings in every aspect of the business operation is sought after. Most college textbooks mention Walmart as an example of a firm following a low-cost strategy. In fact, Walmart takes extreme measures in cutting its costs, including requiring its suppliers to reduce their prices over the length of the contract.[4] Usually, companies that follow a low-cost strategy will also pass those savings on to their customers. Again, Walmart serves as the premier example, as they are absolutely committed to providing products at the lowest prices possible.

Suppose you were to buy your next pair of jeans from Walmart. Chances are you would find jeans that were a good value (low price) and of generic quality (in other words, adequate quality, but not necessarily high quality). For buyers who shop at low-cost firms, the expectation is always low prices, not quality, as the first priority. It is understood that appliances will wear out sooner, clothes will not be the top name brands, and foods may be packed in large containers, offered by generic suppliers. For the consumer who is interested in better quality, shopping at a firm that follows a differentiation strategy is a better option.

A differentiation strategy puts the emphasis on better quality, not low prices. Companies following this strategy seek to provide products and services that are unique and different from those of their competitors. Following this strategy means that more resources can be expended on adding value to the product. As a result, the goal is a quality product, not a cheap product at a low cost. Now suppose you decide to buy that next pair of jeans at Nordstrom's, a high-end retailer. The product line would offer better name brands, better quality, and, of course, higher prices.

Sometimes a firm seeks to compete in just a narrow market segment, or niche. Instead of offering clothes for all ages, men and women, and all sizes, they choose to focus on a select market. For example, Tall and Big Men clothing stores focus on large sizes of clothing for men. Some stores offer only maternity wear. Others specialize in clothes for babies or children. These companies are following the third strategy discussed by Porter, the focus strategy. The focus strategy looks at a narrow market and zeroes in on those customers only. These companies realize they do not want to be all things to all people, so they focus on that part of the market they can serve best. Another example is the quick oil change companies that change your oil and check fluids in less than thirty minutes. They are focus companies because they do not perform all auto repairs, just those related to oil and fluids.

There is a fourth strategy that some companies may follow, although it is not universally recommended. This occurs when a company tries to pursue low-cost and differentiation strategies simultaneously. Porter calls this strategy "stuck in the middle." Simple logic tells us that this strategy is hard to pursue because it suggests that quality and low cost can exist in the same product. Take the example of Ruth's Chris Steak House (a differentiator) and McDonald's (a low-cost provider). Both companies sell beef, but Ruth's Chris Steakhouse cannot sell its products at low prices like McDonald's, that is simply not the goal. Of course, this example is a bit extreme, but it does illustrate what *stuck in the middle* means: two strategies going in divergent directions cannot be used at the same time. It is better for a company to pick one or the other.

The Four Classical Management Functions

Pick up any college-level Principles of Management textbook and you will find it outlined according to the four functions of planning, organizing, directing, and controlling. Indeed, this describes what managers do, although these functions are not necessarily sequential and segregated from each other, as many new students might believe. Instead, the functions are tightly interwoven, often overlapping and sometimes hard to distinguish from each other.

The four classical management functions can be traced back to the thinking of the French engineer Henry Fayol (1841–1925). As a young manager of a large company, Fayol began to see that technical competence and managerial competence were two separate skill sets. In fact, he felt that a good manager with mediocre technical skills was more beneficial to an organization than a mediocre manager with good technical skills.[5] The teaching of management skills was not a practice during Fayol's time because the discipline of management as a field of study was virtually unheard of. Fayol gets the credit for getting the ball rolling in that direction, though. He sensed a need for a theory of management to be developed, and as a result spent a great deal of time writing and presenting papers on the fundamental principles of management. Today, his fundamentals are the backbone to much of the field of modern-day management.

We will look again at those four functions originally proposed by Fayol.[6] To illustrate the functions, we will envision a fictitious pizza restaurant chain[7] that is about to go through some major changes. For many years, the company offered one style of pizza dough, a very thin yet tasty crust that launched the company into a regional chain, well known particularly in the Midwest. Of course, a variety of toppings could be ordered with each pizza, but the crust was the same for all of them: very thin. And that is where we will begin with the four classical management functions.

Planning

This part of management is involved with setting goals and deciding how to achieve those goals. At some point in the company's history, it realized that if it were going to ever grow and be a nationwide chain, it would need to make a dramatic change in its menu. Consequently, the decision was made to add a new style of pizza dough to its menu, one that was thick—in other words, one that was the opposite of the current thin dough. The plan was to have two styles of dough for the customer to choose from, thin or thick.

Obviously, this was going to take a lot of planning. Goals and subgoals would need to be set. Timetables and budgets would have to be managed. In fact, if the company grew as hoped, the whole growth process would have to be planned as well. In short, this was a big undertaking, but one that offered a huge payoff if successful.

Organizing

So how do you organize such a big task? With a lot of planning, of course! Planning and organizing are very much intertwined. However, there is enough difference in

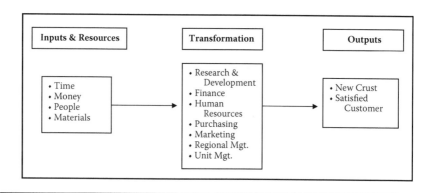

Figure 8.1   ITO Model at a pizza chain developing a new type of crust.

the two processes that we can separate them out for a moment. Organizing deals more with gathering the resources needed and allocating them in a controlled manner. Basically, there were at least four major resources that needed to be organized: time, money, people, and materials. Within these four broad categories, various departments in the company would be affected. Some of these departments included research and development (to invent the new pizza crust), finance (to arrange the extra funds needed to develop the product), purchasing (to buy the new equipment needed to make the crust in the stores), human resources (to arrange training on how to make the new crust), marketing (to advertise this new product), regional managers (to coordinate the implementation of the new product into the restaurants), and unit managers (to make and sell the product). A look back at the Input–Transformation–Output Model in Chapter 3 will remind you that the resources and departments just described are found in the major components of the model shown in Figure 8.1.

Directing

Leading the company through the development of this new product will be a series of upper-level executives and mid- to lower-level managers. Directing then can be thought of as the leadership that must accompany the desired changes in the organization. Upper management is responsible for seeing the need for a new change in the menu philosophy. They must also sell the idea to the rest of the organization, which may think that a new pizza crust is totally unnecessary. You can almost hear the older managers murmuring among themselves: "We've been selling this same style of pizza for many years now. Why change it at this point?" Indeed, much of what directing involves at the higher levels of the organization is selling the skeptics at the lower levels of the organization.

Somewhere in the mid-management area of the company, the hard work of developing and implementation will need to be conducted. Research and development (R&D) will need to plan how it will organize its efforts to develop the new crust. Regional managers will need to prepare their store managers for the big changes ahead. Marketing managers will work on promoting the new product while finance managers figure out how to pay for it all.

The directing will continue at the lowest levels of the organization as store managers teach their pizza cooks how to make the new product. Servers will be taught how to use suggestive selling, enticing customers to try the product. The directing process is always occurring, at all levels of the company. Without successful leadership, the effort to promote the new pizza dough may fail.

Controlling

Throughout the entire process of envisioning, developing, and selling this new product, a separate process, called controlling, will need to take place that monitors the efficient use of resources. This process seeks to answer two major questions: (1) are we reaching our goals, and (2) are we being wise in our use of resources? Above all, the goal of controlling is to introduce the new product, in a timely fashion, at a cost that does not bankrupt the company.

The resources mentioned earlier in the transformation process will need to be monitored. For example, time is a limited resource. The company will most likely have certain deadlines that must be met. Above all, the product will need to be introduced to the market in a reasonable time period. That time period must be long enough to ensure adequate developing and testing of the product so that it is of the highest quality possible, but short enough so that the competition does not step in and introduce the same product first. Delays in product introduction can also be expensive, as each day lost past the introduction date becomes an inefficient use of cash. The use of budgets and timetables are important aspects of the controlling process. Without them, there is no way of knowing if the company is meeting its goals in a timely and efficient manner.

Organization Structure

Most managers will think of the company organizational chart when you mention organizational structure to them. This is true to an extent. A visual representation of an organization's structure can be seen through its organizational chart. However, different types of structures are good for different types of company goals—there is no one-size-fits-all formula for organizational structure. Structures fall into four basic categories: functional, product, geographic, and matrix. We will look at each one in more detail.

Functional

Many organizations are set up exclusively by functional departments. There is an accounting department, finance department, operations department, marketing

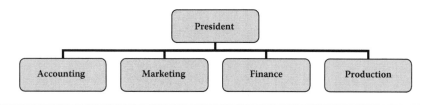

Figure 8.2   A functional organization.

department, and so on. These departments serve the entire organization. Those organizations look similar to Figure 8.2.

Functional department organizations are good at grouping expertise into one department, thus creating opportunities for more new ideas to be developed relating to innovation and efficiency. However, on the downside, functional organizations are not set up according to product lines; therefore it is difficult to isolate costs according to a particular product.[8] In addition, functional departments are notorious for infighting among departments; the classic example cited in many articles is what can occur between marketing and production. Typically, marketing wants new products with lots of features, regardless of the cost. Production desires standard production items with cost efficiencies as a top goal. Needless to say, these two have been going at it for years.

Product

Some organizations are set up by product line. Functional departments also exist, but they are set up under each product line. The goal is to view the business as a set of different businesses with each mini-business representing a product line. An organization set up on this basis may look like Figure 8.3.

An organization set up by product lines offers several advantages. First, there is more financial accountability for the success or failure of that product line because it can be set up as a profit center.[8] A second advantage is that managers can learn the entire scope of a business by being involved in the running of a product line,[9] something that cannot be accomplished if they work in a functional area most of their careers. On the downside, organizations set up on this basis are generally not as efficient in terms of operating costs. This is because functional departments need to be created for each product line.

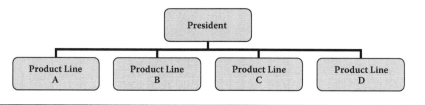

Figure 8.3   A functional organization.

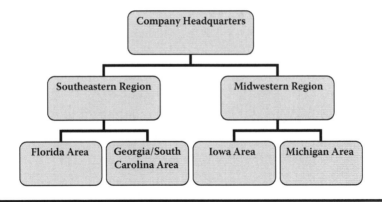

Figure 8.4   An organization set up by geographic region.

Geographic

Some companies will set their structure up according to a geographical framework. For example, there may be an area and regional division, with several areas comprising a single region. If the company is located in several countries, there may be a domestic and an international division. There are numerous ways companies can be set up by geographic area, as seen in Figure 8.4.

Organizations set up by geographic region offer the advantage of being able to adapt better to the local regions they are serving without formal approvals from company headquarters. This is common in the food service industry, where local tastes can be accommodated in different regions. Imagine serving grits in the Northeast? Or scrapple in the South? In geographic arrangements, this kind of mistake can be avoided. The geographic organization does have the disadvantage of being somewhat costly to administer, and there is also the problem that one region of the company may try to upstage another region.[10] However, for large organizations that cover a wide geographic area, this arrangement is suitable.

Combinations of the above organizational structures can occur as well; in fact, they are quite common. Take a large hotel chain, for example. The company may be set up in a geographic configuration, with regions of hotels in Europe, Asia, and North America. However, each hotel may be set up on a functional level, with housekeeping, front desk, food service, and security all operating as separate departments.

Alternate Organizational Structures

The vanishing boundary between manufacturing and services has blurred some of the characteristics of the traditional organizational forms described above. Newer organizational forms have appeared that have the characteristic of being able to adapt better, both to changes in the company and changes in the business environment.

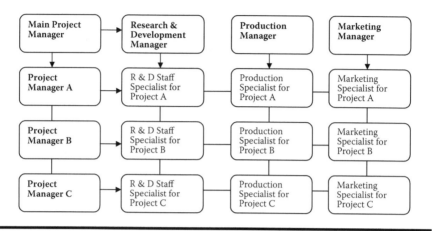

Figure 8.5   A matrix organization.

The Matrix Organization

Imagine combining a functional and divisional structure together into one format; the result would be a structure that looks like a matrix. Some companies actually follow this arrangement in selected areas of their infrastructure. Temporary projects and even some permanent product lines may be set up in a matrix organization. Figure 8.5 shows what a matrix format might look like.

Companies that are extensively involved in projects or are highly geared toward technological innovation may be well suited to using a matrix format. This structure offers the advantage of putting specialists on specific projects, allowing managers to be freed up to work on other duties. On the downside, there is the problem of one employee reporting to two supervisors, a situation that can create confusion and divided loyalties.

The Horizontal Organization

Organizations can be tall or flat. Tall organizations have many levels of management tiers. Tall organizations were typical of many businesses prior to the 1980s. Then a wave of downsizings began, which eliminated many middle managers from the company. More noble terms were chosen for this type of downsizing, including the very popular *re-engineering*. The goal of re-engineering was to reorganize the company so it would function more efficiently, and at the same time, better serve customers.[11] No doubt, re-engineering has accomplished this feat in many companies that were heavy with too many managers and inefficient in responding quickly to customer needs. Countless stories are written in business textbooks about companies that cut their number of employees and responded more quickly to customer demands. Some of these cuts in staff occurred among hourly employees. Other

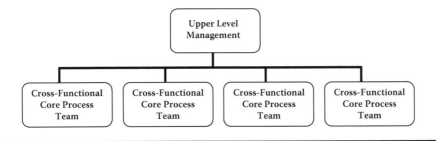

Figure 8.6   A horizontal organization.

cuts were more far-reaching as entire levels of the company, usually in middle management, were eliminated.

The impact of re-engineering has also led to the creation of a new organizational structure, the horizontal organization. Traditional vertically oriented organizations were built around functional units such as production, marketing, and finance. Horizontal organizations are flat and organized around core processes.[12] A horizontal organization may look something like that shown in Figure 8.6.

Replacing the traditional functional departments are cross-functional process teams.[13] These teams are designed to address the specific needs of the customer as opposed to running an internal process of the organization. The desired result is an expedited process transaction with the customer. For example, instead of taking weeks to process an insurance policy or a mortgage, the goal is to complete the transaction in a matter of days. Streamlining the transaction process requires a careful analysis of the job, a process human resource management specialists like to call job analysis.

Of course, a lesser form of the horizontal structure is simply a flatter organization. De-layering the organization compresses the hierarchy and widens the organization chart. The result is a flatter organization that is more flexible in responding to change. Translated, management is able to make decisions faster because there are not as many levels of approval needed. Because fewer personnel are on the payroll, particularly in middle management, a savings in labor cost is also realized.

The Virtual Organization

There appear to be at least two notions of what the virtual organization is.[14] The first is a business operating with a small core of employees, while outsourcing much of the other activities involved in its infrastructure. For example, a business may perform the tasks that it does best, sometimes called its core competencies, while contracting other businesses to perform those tasks that are not in its expertise. One Smooth Stone, an event-planning company, uses this approach as it plans out major business meetings for its clients. Their core competency is in the planning end, accomplished with just seventeen employees. All other functions are contracted out.[15] This type of virtual organization has been given an intriguing name, the *inter-organizational virtual organization*.[16] Figure 8.7 shows what this type of

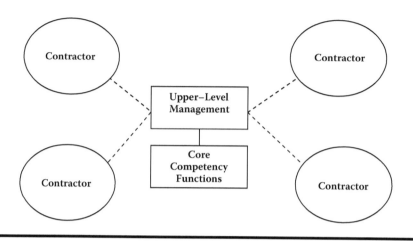

Figure 8.7 A simple depiction of an inter-organizational virtual organization.

virtual organization may look like. The dashed lines indicate a contractual relationship between the contractor and upper management. The solid line between upper management and the core competency functions indicates that both groups are in the same organization. Of course, the number of contractors can increase or decrease, depending on the number of projects the company is managing at the moment. Thus, virtual organizations are constantly in a state of flux.

Another viewpoint holds that the virtual organization is a business that performs most of its own functions but is linked over a wide geographic area via computer networks. The organization may still hold a traditional structure, but in terms of communications and performing various processes it is linked through smaller virtual teams. These organizations may have virtual teams working on advertising campaigns, product development, or a number of other common business functions. Instead of working in a department in one building, virtual teams are linked via computer networks over a geographic area. This type of organization has become more widely used with the increase in videoconferencing technology and the wide geographic distances between supply chain participants. Boeing used this type of organization in developing their 787 Dreamliner. While the Dreamliner project has experienced delays because of the difficulty in coordinating among participants, it does represent an innovative way of organizing. Future projects should benefit from the experience curve and result in smoother implementations.

Several advantages are available using virtual teams. First, the expertise that is closest to the point of action can be tapped into. Field personnel can work closely with the home office on a real-time or close to real-time basis. This advantage is especially important when developing new products or advertising campaigns. A second advantage is that the best talent in the company can be utilized for specific projects, regardless of where those personnel are located.[17] Larger companies that cover a vast geographic span are especially able to utilize these types of virtual teams.

Depicting a visual chart of a "typical" intra-organizational virtual organization is not possible because many organizations of any size already have elements of virtual communications in place. However, one commonality that may exist for specific project management is the matrix format. This is because virtual teams are often composed of temporary members who work on a specific project for a set length of time. When the project is completed, that virtual team is disbanded. Consulting firms use this in assembling teams of functional and industry experts for large consulting projects.

Trends in Organizational Structures

In addition to the alternate forms discussed above, a number of trends have been occurring in organizational structures, which we look at next.

Moving from Centralization to Decentralization

Centralization deals with where the major decision making takes place in the running of the organization. Highly centralized companies have major decisions made at the upper levels of management, or the home office. Most restaurant chains are highly centralized, as major decisions such as menus, advertising campaigns, logos, store design, and prices are made at corporate headquarters. Decentralized companies push the major decisions to the lowest levels of the organization possible. Consider a major food service contract company like Sodexo or ARAMARK. These companies are responsible for running corporate, hospital, and college cafeterias all over the country. Major decisions concerning menus and how to set up the local facility are made by the food service managers at the location where the account is administered.

In general, businesses must learn to adapt in changing times, which means being flexible and responsive to the customer. As a result, there appears to be a trend toward decentralization.[18] The blurring of the service–manufacturing boundary is showing us that managers at the lowest levels of the organization are closest to the action and are often the best equipped to make decisions that require adapting to customer needs. This is especially important to note in larger companies that have tended to be more centralized in their decision making. One of the authors remembers hearing a student lament the fact that fashion decisions were set by her company (Sears) in Chicago, and she was a buyer in Memphis, Tennessee, where fashions could be quite different. The top-down, centralized approach of the company did not work well in her opinion because local fashion trends were not being accounted for. While centralized buying may not work well for style or regional-dependent items, some companies are moving toward centralization of buying staple items in order to get leverage from placing larger orders with suppliers.

Moving from Vertical Structures to Horizontal Structures

Tall, vertical organizational structures are typical of highly centralized companies. It is no surprise then that as decentralization becomes more popular, horizontal structures, which work better in decentralized environments, are becoming more popular.

A related topic concerns span of control, the number of employees who report to a manager. Simple logic tells us that as organizations become more horizontal, a single manager will take on more employees to supervise, thus increasing the span of control. Increasing span of control has good and bad points. On the positive side, the company saves money because less labor cost is needed because fewer managers are employed. Also, employees must be given more authority to make decisions, a process we in the management field like to call empowerment. On the downside, too wide a span of control can cause lapses in supervision, because a manager can only keep up with so many employees. Nonetheless, spans of control are increasing in today's economy, and so is employee empowerment.

Horizontal structures facilitate the scalability of organizations. They can be more easily formed, managed, modified, and dissolved as the need arises to increase or decrease their size and scope.

Moving from Autocratic Managers to More Empowered Employees

Although not directly an organizational structure concept, a discussion on the movement toward employee empowerment is appropriate here. Autocratic management is a style of leadership that can be effective in some environments. As management researchers Hersey and Blanchard have noted, the style of leadership needed may vary according to how ready the employees are to perform the task. Employees who are unwilling and have low abilities will need a more directive style of supervision.[19] Most readers of this book have probably made the observation that some employees seem to need more direction than others.

Sometimes a business or organization will adopt an autocratic style because it fits with the culture of the industry and seems to work well. Branches of the military have traditionally adopted this approach. Nonetheless, the trend in many organizations is to move toward more employee empowerment and away from autocratic styles of leadership. From a structural perspective, this trend seems logical. We have mentioned that managerial spans of control are increasing and organizations are getting flatter. The result is that managers will expect their employees to perform more of the "thinking" aspects of their jobs, roles traditionally taken care of by managers. Autocratic and micro-managing styles of leadership are thus replaced by participative styles of leadership, where employees have more say in the decision making of the operation.

One example of a successful service operation that has adopted this approach is the Ritz-Carlton Hotel Company. "The Ritz" goes to extreme measures to ensure that guests are well satisfied. As a five-star hotel, it is expected to deliver only the best in service and food. Employee empowerment is one way the hotel accomplishes this goal. Each employee is empowered to spend up to $2,000 without managerial approval to fix a customer problem.[20] That may sound extreme, but one thing service industries have taught us is that it is more cost-efficient to keep a current customer than it is to find a new one.

Moving from Job Specialization to Higher Skill Variety

Thanks to the work of Frederick Taylor (1856–1915), a great deal of job specialization has permeated our working landscape. Taylor, if you remember, was the engineer who pioneered the principles of scientific management. His identification with the working class was impressive but misunderstood in his later years, as critics from labor unions thought of him as an exploiter of manual labor.[21] His contribution to the modern economy is the work philosophy of finding the one best way to perform a task. Usually, this involves breaking a job down into its component parts and analyzing the most efficient manner to perform that job. Selecting the best worker for the job, training the worker in the exact methods, and giving the worker the right equipment were important for the highest productivity. Taylor's hope was that both the worker and management would prosper under this approach.[22]

Today, we see evidence of scientific management in both manufacturing and services. Fast-food restaurants have rigid methods on flipping burgers, assembling sandwiches, frying food items, and serving the customer. The rigidness of job specialization has resulted in faster production and a standardized product that can be made to taste the same across a wide geographic area. Indeed, the fast-food industry owes much of its success to Frederick Taylor. But job specialization does have its drawbacks. For one, the work, over a period of time, can become boring. Boring work creates its own problems, as workers can become apathetic, which can result in sloppy work, accidents, and employee turnover. Many restaurant managers will attest that as a result of this, much of their work involves training new employees.

Job specialization may be eroding in the future, at least somewhat. We will look at how organizational structure enters into the picture. First, there is the simple math consideration. As companies reduce their numbers of employees, the remaining employees will perform more work. In some ways, this is a negative occurrence as it can result in work overload. However, in some jobs, there could be a surprising positive benefit: the work will become more interesting. This is because the skill variety to perform the job may increase. Two management researchers, Richard Hackman and Greg Oldham, discovered this back in the 1970s. Simply put, when the skill variety of a job goes up, worker motivation will go up also.[23]

Another reason skill variety may be increasing is due to the flattening of organizational structures. As fewer managers are in the hierarchy, more demands are put on those remaining managers. Again, skill variety may increase, which can increase the meaningfulness of that managerial job.[23] So reducing the number of employees and flattening out the structure of the business can have a positive side effect besides saving money—it can be motivating to the worker. In some companies management has purposely decided to broaden jobs and reduce work specialization. Avery-Dennison, Hallmark, and American Express have all made decisions in this direction.[24]

One caveat is that broadening the scope of a job but overworking the employee can be a de-motivator. Managers at all levels need to make sure they are not overworking the managers or employees below them. Otherwise, the gains that could be made by changing the structure of the organization could be lost due to worker apathy and turnover. Also, not all jobs are suitable for enlarging. Fast-food jobs are a case in point, as broadening these jobs could significantly slow down service to the customer.

Moving from Line Managers to Self-Directed Work Teams

The factory floor setting has traditionally been set up using line managers to direct a group of employees. This format is effective and is still in use in a number of work settings today. The underpinnings for this approach have some tie-in with Taylor's scientific management principles. The line manager (sometimes called a supervisor) directs a smaller group of employees using precise work methods that have been carefully created to accomplish the job. When workers deviate from these precise work methods, the supervisor is always there to correct the situation by putting the worker back in line with the required method of performing the job. Supervisors are also responsible for scheduling the work, ordering the supplies, and arranging nonwork items such as vacation time for the employees.

The line manager arrangement is also popular in the service industries, particularly in fast-food restaurants. In these environments, supervisors often lead younger employees, many of whom are working for the first time, into job tasks that require precise methods and timing. For example, the cooking of a food item involves a step-by-step procedure, with critical timing events within the process. In addition, stringent safety and sanitation standards must be maintained, so a strong supervisor who leads a group of inexperienced workers must guide and direct those employees on a close basis.

There are some work situations where a different format is used. In these cases, the work group does not have a line manager one hierarchal level above them who directs their work. Instead, the group does all of its own directing. These groups, or self-managed work teams, perform a complete work process while managing themselves and arranging nonwork-related items such as sick days and vacation time. Self-managed teams require workers to move away from performing only a specialized task to becoming proficient at a number of skills, a concept called multi-skilling.[25] This concept is consistent with the trend discussed above, where skill variety among workers is being increased. Self-managed teams are found primarily in manufacturing environments and can be seen at well-known companies such as Corning, Boeing, Pepsi, and Hewlett-Packard.[26]

Moving from Specialized Departments to Cross-Functional Teams

Specialized departments (sometimes referred to as "silos") are among the more traditional ways to organize employees. The functional organization discussed earlier is an illustration of a specialized department at work. In contrast, cross-functional teams are composed of employees from different departments in the same organization. The matrix organization, also discussed above, illustrates the use of cross-functional teams.

Cross-functional teams have enjoyed a fair amount of popularity in contemporary organizations, and their use appears to be growing. They are not departments in themselves, but groups of employees set up on a temporary basis and assigned to various projects. Some readers might remember the best-selling book *In Search of Excellence* that came out in the 1980s. In their account of excellent companies, Tom Peters and Robert Waterman Jr. documented an early example of a cross-functional team, called a "skunk works."[27] In this somewhat amusing example, a member of a company skunk works was called in to analyze a piece of machinery that was overheating. Engineers within the company had proposed several expensive solutions, but a skunk works member solved the problem by aiming an inexpensive fan at the machine. From skunk works to cross-functional teams, the concept of taking a multidisciplinary approach to solving business problems has become a popular practice.

Moving from Top-Down to Multidirectional Communications

The traditional model of communications in organizations has been referred to as the "top-down" approach. In this arrangement, orders and directions always flow from the manager to the employee. Employees are not expected to question their manager or participate in decision making. The basic mentality is to take the order and run with it. This arrangement works fine in some companies and industries but cannot be expected to be effective in all situations. Top-down communication formats are common in tall, vertical organizations, with more specialized jobs.

Many organizations have found that using multidirectional communication formats is more to their advantage. Multidirectional formats use top-down, upward, and lateral communication pathways. Top-down communications are useful when giving directions, communicating goals, and indoctrinating employees.[28] Upward communications occur when lower-level employees are actively communicating with their immediate supervisor. One of the goals of upward communication is to foster an environment of participative management. Participative management encourages lower-level employees to work with their managers to solve problems related to the unit they work in. Companies have found that using participative management practices can result in better employee motivation, improved product quality, less stress for the employee, higher job satisfaction, and ultimately, higher profits.[29] The Ford Motor Company was one of the first pioneers in the participative management arena. This was a tall task, given that the United Auto Workers union was initially against the move.[30] However, although the company still has its problems competing, it does well in working with its union.

Multidirectional communication also emphasizes lateral communication among peers in the organization. The trend is to encourage this type of communication more than in the past. Cross-functional teams, described earlier, are an example of lateral communication. The advent of e-mail has made lateral communication faster and more effective. Lateral communication is especially useful when seeking advice from other departments or coordinating projects across departments.[31]

The amazing growth of social media groups has increased the speed and volume of communications among linked individuals. Although its use as a business tool for communication is still somewhat spotty, it is almost certain to become more widely used in the future.

Moving from Rigid Policies and Procedures to More Flexibility

There is no doubt that businesses need some degree of formalization. As companies grow, they typically develop policies and procedures to handle recurring situations. In doing so, life is simplified because the employee does not have to stop and think to himself, "What am I supposed to do in this situation?" This need for formalization also allows large companies, particularly in the service arena, to offer a consistent product over a wide geographic area. That is why fast-food giants such as McDonald's and Wendy's maintain their consistency—through rigid policies and procedures.

Sometimes a rigid procedure can backfire and cause a dissatisfied customer not to shop at a store again. Consider this example: a retail store has the policy that all returns must have one manager sign off on the return in the presence of the employee and the customer. This in itself is not a bad policy. But suppose that at one point during the day, all managers are briefly off the premises at a meeting with a prospective supplier. A customer arrives with an item that is defective and wants a refund. Under this rigid policy, the customer is told they must come back several hours later when the manager is on duty. Not a good situation. This example is hypothetical, but it does illustrate many other possibilities where customers are sometimes the victims of a rigid store policy.

Because there are situations like the one described above where some leeway may be appropriate, there is a trend in many companies to extend some discretion to employees in enforcing policies and procedures. Ritz Carlton Hotels is one of many companies that offer flexibility in decision making to their service employees. The movement away from rigid procedures and rules does not mean abandonment of them. Certainly, the advantage of formalization is the consistency in service and quality that results. However, when that rigidity works against the best interests of the company, then relaxing the rules a bit and letting employees make informed decisions is often a better practice.

Moving from Mechanistic Structures to Organic Structures

By now, it may seem as if all these facets of organizational structure change in concert with each other. In other words, as one item changes, so does another. For example, as mentioned earlier, as organizations become flatter, the span of control of the manager automatically increases. The fact is, the parts of organizational structure all do work together, and this was acknowledged over forty years ago when management researchers Tom Burns and G. M. Stalker outlined the concepts of mechanistic and organic organizations.[32] If the terms seem vague, the concepts are clearly described in Table 8.1 and summarize our discussion on this topic so far.

Table 8.1 can be viewed as two end points on a continuum, with mechanistic characteristics on one end and organic characteristics on the other. In practice, most companies operate somewhere along the continuum. Within an organization, it is possible to see that some departments may be mechanistic and others are more organic. This is not uncommon and in fact makes sense given the mission of different departments. For example, the production department (or any department, service, or manufacturer that actually assembles the product) is likely to be more mechanistic. That arrangement works well given that order, precision, and quality are necessary to produce a product such as a food item in a restaurant. Another department in the same organization, such as R&D, would function better under an organic framework. This would be logical, because the creative development of a product idea is not a rigid process.

Policies lead to action in the form of processes, which define what a business does to run its operations on a daily basis. Just as policies need flexibility, processes must also have some flexibility to adapt to the need for customization of products and services to meet customer requirements.

Processes lead to procedures, which formalize the processes, sometimes to the point of rigidity. If not careful, an organization can constrain its ability to deal with change by overemphasis on procedures that do not allow for the exercise of judgment by trained employees.

Table 8.1 Mechanistic versus Organic Organizations

Organizational Structure Element Being Discussed	Mechanistic Organizations Look More Like This	Organic Organizations Look More Like This
Degree of centralization	Highly centralized; main decisions are made by top management	Highly decentralized; main decisions are made at the unit level
Number of management levels	Tall organizational structure with many managerial levels	Flat organizational structure with fewer managerial levels
Power source	In the hands of management	In the hands of employees and managers
Job design	Jobs are more specialized	Jobs are broad in scope and contain more skill variety
Control source	Control is in the hands of line managers	Control is in the hands of self-directed work teams
Coordination of work	Coordinated primarily by functional departments	Coordination accomplished by departments and by cross-functional teams
Communication	Communication is top-down; orders come from the manager to the employee	Communication is multidirectional
Policies and procedures	Rigid policies and procedures exist	More flexibility allowed in procedures

Practices are "the way we do things." These may exist long after the policies, processes, and even the procedures have moved on to meet new conditions. Practices have a strong link to culture, a subject discussed more fully in Chapter 9.

The Role of the Internet in Changing Organizational Structure

A discussion on organizational structure would not be complete without acknowledging the impact the Internet has had on businesses. The Internet has changed many aspects of running a business as well as being an informed consumer. To understand these changes at a more fundamental level, the reader should understand two concepts—one related to economics and the other related to consumer information.[33] The first is a type of expense called a *transaction cost*. The second concept is *information symmetry* and *information asymmetry*.

Transaction Costs

Transaction costs are the costs of doing business that do not actually add value to a product or service but must be paid by the business nonetheless. For example, paying brokerage fees is a transaction cost. Driving to the store to buy an item involves transaction costs in the gas needed to drive your car to the store. The farther away the store is, the higher the transaction cost. Jokes have been made about frugal consumers who drive endless miles to multiple stores, just to save several cents on an item. These frugal consumers do not consider the transaction costs of driving around or the value of their time, which is also important. For a business, a transaction cost could be the start-up cost paid to use a certain vendor, or the one-time costs needed to begin selling your product to another business. In a sense, transaction costs may be viewed as the number of "touches" a product or service passes through on its way to the customer. In today's world, the concept of lean production, or lean processes, and the emphasis on eliminating non-value-added costs could be likened to the transaction cost concept.

So where does the Internet factor into all of this? In short, it has helped to reduce transaction costs. Consider the example of driving around town comparing prices on a new outdoor grill. With the Internet, driving around is no longer necessary because prices can be compared by researching online. Hence, the "driving-around transaction cost" has been eliminated. Now consider what happens when all businesses and consumers use the Internet for researching and shopping—the net effect is a reduction in transaction costs, which is a good thing for consumers, but at the same time it intensifies competition in an industry. When competition intensifies, businesses need to evaluate all their expenditures carefully. One area is the "make or buy" decision. Should our company make the product, or buy it? Assuming that quality is the same whether we make or buy (economists like to make those kinds of assumptions), the final decision will most likely be made on how much it costs to make the item. If it costs more to make the item, then we outsource it. If it costs less to make the item, then we make it.

It is not difficult to see what the trend is in our business environment today—to outsource the making of that product whenever it results in a cost savings. From an organizational structure perspective, this means companies are getting smaller by laying off production employees so the same product they used to make can be manufactured somewhere else (usually overseas), where costs are lower, usually due to cheaper labor.

To summarize, the Internet lowers transaction costs, which accelerates competition, which causes the production jobs to go overseas. Of course, this is a very simplistic (and yet realistic) scenario, and it should be qualified by saying the pressure on low prices is also influenced by global competition.

Information Symmetry and Asymmetry

Now, on to the information symmetry/asymmetry discussion. The concepts here are really quite simple. The basic idea revolves around the availability of information to the parties in a business transaction. When all parties have access to the same information about the details of a transaction, then information symmetry exists. We can continue with the previous example. Suppose that while you are shopping for an outdoor grill, you construct a chart with the features and prices of each grill you are interested in. You also list the various retailers that offer these grills to minimize your driving time when you make your final decision. As mentioned above, the Internet makes this analysis easier by offering the information online (an example of information symmetry) and by reducing the number of miles you have to drive to get this information (an example of lowering your transaction costs). Before we had the Internet, analyzing this cost information was not so simple. It was more difficult for prospective buyers to compare retailers and models, and it was more difficult for retailers to see what the competition was offering and what they were charging for their products. Because information was not readily available to all the potential parties in the transaction, information asymmetry existed. The impact of the Internet has been to move toward information symmetry, which is advantageous for consumers. Again, as in the reduction of transaction costs, moving toward symmetry intensifies competition that can lead to more outsourcing. Organizational structures ultimately change as a result of these processes.

The Integration of Knowledge Management into Organizational Structure

A discussion on new movements in organizational structure also involves noting a recent trend—knowledge management. Knowledge management seeks to use the intellectual resources that already reside in the organization, to a more effective degree. It is the concept of working smarter by working smarter.

Data, Information, Knowledge, Wisdom

Suppose that your company is experiencing a problem with employee turnover. You are concerned because the employees who are leaving seem to be good workers, many of whom are younger managers with college degrees. This is perplexing to you because it appears that although your company is able to attract excellent candidates, it is not able to keep those up-and-coming managers that every company needs. You decide to do some investigating into the problem and begin by noting, by month, the number of employees submitting resignations. After some reading through employee files, you construct the chart shown as Table 8.2.

Table 8.2 Data on Employee Turnover for Managers

Name	Month Left	Age	Years with Company	Reason for Leaving	Performance Level
John	January	23	1	Went to work for another company	Excellent
Shirley	January	25	2	Went to work for another company	Excellent
George	February	62	32	Retired	Good
James	March	28	4	Unknown	Good
Mary	April	29	5	Went to work for another company	Good
Rick	May	33	8	Went back to graduate school	Excellent
Mitchell	June	28	3	Went to work for another company	Excellent
Justin	August	30	4	Unknown	Good
Samantha	September	29	2	Moved out of the area	Good
Thomas	November	24	0.5	Went to work for another company	Still on probation

At this point in your analysis, you have data in front of you. It contains some quantitative and qualitative features. The quantitative features relate to numerical observations, such as how long the managers worked for you and what their ages were when they left. The qualitative information focuses on why they left and how satisfactory their job performance was. At this point, the data is not very useful to you. It needs some interpretation to show the patterns that exist in your company concerning managerial turnover. What is needed next is called information.

Information attaches meaning to the data. Suppose now that as you look at your newly constructed chart, you note the following trends:

- Six of the ten managers, or 60 percent, have purposely left your organization to go to work for another company or to pursue graduate school.
- Of these six managers, the average age was only 28 years old.
- These six managers worked an average of 3.25 years for your company before leaving.
- All managers who left had performance ratings of good to excellent.

The data in the table has now been interpreted to produce statements that reflect information. Information, then, is data with relevance and purpose.[34] What appears to be a problem for your business is the inability to keep good managers. But the obvious next question is, why? Now suppose you meet with other executives in your company, including the director of human resources. During your meeting, you discuss the possible causes of and remedies to the turnover problem. One obvious suggestion is that the compensation may be too low, but the human resources director tells you that a salary survey indicates your company is competitive with other companies in your industry. Another suggestion is that perhaps the culture of the organization does not support younger managers. In other words, maybe the "old guard" mentality is so entrenched in the company that it is difficult or impossible for young, up-and-coming managers to express their ideas about change. After a lengthy discussion about this cause of turnover, it is decided that the six managers should be contacted to see why they left the company.

Some readers may comment at this point that an exit survey should exist as standard practice when a manager leaves. However, this is not always the case. Businesses committed to learning from the past may conduct such surveys, but an organization not committed to learning from the past may overlook such obvious measures. Suppose for now that the six former managers are successfully contacted and exit interviews are conducted. The exit interview seeks the underlying reasons why employees leave the organization. Sure enough, we discover that overall the former managers were satisfied with their pay and compensation but felt frustrated that they were not able to grow professionally at the company. Furthermore, they expressed concern over their inability to make changes in their departments without a lot of approval from upper management. Five of these managers found companies that were more "forward thinking" and open to change. The one manager who left for graduate school also stated that the problem was the inability to change the organization but felt a calling to go into higher education as opposed to staying in the business world.

At this point, your company has now acquired some knowledge that can be useful as upper management seeks to staff the company with young new managers. Therefore, knowledge is information that has been subjected to human reasoning and in-depth analysis. The knowledge that has been acquired is as follows:

- To keep good managers, it is not necessary to increase the compensation. In fact, doing so will raise the cost of the problem, and ultimately not solve the problem.
- To keep good managers, the company must become more user-friendly to its younger managers. Becoming more user-friendly means letting these managers make changes in their departments without requiring excessive administrative approval to make those changes.
- Developing a culture that values younger managers on the part of senior managers is necessary in your company.

So far, it sounds straightforward and within the grasp of management to begin implementing changes using the new "knowledge" that has been acquired. Nevertheless, the problem at this point is that only a few executives within the company actually "know" this. Getting the word out to the rest of the company is important. In short, the company needs to become a learning organization. Knowledge management is the practice of helping companies become learning organizations.

In a sense, knowledge implies an understanding of what exists. Wisdom is an extension of knowledge in a way that makes it possible to use the knowledge in planning for the future. How do we anticipate changes in customer preferences in order to design a product to meet that want? Or how do we anticipate changes in employees and cultures to devise better designed jobs, selection, and training to make the best use of that essential resource? Can wisdom be acquired through education and experience, or is it a skill held only by a limited number of gifted individuals?

Why Knowledge Is Not Transferred

Although it would seem plausible to "get the word out" to the rest of the organization on the matter of managerial turnover and why it may exist, the fact is, the word often does not get out for various reasons. Three reasons why knowledge management may be impeded are:

1. The organizational structure makes it hard to transfer knowledge. This chapter is about organizational structure factors, and structure can hurt the transfer of knowledge under some circumstances. Management researchers tell us that companies with rigid functional structures—that is, departments that may actually compete against each other—are at a disadvantage when it comes to sharing knowledge within the organization.[35] It should come as no surprise, then, that flatter structures, virtual organizations, and companies that use cross-functional teams do a better job of knowledge management.
2. The right technology is not in place to share knowledge. Moving knowledge throughout the organization should be a systematic process, not a random

series of events. Achieving this task will involve a new culture of openness among management and staff. It will also help if the right technology is used to move knowledge through the company. Intranets are excellent vehicles for doing this, especially in larger companies where employees are geographically separated from each other. There are also knowledge management software programs that can aid in this process. Social media networks may have potential to help in this knowledge-acquiring process, although it will probably take adaptations and iterations to make it useful.

Knowledge management may have already saved your life, or at least kept you from becoming very sick. Pharmacists and doctors use knowledge management to make sure certain prescription drugs are not taken together, lest the patient have a serious reaction or even die. Hackensack University Medical Center uses a system to make sure patients do not receive dangerous combinations of medication. The medical center also utilizes a robot to help doctors make rounds of their patients while at home. The device, called Mr. Rounder, can be operated from the doctor's laptop computer, from anywhere. Mr. Rounder can enter a hospital room and use a two-way video to talk to the patient about his or her condition. The robot even wears a white lab coat and stethoscope. Hackensack University Medical Center may be ahead of the curve when it comes to using knowledge management techniques, and the results have been good. According to a recent *Business Week* article, patient mortality rates are down, while productivity and quality of care are up.[36] Electronic medical records, as they become more commonplace, also hold promise as an aid in the diagnosis and treatment of illnesses.

3. There is the fear among some managers and executives that sharing too much of their expertise may compromise their power base. This mind-set could be true in some businesses, especially ones not committed to knowledge management. But in fact, the willingness to share, along with a supportive organizational culture, is often viewed as being the main factor needed for knowledge management to flourish.[37]

One final note about this thing we call knowledge. It is not limited to special problems such as the investigation of employee turnover. It can be anything that is of value to someone in the business. It might be tips on how to perform a task better. Or it could be a series of workshops on how to prevent certain types of operational crises, such as machine failures or safety mishaps. At Ernst & Young, best practices are documented and then shared through a computer application called COIN (community of interest). Other companies actively involved in knowledge management include General Electric, Toyota, Hewlett-Packard, and Buckman Laboratories.[38]

Does Your Business Need a Change in Its Infrastructure?

In this chapter we have provided an overview of the key elements in an organization's infrastructure. It is normal for businesses to make infrastructure changes as the company grows or as the competition in the industry becomes more intense. Infrastructure changes also occur when manufacturing and services become more integrated within a single business. Manufacturing companies are adding more services, which means changes in infrastructure are not only desirable but also inevitable.

So what does it take to make changes in your infrastructure and thus ease the transition of integrating your manufacturing and services functions?

- **Awareness of the need to change.** The first step is to realize that all businesses are, to a degree, in a state of transformation. This means that management should periodically assess whether improvements are needed in the infrastructure.
- **Top management endorsement and continuing support.** Change can occur anywhere in the business, whether in the manufacturing shop, the frontline contacts with the customer, or the internal financial operations at the home office. It is important that top management be visible in its support of change. Change is difficult in most organizations, so when enthusiasm for making changes wanes, employees will look to top management and gauge their degree of wanting to change based on what they see in their superiors. Andrew Grove, former CEO of Intel, explains that sometimes top management may not understand the need to change:

> If you work in one of those industries (high tech) and you are in middle management, you may very well sense the shifting winds on your face before the company as a whole and sometimes before your senior management does. Middle managers—especially those who deal with the outside world, like people in sales—are often the first to realize that what worked before doesn't quite work anymore; that the rules are changing. They usually don't have an easy time explaining it to senior management, so the senior management in a company is sometimes late to realize that the world is changing on them—and the leader is often the last of all to know.[39]

- **A plan to make the change, including the project plan.** Change takes time, and to be successful it should be organized into smaller steps. Making changes should also involve the input of lower-level managers and employees. They will ultimately be affected by the change, so they should have a say in how to make the proposed changes work.

- **The resources, or monetary support, to complete the change.** It takes money to make the required changes. Money buys new equipment, new employees, and additional training. There was a saying going around where one of the authors had worked that went something like this: "You can change anything you want, just as long as it doesn't cost anything." Needless to say, that particular organization was not very innovative or open to change.
- **The energy to carry out the plan (probably over a multiyear period).** Some changes can be carried out quickly, but most changes in infrastructure will involve several years. Be prepared for a longer time span to work with, and do not get discouraged if progress is slow. It is better to make slower progress and be moving than to be standing still and ready to topple over.

Notes

1. Parnell, J., *Strategic Management: Theory and Practice* (4th ed.), Sage, Thousand Oaks, CA, 2013.
2. Robbins, S. and Coulter, M., *Management* (9th ed.), Pearson-Prentice Hall, Upper Saddle River, NJ, 2007.
3. Porter, M., *Competitive Strategy: Techniques for Analyzing Industries and Competitors*, Free Press, New York, 1980.
4. Fishman, C., *The Wal-Mart Effect: How the World's Most Powerful Company Really Works—And How It's Transforming the American Economy*, Penguin Press, London, UK, 2006.
5. Wren, D., *The Evolution of Management Thought* (3rd ed.), John Wiley & Sons, New York, 1987.
6. Fayol, H., *Industrial and General Administration*, Dunrod, Paris, 1916.
7. Actually, this example will be "loosely" based on the Pizza Hut restaurant chain during the early 1970s. However, the example is meant to be more universal in that it illustrates the four classical management functions within the context of introducing a new product.
8. Lester, D. and Parnell, J., *Organizational Theory: A Strategic Perspective*, Atomic Dog, Cincinnati, 2007.
9. Wright, P., Kroll, M., and Parnell, J., *Strategic Management: Concepts*, Prentice Hall, Upper Saddle River, NJ, 1998.
10. Lester, D., and Parnell, J., *Organizational Theory: A Strategic Perspective*, Atomic Dog, Mason, OH, 2007.
11. Hammer, M. and Champy, J., *Reengineering the Corporation: A Manifesto for Business Revolution*, Harper Collins, New York, 1993.
12. Kreitner, R. and Kinicki, A., *Organizational Behavior*, McGraw-Hill Irwin, New York, 2007, p. 551.
13. Ostroff, F., *The Horizontal Organization*, Oxford University Press, New York, 1999.
14. Kasper-Fuehrer, E. and Ashkanasy, N., The interorganizational virtual organization: Defining a Weberian ideal, *International Studies of Management & Organization*, 33(4), 34–64, 2004.

15. Publisher McGraw-Hill Irwin provides an excellent video series to accompany its management textbooks. *One Smooth Stone* was recently featured in one of these videos.

16. Kasper-Fuehrer, E. and Ashkanasy, N., The interorganizational virtual organization: Defining a Weberian ideal, *International Studies of Management & Organization,* 33(4), 34–64, 2004.

17. Kerber, K. and Buono, A., Leadership challenges in global virtual teams: Lessons from the field, *S.A.M. Advanced Management Journal,* 69(4), 4–10, 2004.

18. Robbins, S. and Coulter, M., *Management* (9th ed.), Pearson-Prentice Hall, Upper Saddle River, NJ, 2007.

19. Hersey, P. and Blanchard, K., *Management of Organizational Behavior* (5th ed.), Prentice Hall, Englewood Cliffs, NJ, 1988.

20. Mosely, T., Are you being served? The route to good customer care, *Consumer Policy Review,* 12(6), 223–228, 2002.

21. Wren, D., *The Evolution of Management Thought* (3rd ed.), John Wiley & Sons, New York, 1987.

22. Taylor, F., *Principles of Scientific Management,* Harper, New York, 1911.

23. Hackman, R. and Oldham, G., Motivation through the design of work: Test of a theory, *Organizational Behavior and Human Performance,* 16(2), 250–279, 1976.

24. Robbins, S. and Coulter, M., *Management* (9th ed.), Pearson-Prentice Hall, Upper Saddle River, NJ, 2007.

25. Newstrom, J., *Organizational Behavior: Human Behavior at Work,* McGraw-Hill Irwin, New York, 2007.

26. Robbins, S. and Coulter, M., *Management* (9th ed.), Pearson-Prentice Hall, Upper Saddle River, NJ, 2007.

27. Peters, T. and Waterman, R., Jr., *In Search of Excellence: Lessons from America's Best-Run Companies,* Warner, New York, 1988.

28. Jennings, D., *Effective Supervision: Frontline Management for the 90s,* West, St. Paul, MN, 1993.

29. See Chandler, P. and Schraeder, M., Will the TEAM work for employees and managers? *The Journal for Quality & Participation,* 26(3), 31–37, 2003; Crandall, W. and Parnell, J., On the relationship between propensity for participative management and intentions to leave: Re-opening the case for participation, *Mid-Atlantic Journal of Business,* 30(2), 197–209, 1994; Gonring, M., Communication makes employee involvement work, *Public Relations Journal,* November, 39–40, 1991.

30. Carrell, M., Jennings, D., and Heavrin, C., *Organizational Behavior,* Atomic Dog, Cincinnati, 2006.

31. Jennings, D., *Effective Supervision: Frontline Management for the 90s,* West, St. Paul, MN, 1993.

32. Burns, T. and Stalker, G.M., *The Management of Innovation,* Tavistock, London, 1961.

33. See Lester, D. and Parnell, J., *Organizational Theory: A Strategic Perspective,* Atomic Dog, Cincinnati, 2007, for an expanded explanation of these concepts within the framework of organizational structure and strategic management.

34. Lee, J., Sr., Knowledge management: The intellectual revolution, *IIE Solutions,* 32(10), 34–37, 2000.

35. Mohamed, M., Stankosky, M., and Murray, A., Applying knowledge management principles to enhance cross-functional team performance, *Journal of Knowledge Management,* 8(3), 127–142, 2004.

36. Mullaney, T. and Weintraub, A., The digital hospital, *Business Week,* March 28, 2005, p. 77. See also Kreitner, R. and Kinicki, A., *Organizational Behavior,* McGraw-Hill Irwin, New York, 2007, p. 379, for an excellent summary on Hackensack University Medical Center and its use of knowledge management techniques.
37. Lee, J., Sr., Knowledge management: The intellectual revolution, *IIE Solutions,* 32(10), 34–37, 2000.
38. Robbins, S. and Coulter, M., *Management* (9th ed.), Pearson-Prentice Hall, Upper Saddle River, NJ, 2007.
39. Grove, A.S. *Only the Paranoid Survive: How to Exploit the Crisis Points That Challenge Every Company,* Doubleday, New York, pp. 21–22, 1996.

Chapter 9

Understanding Organizational Culture—The Elusive Key to Change

Introduction

Every organization has a culture. Think about that a moment. The authors, like you, have been to many different types of businesses and government agencies. One trip to the state drivers' license agency was especially memorable.[1] The line was long, the employees were not particularly friendly, and the facility was small and in need of modernization. Getting service at that place was a frustrating experience. What a different experience when we later ate at a nice restaurant. The service was friendly, the food arrived quickly, and the facility was very modern. In short, the organizational culture in those two places was very different.

Of course, you might be thinking at this point, "Wait a second, how can you compare a state agency with a modern restaurant?" Not a problem, remember the Input–Transformation–Output (ITO) Model? In both organizations this model was at work, taking resources and other inputs, transforming them into products and services, and then distributing them back to the customer. However, there are differences. State agencies are funded differently from restaurants. Agencies depend on tax revenue to run their divisions, and restaurants depend on private investors and the revenue from customers to run their businesses. State agencies must be efficient; offering extra frills to the customer means that tax rates will have to go up. Restaurants must also be efficient, but to differentiate one restaurant from another, owners or managers spend extra money to enhance décor, food, and entertainment options.

The missions of the state agency and the restaurant are also different. State agencies serve the general public by ensuring that all motorists are properly licensed and qualified to operate a vehicle. A restaurant, on the other hand, serves a small segment of society by providing food, and perhaps an entertaining experience. The processes between the two organizations are also different. State agencies provide a service; a restaurant is both a manufacturer of a product and a service provider. As a result, their external facilities will look different. State agencies rely more on information systems to maintain the correct data for its citizens. A restaurant relies more on production equipment because food must be stored, prepped, cooked, and then served in an appetizing manner. One final difference should be acknowledged between the two organizations, and perhaps this one is the most obvious—the profit motive. The restaurant must make a profit to survive; the state agency's existence is more certain, at least as long as its services are required.

To summarize, it appears the state agency and the restaurant share one thing in common, the ITO Model, and that is about it. Everything else is different, including one item we have not mentioned yet: the organizational culture. Both have a culture, but they are very different from each other. Organizational culture is intriguing because it is easy to see, yet we seldom stop to think about it. This chapter is about organizational culture. We look at what it is and why it is so important. To help you understand it better, the different components of culture will be examined. We also look at different types of culture and the role ethics plays in shaping an organization's culture. Finally, how you can change your company's culture is discussed. Changing cultures is often an essential part of implementing improvement programs. First, let us look at what organizational culture is.

What Is Organizational Culture?

Organizational culture is the belief system that members of an organization share. For example, members of a church, synagogue, mosque, or other religious organization share a common belief system. Religious organizations are an obvious place to begin discussing culture because their beliefs separate them from each other. The belief systems of these organizations are written in sacred documents and taught to their members. Each religious group has important people (some called prophets) that played a key role in the founding of that particular religion. The belief system includes a code of conduct for members. Such guidelines are important when members face difficult situations in their lives, and help members know how to act when facing a new situation. In addition, each religious organization has a set of rituals that help members act out their belief system in a formal manner.

Businesses and nonprofit organizations can have their own cultures, just like religious organizations. For example, Southwest Airlines is one of the most talked-about companies in the business media. One reason is that the airline's unique culture is based on having fun at work and incorporating a sense of humor into the workplace. The culture stems from the founder, Herb Kelleher, whose zany and outgoing personality has dazzled admirers from the business world for years. The culture is reflected through the employees, who like to make flying fun for their passengers. Gag announcements are common, such as the now-famous one where passengers who wanted a smoke were encouraged to visit the lounge, located on the wing of the aircraft, where they could also enjoy the movie *Gone with the Wind*.[2]

One of the fundamental beliefs at Southwest is that employees, not customers or shareholders, are the number one priority. Kelleher explains, "The employees come first. If they're happy, satisfied, dedicated, and energetic, they'll take real good care of the customers. When the customers are happy, they come back. And that makes the shareholders happy."[3] Because that is a fundamental belief of the company, the treatment of employees reflects that. Many companies boast that employees are their number one asset, but not all follow through with the rhetoric. For Southwest Airlines, there is much documentation about the positive treatment of its employees. The company consistently made *Fortune Magazine*'s "Best Companies to Work For" list from 1997 to 2000, although it has chosen not to participate in this ranking since 2000. *Fortune* also ranked Southwest as the fifth most admired company in 2004.[4]

Based on this, it sounds like organizational culture is important; but then again, we have talked about only one company. We include discussions on a number of other companies and how culture is important to their success. We also look at how culture is an elusive element of organizational change, one that many managers forget about when they are implementing reforms in their companies. First, let us look at why organizational culture is so important.

Why Is Organizational Culture So Important?

Organizational culture is important, yet it is also one of those "touchy-feely" topics that make some executives feel uncomfortable. One of the early writers on the subject, Linda Smircich, identified four down-to-earth reasons why organizational culture is important.[5] She prefers to call them functions of culture, and we describe each of them in some detail. The functions of culture give the company an identity, help an employee make sense of things, enable employees to be committed to the company, and add stability to the organization.

Organizational Culture Gives the Company an Identity

First, culture gives the company an identity. Many companies demonstrate a unique culture. In the 1980s, IBM was the epitome of a conservative, suit-oriented culture, while its rival, Apple, and especially the Macintosh division, was considered the bad boy of the computer industry because of the outlandish "insanely great" slogan[6] that permeated the company culture and its darling new product, the Macintosh.

Members of organizations with unique cultures take pride in being part of that culture. In essence, it is similar to how we felt when we were part of the "in group" in high school or college. This simple behavioral quirk is important for management to remember because employees want to be part of groups with which they derive an identity. Consequently, it is important that employees take pride in their company's work culture.

A number of companies are identified with their cultures. We begin with an easy one, Microsoft. If we were to sum up the culture in two words, we might say "really smart." That would be true, as prospective employees go through grueling interviews that include such questions as "Why are manhole covers round?" or "Given a gold bar that can be cut exactly twice and a contractor who must be paid one seventh of a gold bar a day for seven days, what do you do?"[7] Of course, the people at Microsoft are smart, and they know they are smart, and one of the cultural beliefs, according to author Mark Gimein, is that they know they are different from other people. Hard work, long hours, and innovation are corporate values. One of the biggest values is the push to ship a product to the market successfully.[8] Every team person who gets his or her product to market earns the SHIP-IT plaque. The plaque is also an example of a cultural artifact, which is discussed later in the chapter. With Microsoft, it should be easy to see how the culture helps employees derive an identity.

In a manner similar to Microsoft, two younger companies are also taking in an abundance of high-tech talent—Google and Yahoo! In the past, software engineers traditionally flocked to places like Microsoft or Sun Microsystems. Both Google and Yahoo! have done well recruiting; in fact, some of the recruits have come from Microsoft.[9] Like Microsoft, the talent that goes to the Internet search companies is very smart and is interested in working on cutting-edge technical problems. Companies like these are not just a place to work but a chance to be with the best minds in the business and work on real problems. Culture drives these companies to seek employees who want to make a difference, particularly with problems that impact the Internet.[10]

Another company with a unique culture is REI (Recreation Equipment). The Sumner, Washington-based retailer sells high-quality outdoor recreation equipment for the serious backpacker. It recruits many of its employees from its customers, who value the mission of the company and its products. Love of the outdoors and a concern for the environment are two fundamental values of the corporate culture. Conveniently, they are also part of their customers' personal culture, which makes them good recruits. As a result, about half of the sales force consists of customers.[11] At REI, culture is both a recruiter and an identity for employees.

Another company that links its identity through its employees is Hot Topic. The central values that come through this culture are music and young fashion. Customers are young and mostly female, stores are in malls, and the merchandise is clothing that is identified with music and entertainment promotions. Executives must live the corporate culture by attending concerts and movies, and by staying tuned to MTV. Comments Cindy Levitt, Hot Topic VP, "For us, Hot Topic is a corporate culture. We must live the lifestyle."[12]

Apple is another company that exhibits a strong culture that was derived from its founder and former CEO, Steve Jobs. He was recognized as an astute innovator of products and their design to appeal to a wide market of users. With his recent death, it will be interesting to see if the culture changes.

Organizational Culture Helps Employees Make Sense of Things

We all want our worlds to make sense to us. When something good or bad happens, it helps if we understand a reason for it. Religions have pondered this dilemma for centuries. Companies also offer their spin on the meaning of life, particularly business life, through their organizational cultures. In turn, employees can understand why their company does things the way it does.[13]

While in college, one of the authors[14] worked for a company that sold books door to door. The work was very difficult, and rejection was common. However, the culture of the company was to treat disappointments and rejections as a way to become a better salesperson, and to grow on a personal level as well. Student salespersons were reminded that each rejection meant they were one step closer to a sale. In addition, hard work and perseverance would eventually pay off. These values permeated the company and helped college students make sense of the disappointments that awaited them in the real world.

The Boston Consulting Group (BCG) is a think tank committed to making sense of the business environment. The organizational culture can be seen in its mission statement: "BCG aims to help the world's best organizations make decisive improvements in their direction and performance by sparking breakthrough business ideas." Three value statements solidify this mission to make meaning of the complicated business world:

- We continually strive to generate deep insight into what drives value creation and competitive advantage in our clients' businesses and the economy as a whole.
- We work collaboratively with clients to convert insight into strategy that will have a substantial positive impact on performance.
- Consistently delivering impact earns the trust that is the foundation of lasting relationships. These relationships serve as a platform for still-deeper insight and more significant impact.[15]

This illustrates a common thread in organizational culture; the mission statement of the organization serves as a basis for the culture. If the two are not harmonious with each other, then something is wrong. Either management is not communicating the mission well or the employees have rejected it altogether. The result is going to be a culture with an identity complex. At BCG, this is not a problem; mission and culture do go well together. *Fortune Magazine* ranked it second in its "100 Best Places to Work for 2012" issue. According to their report, "...the global consultancy invests 100-plus hours and thousands of dollars to recruit each consultant; once hired, they earn an average of $139,000 a year."[16]

If culture is to help employees understand why things happen the way they do, then it must also enable employees to understand why the company does things the way it does. Management writers Robert Kreitner and Angelo Kinicki discuss the need for all employees to understand the company's mission and culture. They discuss how in 1971, Southwest Airlines originally had a vision to compete with ground transportation. Their strategy consisted of offering low fares, short flights, high frequency of flights, and point-to-point service (a fancy way of saying "no hubs" like the big airlines use). Translated, this means a plane must be in the air more than it is on the ground. To accomplish this, flights that land must be quickly serviced and put back into the air, sometimes in as little as twenty minutes.[17] This means that all employees, sometimes even the pilots, must help clean the plane so it can be put back in service. The Southwest Airlines culture is to communicate to all employees the importance of fast turnaround. To the employees, the work they do makes sense, because they understand the company's mission and culture.

Organizational Culture Enables Employees to Be Committed to the Company

Culture helps an employee stay committed to the employer. When employees feel like they can identify with the values of their employer, they are more inclined to want to work for that employer. Usually, the values the company puts forth as being important include acknowledging the dignity and worth of the workforce. Thus, a company with a good organizational culture values its employees.

Let's look at the Container Store. The company sells items that help get your home or office organized. If you visit the Container Store's website, you will come across the cliché, "employees are our greatest asset." This is nothing new; all companies say that. Then, the corporate culture suddenly leaps out in these statements: "One of the Container Store's core business philosophies is that one great person equals three good people in terms of business productivity. So, why not hire only great people? With that said, when it comes to selecting who will join the team, we go to great lengths to look for that one great person to fill a position. Most employees at the Container Store are college educated and most were customers first."[18]

The statement "one great person equals three good people" puts a quantitative dimension to a qualitative subject. Obviously, the company has thought this through and realizes one of the key elements of organizational culture—hire only those applicants who will perpetuate the culture you want in your company.

Furthermore, customers make good prospective employees because they are already interested in what the company sells—organization. Note how the Container Store and REI share a similar idea on recruitment by looking at their customers as potential employees. From an organizational culture standpoint, it makes perfect sense.

Does the Container Store actually have committed employees? The answer appears to be yes. The company has been consistently in the Fortune 100 Best Companies to Work For, and in 2007, they finished an impressive fourth.[19] Furthermore, voluntary turnover, a key indicator of the success of the culture, was at a low 19 percent. According to the 2012 *Fortune* report, the "Employees here heap praise on management for avoiding layoffs during the recession and for an attentiveness to well-being that includes handing out cold water at distribution centers during the summer months."[20]

As we have seen with the Container Store, if a company proclaims that employees are a great asset, that should translate into excellent human resource practices. This, in turn, creates a positive organizational culture because employees feel they really are important (as opposed to just hearing that they are important). One company that is widely recognized as being one of the best in its industry area (software design) and in the area of treating its employees well is SAS in Cary, North Carolina. The culture created by this company has made it one of the most talked-about places to work, and it boasts a 97 percent retention rate of its employees.[21]

When management writers talk about the SAS culture, they inevitably expound on the great benefits the company offers, and these are important to mention:

- On-site (or in SAS talk, on-campus) medical facilities
- On-campus swimming pool, exercise room, and basketball courts
- On-campus massages, dry cleaning, auto detailing, and even haircuts
- On-campus Montessori day care center.[22]

However, to stop there would be missing an important part of the culture—the ability to manage highly creative employees. Managerial Comment 9.1 explains the philosophy behind fostering creativity at SAS.

Managerial Comment 9.1
Fostering Creativity at SAS

Creative people … crave the feeling of accomplishment that comes from cracking a riddle, be it technological, artistic, social, or logistical. Though all people chafe under what they see as bureaucratic obstructionism, creative people actively hate it, viewing it … as the enemy of good work.[23]

The application of this approach to creativity is to enable employees by keeping them intellectually engaged and removing such distractions as unnecessary meetings. It also means removing distractions that can occur off the job. That is why SAS has an on-site medical facility, so employees can return to work faster, rather than trekking across town and waiting for hours to see a doctor. The basketball courts, gym, and swimming pool also clear the employees' lives of distractions because they do not have to travel off campus to take an exercise break.

One final note about the SAS culture. Much has been written about the company's thirty-five-hour workweek. In truth, the creative software developers probably work a little more than that, but they certainly are not encouraged to work the seventy-hour weeks known throughout Silicon Valley. A company proverb states, "After eight hours, you're probably just adding bugs."[24]

Commitment can also be found in the health care industry. The Mayo Clinic offers a glimpse of how a positive culture can be cultivated in the industry. Perhaps the most revealing aspect of the culture is the team approach to patient care. Only the top physicians are recruited to work at the Mayo Clinic, which is not just one facility, but over 58,000 employees in various clinics and hospitals across the United States. Doctors who want to be stars, want to work independently, lack interpersonal skills, or want to maximize their income need not apply. What most hospitals desire is a team player who derives satisfaction from seeing a complicated medical case successfully treated because a number of professionals were involved.[25]

To fit into this type of culture, hospitals seek a certain type of individual, one who can perpetuate the culture, not fight it. That is why the selection process at the Mayo Clinic, and other organizations we talk about in this chapter, must be stringent. Once the right person is on board, that employee can work within that culture because that is what he or she is best suited for. As a result, commitment to the organization usually follows. In the case of the Mayo Clinic, *Fortune Magazine* reports that 89 percent of the employees say they are proud to tell others they work at this hospital.[26]

Organizational Culture Helps Add Stability to the Company

Culture adds traditions to the organization, which can add stability to the workplace. As a result, a positive culture can help buffet the company as it goes through rough times. Some cultures have allowed companies to endure for many years. The William Wrigley Jr. Company makes candy and gum and has been around since 1891. Four generations of the Wrigley family led the company from 1891 until 2006, with the latest being CEO Bill Wrigley, a great-grandson of the founder. In 2006, Bill Perez became the first CEO who was not a member of the Wrigley family. In 2008, Mars Inc. acquired the Wrigley Company, and its current president is Martin Radvan, a longtime employee of Mars Inc. The traditions and culture seemed to appeal to the employees as well, as one-third of the employees have been with the company for more than fifteen years.[27]

Nordstrom's, founded in 1901 as a partnership, is another company that has endured over the years by providing excellent customer service, a foundation of their organizational culture. The Seattle-based retail chain translates its culture of superior customer service into high commissions for its store sales personnel and promotion from within. This type of culture also means that employees have some leeway in how to meet the customer's needs. This is an important consideration when formal rules dictate much of employee behavior. In the area of excellent customer service, sometimes an absence of stiff rules is more appropriate.[28] The cultural norm is to "use your good judgment in all situations."[29] As a result, it is the salesperson who makes the decisions on exchanges and refunds, not the supervisor. The formula works at Nordstrom's, as this company has withstood a century of retail turbulence. Today, sales associates can earn commissions of over $100,000 a year, which in a retail store is almost unheard of. Culture, then, adds stability to the company and income to employees.

Some work cultures are destructive for employees because they can reduce morale and lead to poor productivity, absenteeism, turnover, and even lawsuits. Many discrimination lawsuits result from a culture that tolerates racism, sexual harassment, or a hostile environment of some type. Health care researchers Stephen Crow and Sandra Hartman of the University of New Orleans have reported how "rogue doctors" can contribute to a dysfunctional culture in health care.[30] They describe the rogue doctor as one who exhibits severe ethical and performance problems that can expose a patient to certain risks. Such doctors can be disrespectful toward others, particularly nurses and aides. Some health care institutions look the other way at such behavior, but the result is a culture that tolerates the rogue physician. As the culture solidifies, the result can be regular intimidation of health care staff that is not addressed by management. In such environments, a "zero tolerance" approach to identifying and removing these doctors is being encouraged.[31]

In summary, organizational culture can be an asset or a liability. Learning to identify the elements of culture, both good and bad, is the focus of the next section.

What Are the Components of Organizational Culture?

The Components of Culture

As we have seen, organizational culture emanates from a belief system. This belief system manifests itself in a number of ways. First, it will show in the company's mission statement. The CEO expresses the mission in speeches and written statements. Culture can also be expressed through more indirect means, called cultural artifacts by management researchers who study these types of things. In this section, we look at the components of culture, beginning with values. Next, we look at cultural artifacts in more detail.

Values

The values of a company often come from the founder. These values then become part of the culture. We can look at many well-known companies and see their founder's personality and values expressed through the company. Herb Kelleher's personality and values are obviously ingrained at Southwest. Ray Kroc, although technically not the founder of McDonald's (two brothers actually owned the restaurant; Kroc just built the empire), was instrumental in its success and growth as a chain by instilling the values of quality, service, cleanliness, and value.[32]

With all the hype about Southwest Airlines' culture, one wonders whether there is a company that bases its values and culture on Southwest. Actually, there is. HomeBanc Mortgage, in Atlanta, Georgia, borrows some of its culture from Southwest Airlines as well as the United States Marine Corps and the Walt Disney Company. Like at Southwest, employees come first, not customers. The value here is that satisfied employees will take good care of the customers. Another value is the emphasis placed on character. "The director of human resources is a minister, and internal meetings are often opened with a prayer. No experience is necessary, but good character is."[33]

Values appear in a number of ways. At the Midwest-based accounting firm of Plante and Moran, values translate into a lower than usual turnover rate. The atmosphere, which has been informally labeled "jerk-free," thrives on discouraging bad supervisory styles and instead puts an emphasis on the Golden Rule.[34] As a result, the turnover rate (8 to 15 percent over the previous decade) is about half that in other similarly sized accounting firms.[35]

As described in the UPS case in Chapter 5, Jim Casey, founder, instilled values in the company that remain strong after a century of operation. A quick way to discern an organization's values is to look at its mission statement. If values appear anywhere, it should be there. The mission should express the organization's reason for existence, as well as a list of key values that are important.

Artifacts: The Display of Organizational Culture

When the culture is visible, it translates into "how things are done around here."[36] Looking at the physical manifestations of culture or the artifacts is important to understanding what the culture is about. In other words, artifacts help reveal what the values are of the organization. Organizational culture researchers have identified various types of artifacts.[37] These include stories, language, symbols, ceremonies, rituals, identifiable value systems and behavioral norms, the physical surroundings characterizing the particular culture, and organizational rewards and reward systems.

Stories

These are also referred to as myths and sagas. The stories reflect the ideals that the company wishes to convey. Stories are often told during orientation and training sessions to let newcomers know what is valued in the organization. Innovation is often a value reflected in such stories. For many years, one of the most popular stories told in business textbooks was the invention of those sticky notes that we take for granted today. The author remembers the most common version of the story going something like this: a 3M inventor was looking for a way to mark his hymnal in church with pieces of paper that would stick to the page, but not stick so well that it would tear the page. Although the story is true, credit should be given where credit is due. Actually, Post-It Notes were the result of two inventors in the 3M Company. Dr. Spencer Silver discovered an adhesive in 1968 that was very thin but did not stick very well. For several years, he tried to find a use for the discovery, to no avail. Another 3M inventor, Art Fry, needed a way to mark his hymnal in church with something that would stick, but not stick so well as to tear the delicate pages of the hymnal. Fry worked on the project and eventually came up with the sticky notes that we are familiar with today.[38] The 3M company uses the term *bootlegging* to describe their practice of letting inventors/scientists spend 15 percent of their work time on any product of their choosing.[39]

Stories serve several purposes in perpetuating a company's culture. First, they involve the listener's emotions, as most everybody likes a good story and can relate to some degree to the events going on. Second, they show listeners how the company dealt with past mistakes and managed to overcome them and forge on. Third, stories communicate the traditions of the company. Finally, stories help listeners become cohesive with their organization by showing them how they can change to become like the organization.[40] This is why stories are often included as part of a company's orientation and training; as a way to show new employees the values the organization embraces.

Language

A more formal way that management researchers refer to language is "language systems and metaphors."[41] Vocabulary and expressions used among employees can convey a sense of the culture of an organization. First, consider vocabulary. There is something in everyone that enjoys using a specialized language that is indiscernible to an outside audience. When we use that language, there is a sense of pride that comes with the type of work we do and, perhaps, the organization we work with. Sometimes, that language uses words or expressions that mean something different from what the average person thinks when using that same word or expression. For example, in the restaurant industry, the following words or phrases take on specialized meanings:

- "86." This is understood to mean that the chef (or the restaurant) is out of that menu item. For example, "86 the crab cakes on the menu tonight."
- "In the weeds." An expression used among staff, particularly in the kitchen, to mean that they are very busy. For example, "We can't talk right now. We're in the weeds."
- "Slammed." Similar to "in the weeds," but conveys more of a sense that things got busy all of a sudden. For example, the tour bus came in, the tickets started jamming up in the window, and "we got slammed."
- "Give me an echo on that." Repeat the order, please.
- "Run it through the garden." Load that sandwich with everything: lettuce, tomatoes, onions, etc.
- "Give me a 31 over easy with bacon and a 21 scrambled." This is breakfast talk among cooks and staff. Translated, the first order gets three eggs over easy with bacon. The second order gets two scrambled eggs, no meat.
- "I need a 62 scrambled, a 21 poached with ham, and a Western." A variation of the previous example. In this case, two of the customers want exactly the same thing, three scrambled eggs with no meat. The second customer wants two eggs poached with ham. The third customer wants a Western omelet.[42]

No doubt, whatever industry the reader is working in, a vocabulary has emerged that may sound foreign to those outside the industry. With this vocabulary comes a sense of pride and ownership in one's line of work. Having pride in one's work is one aspect of a culture.

Language can also express culture through the use of phrases, mottos, and slogans. A favorite expression can go a long way in perpetuating certain values. Some are memorable and descriptive, such as "Good to the last drop" (Maxwell House Coffee) and "It runs like a Deere" (John Deere Equipment). Others are less direct, and therefore more difficult to relate to a specific company, such as "Go further" (Ford Motor Company) or "Your world, delivered" (AT&T). In recent years, companies appear to change slogans more often and make them shorter.

Symbols, Ceremonies, and Rituals

This category of artifacts is comprised of two subcategories, with symbols focusing on physical items, and ceremonies and rituals focusing on processes.

We begin with symbols. Some companies have found ways to use a material item as a symbol. WorldNow is an Internet technology company with the company value of "drilling down to solve problems" for their customers. To help support this culture, a used drill was purchased (for two dollars) at a local thrift shop. The drill is awarded for one month at a time to an employee who displays the quality of working hard (drilling down) to solve a problem. After the month is over, another employee who has performed great work receives the drill. The drill must be personalized in some way by the employee before presenting it to the next employee. One employee was innovative and put a Bart Simpson trigger on the drill. Another employee put an antenna on it to show it was "wireless."[43] Another symbol was used by a dean at a major university who wanted faculty to have an open-door policy. To illustrate his point, he provided each faculty member with a rubber doorstop. The doorstops became a powerful symbol that exemplified the cultural value of working hard to meet the needs of students.[44]

Ceremonies and rituals are also important aspects of organizational culture. These events often celebrate an accomplishment of a key employee or division of the company. Perhaps the best-known examples are annual sales meetings, held by many companies, where top achievers are recognized and rewarded for their work. In education, the commencement exercise celebrates the graduation of students. When an employee retires, there is usually some recognition for his or her work in the form of a party and material gifts. Some of these ceremonies may include personal comments of friends and colleagues. These types of events serve several purposes: (1) they acknowledge the worth of the employee, (2) they celebrate the hard work of the employee, and (3) they encourage younger employees to identify with the accomplishments of the recognized employee. Thus the values of the company—its culture—are perpetuated through these ceremonies.

Some ceremonies serve a different purpose—they seek to unite the members of the organization into a more cohesive unit. These types of events seek to bring employees together and help them feel good about themselves and each other. In this way, employees build commitment to the company.

Identifiable Value Systems and Behavioral Norms

These can be fun to observe as you visit various organizations. The dress code is one behavioral norm that is tied directly to the company's value system. Conservative dress codes reflect the value in the organization that one should look professional at all times. More lenient dress codes, such as those followed in a university setting, emphasize the importance of being comfortable and less formal, particularly when among students. Although no one dress code is superior to another, the code should match the type of organization. A casually dressed executive among a conservative board of directors would not be appropriate. Neither would a professor in a three-piece suit teaching an art history class unless, of course, that is the culture the university wants to convey. There is a trend toward more informal dress in business settings. Executives frequently appear without ties on television interviews.

Another behavioral norm (that could also fall under the category of language) is the use, or absence of the use of profanity in the workplace. The use of profanity is a reflection of the professional atmosphere of the workplace.[46] The atmosphere of the workplace is perceived to be less professional if profanity is allowed. Some organizations have taken steps to clean up the language used in the workplace, and the first place they look is to make sure management is setting a good example. Some companies have realized that profanity be a form of harassment, which could result in lawsuits, and written policies about the use of inappropriate language have been encouraged.[47] Regardless of a company's policy on profanity, inferences can be made about the culture of that company based on the kind of words that are allowed or not allowed in conversations.

The Physical Surroundings Characterizing a Culture

The buildings, offices, furniture, and even the landscaping help convey the organizational culture. On a personal level, the type of car and housing a person chooses may provide insight into his or her belief systems. Exotic, expensive, and fast cars convey a much different mind-set than more conservative cars that are fuel efficient and less expensive to own and operate. The owner of a Jaguar and the owner of a Prius are probably very different types of people. The way people decorate their homes and the type of dwelling they live in convey an impression of the belief system of the occupants. A large home with all the latest technological gadgets shows what is important to the person who lives in that home, and hence a glimpse of who that person is. Then there are people who save everything and live in a state of clutter. This tells us that one of the values of that person is to be frugal and not throw anything away. People who grew up during the Great Depression often illustrate this behavior. The point is not to be critical of the person with the cluttered house, but to see the values of the person conveyed through how he or she manages the home.

We can apply this discussion to the way an organization sets up its buildings and offices. Some organizations have simple, inexpensive looking home offices. Walmart corporate headquarters are far from elaborate or expensive looking. The surroundings are frugal compared to many other corporate offices. Walmart's retail stores also convey this simple, inexpensive approach to business—a large-box, simple store structure, and an eye for efficiency. These are the values that permeate the company—the physical structures illustrate the company values.

Other companies prefer more flamboyant physical surroundings. Oakley, maker of products for athletes and probably most well known for its sunglasses, illustrates this. The company headquarters in Foothill Ranch, California, resembles a spaceship, a bunker, and a futuristic fortress, all in one. The building is intentionally designed to reflect the values stated in the company's innovation philosophy found on the company website. Managerial Comment 9.2 states these values.

Managerial Comment 9.2   Innovation at Oakley

1. Redefine products by redefining what is physically possible. Nothing is impossible—we know that from experience. But new ideas demand new technology, and some are insistent. CAD/CAM engineering lets us experiment and explore.
2. Reject conventional ideas, and when necessary reinvent from scratch. That means sending out hunter-gatherers to roam the world for inspirations, and locking ourselves in a design bunker until even the raw materials get the science treatment.
3. Erase the line between form and function by elevating physics to the level of art. We think they should fuel each other until they reach a critical mass called innovation.
4. Transcend performance innovations into solutions for everyday applications. By serving the demands of professional athletes, we create innovations that serve all.
5. Deliver the unexpected. That's what invention is all about. Science and art come together, producing a new formula of performance and style. And the occasional nasty explosion.[48]

Some physical surroundings may be offbeat, sophisticated, and innovative. At Google, employees can race their remote-controlled blimps through their large offices. Digital toilets are also in place, with adjustable seat temperatures that can be controlled via remote control.[49] Such surroundings help convey a culture of creativity, high technology, and fun.

The arrangement of offices is also an indicator of the company values. Cubicles and open office arrangements do save money, but they also promote better communication among employees. This value is important when cross-functional teams are part of the organizational structure. Individual offices limit the type of face-to-face communication that can occur. Such offices are status symbols for some people; for others, they provide a necessary retreat when quiet and concentration are required. Some companies incorporate the benefits of both arrangements by having open office areas and individual offices for those who need time away from the noise.

Organizational Rewards and Reward Systems

In general, rewards tend to motivate employees. To perpetuate a desired culture, an organization will reward the types of behavior it is seeking from its members.[50]

For example, Timberland, an outdoor boot and gear retailer, takes the environment seriously and wants its employees to do the same. Employees who buy a hybrid car can receive a $3,000 credit.[51] The company also promotes a culture of community volunteerism and pays up to forty hours each year if employees do volunteer work in their communities.[52]

Most companies want to perpetuate a culture of excellent performance from their employees. As a result, most companies also have pay-for-performance compensation programs to encourage this behavior. Across-the-board pay raises should be a thing of the past for any organization seeking to motivate employees. Such programs, which still hang on in some organizations, only reward seniority, not performance. Consequently, these programs reward mediocre employees, who are glad to have a job, and drive off excellent employees, who feel offended by such out-of-date pay programs.

It is important to select performance criteria carefully; otherwise, the wrong cultural values may be encouraged. One of the authors traveled to a conference recently and received a form letter and survey from the hotel manager a day before the stay was over. The survey was worded so that a "1" was a "poor" indicator and a "7" was an "excellent" indicator, with the numbers in between indicating the degree of each. Each question in the survey addressed an operational aspect of the hotel. The letter thanked the guest(s) for their stay and asked that they complete the survey. In the letter there were several appeals to contact the manager or designee if there was a reason a score of "7" could not be entered with each question. In other words, that manager wanted to see "7s" because another party was going to be reviewing those surveys, perhaps his or her supervisor. It did not seem to be a genuine method of capturing data about where improvement was needed because the instruction was to contact the manager before scoring anything lower than "7."[53] This practice violates common survey research design. This performance criteria and culture were encouraging perfect surveys, not excellent customer service.

In his research of disastrous reward systems, *Management Review* writer Dean Spitzer found the following examples:

- A pizza delivery company rewarded its drivers for speedy, on-time deliveries. Unfortunately, the drivers were also causing numerous accidents.
- An insurance agency rewarded its sales agents for the number of telephone calls made (as opposed to the number of sales), and what they got from this reward system was a high telephone bill and few sales.
- A manufacturing company rewarded its maintenance mechanics based on the amount of time they were spending making physical repairs. Time spent on problem solving and preventative maintenance was not rewarded. The result? There was little problem solving and preventative maintenance being performed.

■ A freight company came up with the idea of basing rewards on the number of containers shipped. The amount in the container did not matter, just that the container was shipped. A check on the capacity of the containers shipped revealed that only 45 percent of them were full when shipped, thereby encouraging a culture of waste and high shipment numbers, and eroding profits for the firm.[54]

The bottom line is that the monetary reward system must match the desired cultural behaviors.

Of course, other rewards can be dispensed that are not monetary in nature, and these rewards can also perpetuate the organizational culture. One such reward is recognition. That is why many companies have annual sales meetings so they can identify and formally recognize their top salespeople. There may be some monetary rewards given out as well, but the formal recognition serves not only to honor the exceptional employee but also to encourage others that they will be rewarded if they perform well.

What Types of Organizational Culture Are There?

Business researchers have identified various types of organizational cultures. Some cultures are conservative; others are more innovative and free flowing. Some cultures thrive on providing excellent customer service, and others seem not to care whether the customer is treated well. This divide is generally seen more when comparing for-profit organizations, such as a private business, with government agencies that must provide a service, regardless of the number of customers.

Types of Cultures

One of the best-known classifications of organizational culture focuses on four distinct groups: adhocracy, market, clan, and hierarchy cultures.[55]

■ Adhocracy. The main characteristics of this culture are the striving for a spirit of entrepreneurship, innovation, and risk taking. These businesses are always developing new products. There is an emphasis on growing the company and acquiring new resources, so these cultures are externally focused. The type of person that does well in this culture is one who is flexible, creative, and adaptable.

■ Market. Like the adhocracy, this culture is externally focused in that it is seeking to grow and be competitive in the market. The main characteristics of market cultures are their enthusiasm for competitiveness and wanting to lead the industry in which they operate. As a result, this culture is also very goal oriented. Individuals who do well in this culture are achievement oriented.

■ Clan. As the name implies, this culture is more like an extended family. The culture emphasizes harmonious relationships among members, solid mentoring of newer employees, and working in groups or teams. Unlike the adhocracy and the market culture, the clan culture is internally focused in that it seeks to run well as a cohesive unit. The type of person who does well in this culture desires to be part of the team and seeks the good of the company. Individual stars who are performance-driven fit better in a market culture than a clan culture.

■ Hierarchy. The hierarchy culture thrives on rules, policies, procedures, and order. Hierarchy cultures are internally focused in that the goal is to make the operation predictable and smooth running. These cultures do best in slow-changing environments. Innovation is not one of the strengths of the hierarchy. Individuals who function best in this type of culture like order, efficiency, and predictability.

Aligning Culture and Strategy

Robert Quinn and John Rohrbaugh feel that all four cultures are useful in the right environment, as each culture has its own unique strengths and weaknesses. However, when the culture and the business environment are mismatched, problems can occur that will hinder the long-term survival of the business.[56] The problem for management, then, is what to do when the culture does not match the environment and, hence, the business strategy.

Another consideration when aligning cultures with strategy is determining which culture produces the best business performance. Using the measures of profitability, size, growth rate, and market share, Rohit Deshpandé and his colleagues found that market cultures perform best, followed by adhocracy, clan, and hierarchy cultures.[57] Following this line of thinking, it comes as no surprise that many government agencies exhibit hierarchy cultures, an indicator that efficiency and predictable service to the public is needed over growth and innovation strategies. Likewise, highly competitive, for-profit companies often exhibit market and adhocracy cultures, a necessity in the types of business environments in which they operate.

What happens when a hierarchy culture exists in a company that is seeking to become more competitive? Or a clan culture exists when a hierarchy culture is needed? Mismatches in culture are common and can often be detected when new technologies are introduced into the workplace. The following section discusses what companies can do to change their organizational culture.

Changing Organizational Culture

To make changes and improvements in an organization, three elements need to be addressed: technology, structure, and organizational culture. This chapter describes

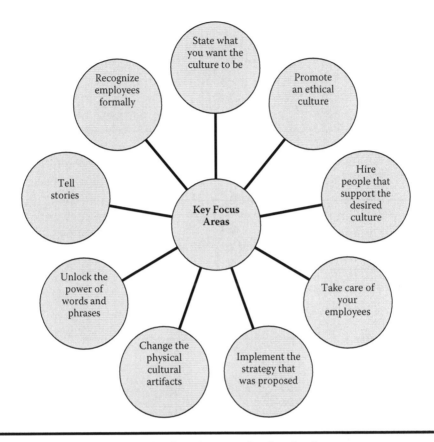

Figure 9.1   Key focus areas in changing organizational culture.

the power of culture as a key to change. Many companies do well at changing the technology when needed; they may be less adept at changing the structure. Changing the culture is often overlooked. It is that elusive element often forgotten in the change process.

The good news is that culture can be changed, but it is often a slower process than changing the technology or structure of the organization. Changing organizational culture involves addressing a number of different focus areas, as shown in Figure 9.1 and discussed in the next section.

State What You Want the Culture to Be

Organizational leaders need to carefully state their vision for the company culture. Such statements will appear in several formats. Because culture is a belief system followed by actions throughout the organization, it makes sense to state those beliefs in the company mission and values statements. Putting these statements on the company website is standard practice now. However, there are other places where mission statements can appear: on business cards and company letterhead, in lobby headquarters, even on coffee mugs. For these statements to be truly effective, employees should be able to not only remember the statements but also relate them to their jobs. Many organizations have a short mission statement, which can be recited, followed by a longer list of values, which can be referred to.

The values of the desired culture should also be verbally repeated by key company spokespersons, such as the CEO, president, or vice presidents. These values can be stated in company literature as well as advertising. At the University of North Carolina at Pembroke, there is a key value statement that we[58] are very proud of: "where learning gets personal." Classes are purposely kept small so students and professors can have more contact with each other. This one value appears in university advertising and is frequently mentioned by the chancellor of the university. Although the university holds other values, this particular value is the one that is promoted the most, and hence permeates the culture of the institution.

Promote an Ethical Culture

Although it may seem obvious in this day and time that a company would want an ethical culture, wanting such a culture and promoting it are two different things. Promoting an ethical culture involves extra work on the part of management in that ethical behavior must be communicated, nurtured, encouraged, and expected. Not promoting such a culture leaves management only hoping that the company will be ethical.[59]

Management writers Stephen Robbins and Mary Coulter offer five excellent guidelines for promoting an ethical culture.[60] First, managers and supervisors need to be visible role models for the rest of the employees in the company. Think about that! If employees are not sure how to act in a particular situation, to whom do they look for guidance? If managers act in an ethical manner, there is a better chance the employees will, too. By the same token, unethical behavior on the part of managers and executives will also be imitated. Management textbooks are full of cases where unethical behavior permeated a company like cancer—the process metastasizing when lower-level employees imitate the unethical behavior of their supervisors.

The second guideline is to communicate that ethical expectations be part of the culture. These guidelines can be expressed in a number of ways: through written means, such as company documents and its website, and through verbal statements by management. Codes of ethics or conduct are also used by many organizations. These codes are usually a series of statements that address ethical areas including complying with all laws and regulations, avoiding conflicts of interest, maintaining the security of company information, avoiding gifts of a certain monetary amount or lavish entertainment from clients and suppliers, and maintaining proper behavior while in a public setting.[61]

The third guideline is to provide ethics training to employees. This practice is usually offered within the format of brief company workshops and training sessions. Likely subjects covered in training include those mentioned above. In addition, sexual harassment has been a frequent topic of ethical training in more recent years.

The fourth guideline is to reward ethical acts and punish unethical ones. The practice of punishing unethical acts is frequently overlooked in organizations. A simple example might help illustrate this point. When a supervisor makes a derogatory or humiliating statement to an employee, the practice may continue if another manager or another employee does not challenge it. The danger, then, is that this type of statement becomes an established part of the organizational culture, no matter how uncomfortable the remarks may make the person to whom they are directed. Almost any undesirable behavior on the part of an employee can be perpetuated and become an undesirable part of the company culture if left unchecked.

The fifth guideline is to provide mechanisms whereby employees can discuss and report ethical dilemmas. Some companies have an ethics officer who oversees this function. Ethics hotlines within the company may also be implemented. Ethics committees can also make recommendations to management on how to handle specific ethical situations.[62]

Hire the Kind of People You Want to See Perpetuate Your Desired Culture

As we noted from the examples of the Mayo Clinic and REI, it makes sense to hire not just good potential employees but ones who share the culture and vision of the organization as well. This distinction should be made because even an excellent new employee in the wrong culture can be a problem to the company.

Colleges and universities see this often when they hire new professors. Table 9.1 illustrates potential scenarios that can occur. The table gives an overview of how well a potential professor will "fit" into a particular college or university. By fit, we mean the ability to perform well on the job and to collaborate effectively with other personnel in the institution. The table indicates that a research-oriented professor will be an ideal fit with a research-oriented university. These types of professors enjoy conducting research, attending professional meetings to present their research, and publishing their research in journals and books. Because research is their passion, they may see teaching as just another part of their job, but not a part they are particularly enthusiastic about. As a result, research universities usually allow these professors more time to research and less time to teach, a practice that fits both the potential professor and the mission of the university quite well.

As we move across the row, the next scenario depicts a research-oriented professor in a teaching-oriented college. This situation is not a good fit. The culture of the college supports teaching, not research, but the strengths of the professor are in

Table 9.1 Potential Scenarios When Hiring a Professor

	What the College or University Wants		
Strength of the Potential Professor	*Research Oriented*[a]	*Teaching Oriented*[b]	*Both Research and Teaching Oriented*[c]
Research oriented	Ideal fit	Not a good fit	A somewhat good fit
Teaching oriented	Not a good fit	Ideal fit	A somewhat good fit
A combination of research and teaching oriented	A somewhat good fit	A somewhat good fit	Ideal fit

[a] Large universities have as part of their mission the goal of generating primary research to add to the body of knowledge of an academic discipline.

[b] Small colleges typically focus on teaching undergraduates as their primary mission.

[c] Medium-sized colleges and universities stress both; however, research is not usually as intensive as at the large university.

research, not in teaching full time. The college will usually assign professors a heavy teaching load, leaving little time for research. The research-oriented professor will be frustrated in this environment and can either leave the institution or stay and try to change the culture so it is more accepting of research. Leaving is usually the better alternative because it will be difficult, if not impossible, to change the culture of the college to a research agenda. Such teaching-oriented colleges will do best if they make sure they are hiring professors who are passionate about teaching. This is not to say the potential professor cannot also be a good researcher, but between teaching and research, the best potential fit will be a professor who favors teaching over research.

Moving across to the far right of the table, an interesting and flexible institution emerges, the university that stresses some degree of teaching and research. Teaching- and research-oriented professors can be a suitable fit in most of these universities, because the cultures of these schools support both. However, the intensity of research will not be as great as at the larger research university discussed earlier. Still, a research-oriented professor will be at home in this environment. What is required, however, is that professors be fairly competent at both teaching and research, not just one area over the other. As a result, the best professors in these environments find they need to balance the time demands of research with the student-centered demands of teaching. This can be a difficult task for some and can result in some professors leaving the university. For example, teaching-oriented professors who do no research will find they are not eligible for promotion or tenure.

Likewise, research-oriented professors who do not perform well in the classroom may also find that the university does not want to keep them.

The rest of the table depicts the other combinations of possible scenarios. It shows that the fit of the potential employee to the organization is an important factor in encouraging the desired culture. Teaching-oriented professors will find they do best at teaching-oriented colleges. Research-oriented professors will prefer the larger universities. Those who choose universities with a hybrid of both teaching and research cultures will have to become proficient at both if they want to continue in this type of school.

Take Care of Your Employees, and They Will Take Care of Your Customers

The examples of Southwest Airlines and top-end retailer Nordstrom's illustrate how taking excellent care of your employees can result in excellent care to the customers. A sobering reality, though, is that when employees are unhappy with their employer, this discontentment may come across to the customer. Most of us have experienced a service worker who "grumbled" about the boss or the company. Such comments may typically be something like, "They need to hire more people around here," or "These procedures they (the company) require are useless."

These types of comments often come from employees who either do not feel valued by their company or feel the company does not care about their work situation. Why else would they complain about not having enough help? They may be frustrated by the level of service they are providing because the company does not give them sufficient resources to work with. Frontline employees, those who deal directly with the consumer, are generally observant of the feelings of their customers. Most want to serve the customers in a way that makes them happy; if the customers are upset with the service, the employee's job will not be pleasant either.

An organizational culture that puts employees first will go a long way in helping the attitude of the frontline employee. When they feel good about the job and their attitude is good, they perform better in their roles. Resources will be available (more help will be provided when business gets busy) and needless restrictions will be lifted. As we learned in Chapter 8 with the Ritz-Carlton Hotel chain, frontline employees need the freedom to remedy customer problems without undue restrictions.

Implement the Strategy Proposed

This might seem obvious, but many changes proposed by management only get partially implemented, which means that ultimately they are never fully implemented. It is impossible to formulate a distinct culture if the business strategies of the firm are not implemented. Managerial Comment 9.3 elaborates on this point.

Managerial Comment 9.3
The Strategy–Culture Linkage

An effective culture is closely related to business strategy. Indeed, a culture cannot be crafted until an organization has first developed its business strategy. The first criterion for using culture as a leadership tool is that it must be strategically relevant.[63]

Most readers should be able to relate to the following scenario. Your organization has proposed a significant change in operations. To accomplish that change, a major shift in how employees think about their work will be necessary. If management expects this change to take place, then actions will follow. Management will follow up with announcements, actions, and implementations, causing the change to take place, and ultimately we will suppose the change is successful. It will follow that the employees involved will also have to change their mind-set about their work and support the change.

To show that this scenario is not just an abstract example, we will now suppose that management only announced that a change would take place; they intend to implement an improvement program to be more competitive and ensure the company's continued prosperity. Employees may be skeptical of the benefit of the change at this point. Some of the "old-timers" (more senior employees who do not like change in general and want things to stay the same because it makes their job easier) may voice their opposition to the proposed change. Suppose that a small part of the implementation process begins but then is halted because of a shortage of employees in a certain sector of the unit. Perhaps there is sickness, absenteeism, or vacations. This shortage stops the momentum of the change. Management hopes that in several weeks things will get back to "normal" and change efforts can be renewed. Then another demand develops that takes management off its focus on the improvement program. Perhaps it is preparation for a key inspection or audit, or maybe a series of critical reports are due. Whatever the reason, momentum shifts away from the changes, those changes are not implemented, and the culture stays the same. The old-timers comment that "we knew it wasn't really going to happen. Management always talks about making changes, but it never happens."

Most readers can probably remember when proposals were made by management for change (maybe it was you and your management team that proposed the change), but because the implementation was weak and incomplete, those changes never occurred. When this happens, the organizational culture does not have a chance to change either, and the status quo becomes even more engrained. To remedy this, management must mean what they say when changes are proposed. Make sure the strategy proposed is enacted. Otherwise, the organizational culture will expect to "hear" of changes that will never be implemented.

Change the Physical Cultural Artifacts

Physical cultural artifacts is a term that encompasses a number of areas, including the layout and décor of the offices, the design of the building, symbols and mascots, and almost any item that takes up space. Although this description may sound a bit facetious, it means to convey the degree of creativity that is available to management.

Several examples illustrate the power of changing artifacts. Consider the impact a face-lift in décor has on the motivation of personnel. Redesigning offices, changing colors, and providing new furniture are all visible indicators that a change is in the works. Of course, change for the sake of change does have its merits, but changing items in an office should also lead to a higher level of functionality and efficiency. The Crummer Graduate School of Business at Rollins College made a number of changes of these types, including providing breakout rooms for small group discussions (to enhance face-to-face interaction and creativity), upgrading the building for wireless Internet service, designing tiered classrooms, providing practice presentation rooms for student projects, and installing plasma monitors in group study rooms.[64]

Many organizations change their physical surroundings to blur the level of hierarchy that exists among management and hourly employees. One such change includes providing "rankless" parking lots, where all employees park on a first-come, best-place allotment, regardless of whether they are in management or the hourly ranks. A variation on this theme is to provide a common food-service area for all employees, leaving the executive dining room for special occasions only.

A physical artifact can also be a highly symbolic item, such as the old drill used at WorldNow to convey the depth that employees will go to meet customer needs. Physical artifacts can also be used during speeches. A favorite tactic by some speakers who crusade against excessive bureaucracy is to show physically the amount of paperwork required to perform certain tasks. President Reagan sometimes produced stacks of paper during his speeches to illustrate the endless requirements that the federal government required of its constituents. Other speakers have thrown papers and forms in the air and let them fall to the floor to show that such paperwork is not necessary. All kinds of props can be produced during a speech. The point is to use the prop (artifact) to convey some value that the speaker wishes to emphasize. Such displays during a speech can be very powerful in communicating a message to its listeners.

Tell Stories

One of the key areas to communicate cultural values is in the orientation and training of new employees. This is an excellent time to relate personal stories about former or existing organizational members who exemplify the qualities of an outstanding employee. Orientation sessions should cover a history of the company, including the personality characteristics of the founders. These types of stories often show how ordinary people overcame great obstacles to establish their company.

At the Crummer Graduate School of Business, students are told stories about previous students who achieved extraordinary goals. A sampling of this type of stories[65] is listed here:

- Students in Free Enterprise (SIFE) is an organization that pits student teams against each other in regional and national competitions. SIFE teams at the Crummer School have won several regional competitions and one national competition. Stories about SIFE are important because they perpetuate a culture of competition and excellence.
- Students are told stories about how consulting projects have resulted in profitable changes at the project companies. These types of stories are useful to stress a culture of problem solving.
- Students learn that international projects provide career offers for many graduates. Many colleges and universities try to promote a culture of international awareness. Stories like these help to motivate students to travel overseas, many for the first time.

Recognize Employees Formally

Reward systems should include recognizing employees formally for the type of achievements that are desired in the culture. This is particularly true when the desired behavior has not been reinforced previously through the existing reward system. When management says it wants "something" (e.g., a desired performance level) but does not recognize when that behavior has been achieved, employees will wonder if it is truly important.

Many colleges and universities have found themselves in this type of dilemma. The scenario goes something like this. Faculty spend most of their time teaching, yet are also held to one quantifiable goal—writing a certain number of research articles in a specified time period. The rationale for this goal is actually a good one: faculty who are inquisitive and want to research outside of the classroom will ultimately be better teachers. Research keeps professors from becoming "stale" in the classroom. Perhaps the reader remembers a class where it was obvious that the professor had not changed the lecture notes in quite some time. The result is usually a bored professor and a bored class. Conducting research helps alleviate this boredom and sharpens the mind of the professor, who, by the nature of the job, should always be looking for new things to teach.

A dynamic tension usually exists in the university setting. As illustrated in Table 9.1, many professors enjoy a combination of teaching and research. Some schools emphasized only teaching from their faculty but are now moving to requiring more research. This type of movement is an example of a major culture shift. Such a shift is difficult for faculty who are accustomed to teaching. To make the transition to research, the dean needs to emphasize continually the importance of research. These statements seek to address the values and beliefs held by the professors. To accompany these statements, formal recognition is needed for faculty members who publish research articles. Many schools recognize these professors each time they publish an article. Some who publish regularly receive research awards. Many schools set up a display case in the lobby and show the articles of their publishing faculty. The point is that recognition is made in a formal manner. Only when this occurs consistently will employees begin to understand that this level of behavior is actually desired by the organization.

Conclusion

In this chapter, we have looked at organizational culture as a key component in addressing change in the company. We have referred to it as the "elusive key to change" because of its subtle and hidden nature. Unlike technology and infrastructure, which are more externally oriented, organizational culture resides "beneath the surface" in the form of embedded values and beliefs held by organizational members. Fortunately, culture can change, although it is a slower process than changing technology and infrastructure. Implementing major improvement programs almost always requires a change in the organization culture.

The key to changing organizational culture is to address the values and beliefs of the employees first. Fundamental values, held dearly by many, may have to be changed or at least challenged. Usually, the source of this value change will come from top management, most likely the president or CEO. In addition to changing beliefs, more external cultural elements, called artifacts, need to be addressed. New slogans may need to be implemented, the physical structure may need to be improved, stories will have to be told of employees who achieve outstanding goals, and reward systems will need to acknowledge excellent performance by organizational members formally.

In short, organizational culture is an elusive key to change, but only if it is not addressed. When culture is considered part of the equation for making organizational change, the rest of the process has a better chance at being successful.

Notes

1. The particular state in this example will remain undisclosed because overall, the author liked living there. Also, the author wants you to know that he has also been to drivers' license agencies where the employees were very friendly—even though the lines were long.
2. Carrell, M., Jennings, D., and Heavrin, C., *Organizational Behavior*, Atomic Dog, Cincinnati, 2006.
3. Godsey, K., Slow climb to new heights, *Success*, October, 1996, p. 21.

4. Kreitner, R. and Kinicki, A., *Organizational Behavior* (7th ed.), McGraw-Hill Irwin, New York, 2007.

5. Smircich, L., Concepts of culture and organizational analysis, *Administrative Science Quarterly*, 28, 3, 339, 1983.

6. This phrase was made famous by Apple cofounder Steve Jobs, who was fond of describing the next generation of PCs that would emerge from the company.

7. See Gimein, M., Smart is not enough, *Fortune*, 143, 1, 124, 2001, for the answers to these questions and others.

8. Gimein, M., Smart is not enough, *Fortune*, 143, 1, 124, 2001.

9. Elgin, B., Hof, R., and Greene, J., Revenge of the nerds—again, *Business Week*, August 8, 2005, pp. 28–31.

10. Elgin, B., Hof, R., and Greene, J., Revenge of the nerds—again, *Business Week*, August 8, 2005, pp. 28–31.

11. Arnold, J., Customers as employees, *HR Magazine*, 52, 4, 76, 2007.

12. Cooney, 2003 Retailer of the Year, *License!* January, 20–30, 2004.

13. Kreitner, R. and Kinicki, A., *Organizational Behavior* (10th ed.), McGraw-Hill Irwin, New York, 2012.

14. On a much too trivial and personal note, between his junior and senior years at college William "Rick" Crandall traveled the small towns of southern Texas during the summer of 1977, peddling books door to door. The lessons the company taught are still appreciated to this day.

15. Boston Consulting Group Website. Retrieved 7/5/2007 from http://www.bcg.com/.

16. 100 Best Companies to Work for 2012. Retrieved 6/2/2013 from http://money.cnn.com/magazines/fortune/best-companies/2012/snapshots/22.html

17. Kreitner, R. and Kinicki, A., *Organizational Behavior* (7th ed.), McGraw-Hill Irwin, New York, 2007.

18. Excerpt from the Container Store website. Retrieved 6/12/2013 from http://standfor.containerstore.com/putting-our-employees-first/

19. 100 Best Companies to Work for 2007. Retrieved 7/6/2007 from http://money.cnn.com/magazines/fortune/bestcompanies/2007/snapshots/4.html

20. 100 Best Companies to Work for 2012. Retrieved 6/2/2013 from http://money.cnn.com/magazines/fortune/best-companies/2012/snapshots/22.html

21. Rawe, J., A homey cubicle helps a little, *Time*, 161, 21, 50, 2003.

22. Florida, R. and Goodnight, J., Managing for creativity, *Harvard Business Review*, 83, 7/8, 124, 2005.

23. Florida, R. and Goodnight, J., Managing for creativity, *Harvard Business Review*, 83, 7/8, 124, 2005.

24. Florida, R. and Goodnight, J., Managing for creativity, *Harvard Business Review*, 83, 7/8, 124, 2005.

25. Berry, L., The collaborative organization: Leadership lessons from the Mayo Clinic, *Organizational Dynamics*, August, 2004, pp. 228–241.

26. Levering, R., Moskowitz, M., Levenson, E., Mero, J., Tkaczyk, C., and Boyle, M., And the winners are…, *Fortune*, 153, 1, 63, 2006.

27. Levering, R., Moskowitz, M., Levenson, E., Mero, J., Tkaczyk, C., and Boyle, M., And the winners are…, *Fortune*, 153, 1, 63, 2006.

28. Chatman, J. and Eunyoung, C., Leading by leveraging culture, *California Management Review*, 45, 4, 20, 2003.
29. Spector, R. and McCarthy, P., *The Nordstrom Way*, John Wiley & Sons, New York, 1995.
30. Crow, S. and Hartman, S., Organizational culture: Its impact on employee relations and discipline in health care organizations, *Health Care Manager*, 21, 2, 22, 2002.
31. Crow, S. and Hartman, S., A prescription for the rogue doctor, *Clinical Orthopaedics and Related Research*, 411, 334, 2003.
32. Pepin, J., Burger Meister Ray Kroc, *Time*, 152, 23, 176, 1998.
33. McGarity, M., Happy at HomeBanc, *Mortgage Banking*, 170, 234, 26, 2005.
34. *Fortune Magazine*, 100 Best Places to Work for 2006, 153, 1, 71, 2006.
35. Bufe, B. and Murphy, L., How to keep them once you've got them, *Journal of Accountancy*, 198, 6, 57, 2004.
36. Triandis, H., *Culture and Social Behavior*, McGraw-Hill, New York, 1994.
37. Higgins, J., McAllaster, C., Certo, S., and Gilbert, J., Using cultural artifacts to change and perpetuate strategy, *Journal of Change Management*, 6, 4, 397, 2006.
38. Anonymous, 1978—Post-It Notes—3M, *Manufacturing Engineer*, 83, 5, 36, 2004.
39. Useem, J., Jim McNerney thinks he can turn 3M from a good company into a great one—With a little help from his former employer, General Electric, *Fortune*, 146, 3, 127, 2002.
40. Newstrom, J., *Organizational Behavior: Human Behavior at Work* (12th ed.), McGraw-Hill Irwin, New York, 2007.
41. Higgins, J., McAllaster, C., Certo, S., and Gilbert, J., Using cultural artifacts to change and perpetuate strategy, *Journal of Change Management*, 6, 4, 397, 2006.
42. Just a few of the many expressions the second author used, and heard, in his former career in the food service industry.
43. Canabou, C., Here's the DRILL, *Fast Company*, 43 (February), 58, 2001.
44. Higgins, J., McAllaster, C., Certo, S., and Gilbert, J., Using cultural artifacts to change and perpetuate strategy, *Journal of Change Management*, 6, 4, 397, 2006.
45. Carrell, M., Jennings, D., and Heavrin, C., *Organizational Behavior*, Atomic Dog, Cincinnati, 2006.
46. Anonymous, What to do about swearing in the workplace, *Leadership for the Front Lines*, 413 (November 1), 1, 2001.
47. Wah, L., Profanity in the workplace, *Management Review*, 88, 6, 8, 1999.
48. Excerpt from Oakley company website. Retrieved 6/2/2013 from http://www.oakley.com/innovation/design. The reader is also encouraged to view pictures of the company's corporate headquarters, also available on the website.
49. Elgin, B., Hof, R., and Greene, J., Revenge of the nerds—again, *Business Week*, August 8, 2005, pp. 28–31.
50. Kerr, J. and Slocum, J., Managing corporate culture through reward systems, *Academy of Management Executive*, 19, 4, 130, 2005.
51. Carlson, L., Companies chip in for workers' hybrid cars, *Employee Benefit News*, 19, 12, 73, 2005.
52. *Fortune Magazine*, 100 Best Places to Work for 2006, 153, 1, 71, 2006.
53. The author did not fill out the survey but used it as a classroom example on how not to collect this type of data.

54. Spitzer, D., Power rewards: Rewards that really motivate, *Management Review*, 85, 5, 45, 1996.
55. The original framework was developed by Quinn, R. and Rohrbaugh, J., A competing values approach to organizational effectiveness, *Public Productivity Review*, 5, 122, 1981. We also draw from the later work of Deshpandé, R., Farley, J., and Webster, F., Jr., Corporate culture, customer orientation, and innovativeness in Japanese firms: A quadrant analysis, *Journal of Marketing*, 57, 1, 23, 1993; and Petrock, F., A quadrant of organizational cultures, *Management Decision*, 34, 5, 37, 1996.
56. Quinn, R. and Rohrbaugh, J., A competing values approach to organizational effectiveness, *Public Productivity Review*, 5, 122, 1981.
57. Deshpandé, R., Farley, J., and Webster, F., Jr., Corporate culture, customer orientation, and innovativeness in Japanese firms: A quadrant analysis, *Journal of Marketing*, 57, 1, 23, 1993.
58. William "Rick" Crandall is Professor of Management within the School of Business at the University of North Carolina at Pembroke. He is very proud of the value the university promotes, "where learning gets personal." The phrase is not just rhetoric; the university truly lives up to this value.
59. Egan, M., Infusing ethics into your organization, *Insurance Advocate*, 117, 25, 14, 2006.
60. The discussion that follows is based on Robbins, S. and Coulter, M., *Management* (9th ed.), Pearson Prentice-Hall, Upper Saddle River, NJ, 2007, p. 69.
61. Rotta, C., Rules of behavior, *Internal Auditor*, 64, 3, 33, 2007.
62. Post, J., Lawrence, A., and Weber, J., *Business and Society: Corporate Strategy, Public Policy, Ethics*, McGraw-Hill Irwin, New York, 2002.
63. Chatman, J. and Eunyoung, D., Leading by leveraging culture, *California Management Review*, 45, 4, 20, 2003. Quote is from p. 21.
64. Higgins, J., McAllaster, C., Certo, S., and Gilbert, J., Using cultural artifacts to change and perpetuate strategy, *Journal of Change Management*, 6, 4, 397, 2006.
65. Higgins, J., McAllaster, C., Certo, S., and Gilbert, J., Using cultural artifacts to change and perpetuate strategy, *Journal of Change Management*, 6, 4, 397, 2006.

Chapter 10

Integrated Supply Chains—From Dream to Reality

Introduction

Supply chain management (SCM) is an important concept. Increased competition is forcing businesses to consider how best to gain a competitive advantage or at least to hold their own against global competitors. One way is to work with selected customers and suppliers to build lean and agile supply chains.

Present Status of Supply Chains

The benefits of successful supply chains are direct and measurable—lower product costs, lower inventories, higher quality, faster response times, fewer stockouts, and higher on-time deliveries. Other benefits that are more difficult to measure and verify include increased market share, improved customer satisfaction, increased customer retention, and a more competitive position overall. In speaking of global strategies, Edward Davis and Robert Spekman suggest that "the extended enterprise is really about creating a defensible long-term competitive position through strong supply chain integration, collaborative behaviors, and the deployment of enabling information technology."[1]

If the benefits are so attractive, why are so few companies building supply chains that satisfy not only the critics but also the companies who want to participate in a successful supply chain? An annual survey conducted jointly since 2003 by Computer Sciences Corporation (CSC) and *Supply Chain Management Review* acknowledges progress but observes that the majority of companies are still struggling to implement true partner collaboration practices.[2,3] Davis and Spekman summarize the situation as follows: "We advocate close ties among supply chain partners. Our observation is that most firms are not even close to developing the requisite mind set; they lack the skills and competencies needed and cannot implement the processes that lie at the heart of the extended enterprise."[4]

Another study evaluated progress toward lean supply chains and found that the majority of companies were still more internally oriented than externally oriented on supply chain matters. The researchers defined the lean supply chain as "a set of organizations directly linked by upstream and downstream flows of products, services, finances, and information that collaboratively work to reduce cost and waste by efficiently and effectively pulling what is needed to meet the needs of the individual customer."[5] Tables 10.1 and 10.2 summarize the results of this study.

Why is it so difficult to build lean and agile supply chains? In Chapter 3, we described the various obstacles in trying to link companies with their customers and suppliers in a meaningful way. In this chapter, we describe in more detail the steps required and why it is difficult. We hope to encourage businesses to realize the opportunities that a lean and agile supply chain can provide.

Before we begin our discussion, we would like to call your attention to one of the most comprehensive descriptions of how a company can build a supply chain. In his book, *Supply Chain Architecture: A Blueprint for Networking the Flow of Material, Information, and Cash*, practitioner William T. Walker, CFPIM, CIRM, CSCP, calls on his thirty-plus years of experience with Hewlett-Packard and its spinoff, Agilent Technologies, to provide his readers with a detailed description of the individual components and how they fit together to form a world-class supply chain.[6]

Background of Supply Chains

There have always been supply chains. The early traders bringing silks and incense from the Far East to Europe are some of the best known. The ships that sailed from the colonies to England represented another supply chain. It has only been in recent years that the supply chain has become a popular topic in the management literature. The concept of supply chains has evolved from several different research areas. First, the term *supply chain* suggests a movement of goods or services from an origin to a destination. Second, the supply chain builds on research from facets of organization theory. Finally, systems theory has also contributed to our understanding of supply chains.

Material Flows

Figure 10.1 shows the movement of materials from suppliers to customers along the horizontal axis. The concept of integrating entities is illustrated along the vertical

Table 10.1 Status of Lean Supply Chain Implementation

Survey Variable	Companies Reporting in Each Area (%)				
	Internally Focused	*Little SC Focus*	*Some SC Focus*	*Significant SC Focus*	*Supply Chain Focused*
Managing the demand signal	14	39	14	19	15
State of demand collaboration	13	34	30	19	5
Waste reduction effort	21	31	29	16	4
Value-added activities	13	28	34	22	4
Planning and production process standardization	7	33	41	13	6
Company product standard	12	30	19	28	12
Industry product standard	17	14	43	14	13
Data standard	21	42	19	15	3
People (expendable or valuable asset?)	9	22	27	26	15
Continuous improvement/change culture	3	20	32	27	17
View of team	10	46	19	21	4
Average (%)	13	31	28	20	9

Source: Adapted from Manrodt, K.B., Abott, J., and Vitasek, K., *Understanding the Lean Supply Chain: Beginning the Journey,* 2005 Report on Lean Practices in the Supply Chain, APICS.

axis. Beginning at the bottom, we show separate functional areas that move product in discrete steps. As companies progress in an upward direction, they begin to link internal functions together and, even more progressively, link with their customers and suppliers. At the most advanced stage (the top level), they exist within an integrated supply chain.

Table 10.2 Progress in Developing Lean Supply Chains

Status of Program to Develop a Lean Supply Chain	%	Obstacles to Developing a Lean Supply Chain	%
Just beginning	12.7	Lack of resources	27.3
Some sporadic implementation	47.3	Lack of training	25.7
Formal integrated approach ready	25.3	Lack of top management commitment	21.2
Entire product line has flow	9.0	Other activities (Six Sigma, TQM)	11.0
Lean is a standard procedure	5.7	Hard to apply in my industry	8.5
		No clear benefit of lean	6.3

Source: Adapted from Manrodt, K.B., Abott, J., and Vitasek, K., *Understanding the Lean Supply Chain: Beginning the Journey*, 2005 Report on Lean Practices in the Supply Chain, APICS.

Organization Hierarchy

Supply chains have also evolved as an organizational concept. Historically, organizations have tended toward a vertical hierarchy, with levels in the organization indicating levels of authority and responsibilities. These organizations also tended toward specialization, or the separation of functional areas into departments. The division of labor concept to increase productivity was first applied in production operations and later extended to staff functions, no doubt helped along by Taylor's "functional foreman" recommendations.[7] Because of this specialization, functions within the organization tended to be somewhat isolated from one another. The current term, *functional silos*, describes this condition. As a result, it is more difficult for information to flow horizontally from department to department.

To remedy this, companies have moved to cross-functional teams as a means of stimulating a freer flow of information and faster resolution of problems. This move has led to closer collaboration of functions within a company and to extending collaborative efforts along the supply chain to customers and suppliers. The result has the effect of reducing the rigidness of vertical organizations and facilitating the horizontal flow of information and funds, much as in the flow of goods and services.

Systems Theory

Systems theory has also played a role in the evolution of the supply chain. A system includes individual entities that link together in a progressive way to achieve a final result. It may be a system to circulate blood in a human being or a supply chain that moves products and services from suppliers to customers. We described this linking

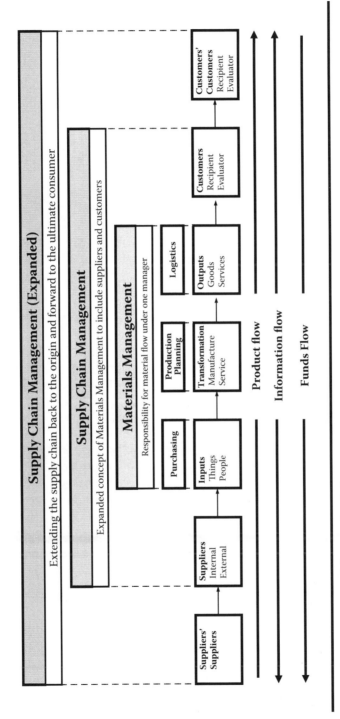

Figure 10.1 Evolution of the supply chain.

in Chapter 3 as the extension of the Input–Transformation–Output (ITO) Model. Systems theory is especially useful in describing the role of information technology (IT) in supply chains. Without the advances in IT, especially in the area of inter-organizational systems (IOS), the modern supply chain would not be possible.

Role of the Supply Chain

Modern supply chains perform three primary functions. They deliver products and services to customers, they collect and assimilate information about customer wants and needs, and they facilitate the flow and equitable distribution of money among the various participants in the supply chain.

In this chapter we assemble the management modules described in Chapter 2 through Chapter 9 into a strategically planned and integrated supply chain. We show how continuous improvement programs fit into this progression from a traditional batch flow operation to a lean and agile organization of the future. First, we explore the implications of moving to an agile and integrated supply chain.

Setting the Stage

Is your supply chain all you expected it to be? On the other hand, are you, like many other managers, finding that the reality is far different from the dream? What did you expect? The *APICS Dictionary* offers the following definitions:

- Supply chain: The global network used to deliver products and services from raw materials to end customers through an engineered flow of information, physical distribution, and cash.
- Supply chain management (SCM): The design, planning, execution, control, and monitoring of supply chain activities with the objective of creating net value, building a competitive infrastructure, leveraging worldwide logistics, synchronizing supply with demand, and measuring performance globally.[8]

Apparently, supply chain management is more than finding a few good vendors who will deliver the right product at the right time in the right quantity and with good quality.

What about demand? Is the customer part of the supply chain? In case there is any doubt, another definition clears up that point. The lean supply chain is "a set of organizations directly linked by upstream and downstream flows of products, services, finances, and information that collaboratively work to reduce cost and waste by efficiently and effectively pulling what is needed to meet the needs of the individual customer."[9]

Does the supply chain have to be lean, too? Yes, and according to some authorities, the supply chain should be agile, adaptable, and aligned.[10] It should also be integrated, innovative, globally optimized, involve inter-organizational collaborative relationships, and share the benefits equitably among the supply chain members.[11] Herbert Kotzab and Andreas Otto describe SCM as "a strategic, co-operation oriented, business process management concept, cutting across organizational boundaries (i.e., integration oriented), which leads to increased results for all supply chain members. SCM demands integration and co-operation beyond the logistics dimension. SCM processes are customer and/or end-user driven."[12]

Does your competitor have a dream supply chain? Probably not. That is, unless your competitor is among the handful of companies that is continuously cited in the literature as having one of the best supply chains around (Apple, Proctor & Gamble, Walmart, and Dell, to name a few). The reality of supply chains is quite different from the dream. Here is what a few researchers were reporting in the early part of the twenty-first century:

- SCM can deliver powerful results—reducing costs, boosting revenues, and increasing customer satisfaction and brand equity by improving on-time delivery and product or service quality. Yet overwhelmingly, senior executives at large companies worldwide believe SCM has failed to live up to the promise during its first two decades.[13]
- Approximately 75 percent of the firms in the United States and the United Kingdom are small and medium-sized enterprises. The authors have concerns that these firms have "grave" problems in their supply chains, because of both legacy systems (hardware and software) and legacy mental models (how they manage their businesses).[14]
- In a comprehensive study of over 600 companies, in which the researchers used eleven variables to assess the status of supply chain implementation, less than one-third of the participants indicated that their efforts had "significant" or full supply chain focus. The remaining two-thirds indicated their program was internally focused or had little supply chain focus.[15]
- In another global study, Computer Sciences Corporation used a five-level scale to place companies' progress toward "full network connectivity," or level 5. Levels 1 and 2 were in the early stages of supply chain evolution. They found that approximately 50 percent of the companies were at levels 1 or 2, with 35 percent at level 3, leaving only 15 percent at levels 4 or 5.[16]
- Supply chains of the future will include reverse logistics, or green supply chains. Businesses will design processes to manage returns, repairs, remanufacturing and redistribution of resalable products, and recycling of disposable materials. This trend is already occurring but will grow in importance.[17]

More recent surveys have tended to focus on the recovery of supply chains from the economic downturn in the United States during the 2008–2012 period. One survey reports that companies see supply chain challenges from increasing pressure from global competition, rising risks of supply chain disruptions, lack of collaboration, and low CEO involvement.[17a]

Another survey reported that visibility and analytics characterized the leaders in building and managing supply chains. Visibility includes having access to customers' sales information, promotional plans, and demand forecasts and to their suppliers' inventory, order lead times, and delivery dates. However, businesses are still not satisfied with the quality of the data they have visibility into.[17b]

An APICS survey found that more companies are beginning to think of supply chains as part of their corporate strategy. While the concept is still new, more than 60 percent of respondents reported that their organizations have adopted their supply chain strategies in the last five years or so.[17c]

IDC reports the following issues for supply chains: complex and extended global supply networks, volatile demand as the new norm, growing regulation, accelerated pace of business, and the rise of the customer.[17d]

There is much more. For example, most supply chains focused on cost reduction, even though some studies suggest that improved customer service and increased revenues will have an even bigger payoff.

Does the gap between reality and dream mean that supply chains are not a good idea? Of course not! It just means that building a really good supply chain is not easy and takes time. What makes it so difficult?

- Supply chains are complex. How many suppliers do you have? Are they all alike? How many products do you have? Are they all alike? How many customers do you have? Are they all alike? How many combinations of supplier/product/customer do you have? Are they all alike? Even in a small company, the complexity is daunting; in a large company, it is overwhelming. Deloitte Touche Tohmatsu discuss the complexity paradoxes of optimization, customer collaboration, innovation, flexibility, and risk.[18]
- If the inherent complexity described above were not enough, most companies are making the relationships even more complex by outsourcing both goods and services. Outsourcing is here to stay, and assimilating it into everyday practice is a challenge for even the most experienced global participant. Thomas Friedman calls outsourcing one of the ten "flatteners" in his best-selling book *The World Is Flat* (a worthwhile read).[19] To make the situation even more subject to change, the current interest is "insourcing" or bringing the outsourced products back to the home country.
- Competition is increasing; this means that every business has to improve itself continually. The supply chain that was innovative just a few years ago is barely adequate today. The order winners of yesterday are only order qualifiers today.
- Variability is rampant along the supply chain. Not only are the relationships complex, they are constantly changing, and change breeds variability. It is difficult to build interpersonal relationships among individuals, but how do you build inter-organizational relationships among companies? It is difficult and takes time, longer than most of us have the patience to devote.

- Another major element of change is the introduction of e-commerce in the equation. That is another whole subject, but the impact on supply chains is indisputable. You probably cannot utilize the identical supply chain for your e-commerce efforts that you use for your regular products.
- Supply chains, as a management concept, are getting more comprehensive. What started as a transactional process is rapidly becoming a strategic process. What started in purchasing or logistics is becoming a cross-functional, horizontal flow process that includes all the functions of a business as well as top management. Concepts like sales and operations planning are definitely in vogue, after years of languishing in the shadows.

Is there any hope to improving supply chain successes? Are businesses doomed to be forever mired in the early stages of supply chain development? The results of surveys indicate slow progress, not lack of positive results. However, the gap is getting wider between those who are getting it right and those who are struggling.[20] Debating whether to have an effective and efficient supply chain is no longer an option; it is an imperative!

Several sources provide direction and encouragement. The consulting group PRTM describes the transformation required to move from a functionally focused supply chain to cross-enterprise collaboration: (1) functional focus, (2) internal integration, (3) external integration, and (4) cross-enterprise collaboration.[21] Cigolini and colleagues offer a similar progression from traditional logistics, with a focus on reducing inventory; to modern logistics, with a shift from cost reduction to improving customer service; to integrated process redesign, which provides a systemic vision of the supply chain; to an industrial organization, which focuses on the strategic alliances among the various members of the same supply chain.[22] On a similar note, Togar Simatupang and colleagues describe the "recursive interplay" of logistics synchronization, information sharing, incentive alignment, and collective learning with an integrated supply chain. There are plenty of answers; it is just that none of them is easy.[23]

Some see a brighter future if managers want to work for it. As Kempainen and Vepsalainen emphasized, "information sharing and coordination are often considered the preconditions for successful supply chains. Our analysis of supply chain practices in industrial supply chains shows that visibility is still limited. The companies have a realistic view on the advantages and risks of information sharing, and so planning information is shared only selectively. Coordination efforts focus on inter-functional operations and relations with selected partners. The companies hesitate to expand their coordination efforts beyond order processing and operational scheduling within the dyadic supplier–buyer relationships."[24] Vijay Kannan and Keah Tan found, in a survey of APICS members, that information sharing was not among the top criteria used by companies in selecting suppliers, yet information sharing was one of the major positive influences on firm performance, in such areas as market share, return on assets, product quality, and competitive position. They suggest increased attention to effective information sharing.[25]

In the interest of promoting improved coordination, Chee Wong and colleagues describe the advantages of loosely coupled supply chains. They categorize tightly coupled supply chains where partners closely connect in a one-to-one relationship, such as through traditional EDI. In contrast, "a loosely coupled supply chain is an extensively integrated but loosely connected supply chain network. These concepts have emerged from the Web services, Web portal and Extended Mark-up Language (XML) technologies." They suggest that a loosely coupled supply chain has a specific niche for short- and medium-term buyer-seller relationships, small and medium-size enterprises, secondary partners, and a large number of partners.[26]

Chuck Poirier and Frank Quinn also offer cautious optimism. In reporting the results of a survey, they comment, "The findings further suggest that those running the business today continue to think of supply chain management as a short-term cost-reduction effort, not worthy of strategic importance. Only 37 percent of the companies indicated that their supply chain initiatives were 'mostly' or 'fully' aligned with corporate strategy. The survey results indicate that collaboration across an extended enterprise is still more theory than practice. Collaboration was the single most pressing need—both internal collaboration and external collaboration with suppliers and customers. Perhaps the most important insight from our survey is that the real business benefit of advanced supply chain management remains largely untapped."[27]

Consider this scenario. Companies report that developing reliable demand forecasts is one of their major challenges. To make better forecasts, a company needs more information from downstream members of the supply chain and the company needs to collaborate with downstream customers and upstream suppliers, to develop a mutually beneficial demand forecast. To obtain better information, partners need effective inter-organizational communications. Improved communications involves more than technology. Not only do you need to have a better way to communicate, such as through electronic IOS, but you need to agree on what information to communicate. Here it gets tricky. Collaboration requires sharing information. What information are you willing to share, and what information are the other members of the supply chain willing to share with you? This is where trust enters the picture, and trust has been largely lacking in most supply chains. Most of those who have studied trust—and the list goes well beyond business researchers to include those in organizational science, industrial psychology, economics, and operations research—have found it a difficult ingredient to integrate into working relationships.[28] Much of the research boils down to the fact that to trust someone or some entity, you have to work with them to verify that the benefits of trust outweigh the risks. At the same time that you are evaluating those parties in the supply chain, they are evaluating you. At some point, you have to make a judgment decision to trigger the upward spiral from trust to collaboration, to better demand forecasts, to improved execution, to increased profits along the supply chain.

So there it is. Communicate, collaborate, trust, and share the benefits equitably with the members of your supply chain; everything will be all right. Just as long as you do the hundreds of other things necessary to make your multiple supply chains become the reality, not the dream.[29]

At the same time, care must be taken that investments in supply chains are monitored closely and overspending does not occur. Suvankar Ghosh and colleagues warn that some companies have invested two to three times the level that would be desirable from a stakeholder perspective.[30]

Supply Chain Models

A number of researchers have proposed models or narratives to describe the evolution to an effective supply chain. We describe several of them and introduce the model that we believe extends the existing models to a more complete state.

Evolution of Supply Chain Models

One approach is to classify SCM into different schools of research. In a comprehensive review of the literature, Cigolini, Cozzi, and Perona developed the following classifications that indicate the progression of supply chain thinking over the past twenty years:

- **Supply chain awareness:** Recognize that there is a continuous chain of functional areas through which materials flow and that extends from suppliers to final distributors.
- **Traditional logistics:** Main objective is to improve supply chain efficiency by reducing inventory levels, whereas little emphasis is given to supply chain effectiveness.
- **Modern logistics:** The focus shifts from mere cost reduction to include also service and quality improvement.
- **Integrated process redesign:** Through quantitative models applied to a systemic vision of the supply chain, studies show how to redesign the entire supply system to obtain more efficient and effective flows of materials and information.
- **Industrial organization:** Focuses on the strategic alliances between the various actors of the same supply chain.[31]

The following eight supply chain management processes are included in the Global Supply Chain Forum (GSCF) framework:

1. Customer relationship management provides the structure for how relationships with customers are developed and maintained.
2. Customer service management provides the firm's face to the customer, a single source of customer information, and the key point of contact for administering the product service agreements.

3. Demand management provides the structure for balancing the customers' requirements with supply chain capabilities, including reducing demand variability and increasing supply chain flexibility.

4. Order fulfillment includes all activities necessary to define customer requirements, design a network, and enable the firm to meet customer requests while minimizing the total delivered costs.

5. Manufacturing flow management includes all activities necessary to obtain, implement, and manage manufacturing flexibility and move products through the phases of the supply chain.

6. Supplier relationship management provides the structure for how relationships with suppliers are developed and maintained.

7. Product development and commercialization provides the structure for developing and bringing to market new products jointly with customers and suppliers.

8. Returns management includes all activities related to returns, reverse logistics, gatekeeping, and avoidance.[32]

The objectives of the Supply Chain Council (SCOR) framework are:

■ **Plan:** Balances aggregate demand and supply to develop a course of action that best meets sourcing, production, and delivery requirements.

■ **Source:** Includes activities related to procuring goods and services to meet planned and actual demand.

■ **Make:** Includes activities related to transforming products into a finished state to meet planned or actual demand.

■ **Deliver:** Provides finished goods and services to meet planned or actual demand, typically including order management, transportation management, and distribution management.

■ **Return:** Deals with returning or receiving returned products for any reason and extends into post-delivery customer support.[33]

Four conceptual perspectives on purchasing versus SCM have been proposed:

1. **Traditionalist:** SCM is a strategic aspect of purchasing, with emphasis on supplier development, and partnerships with first- and second-tier suppliers.

2. **Relabeling:** Simply change the name of purchasing to SCM. The scope of SCM equals purchasing. Other fields have also been relabeled as SCM, such as logistics.

3. **Unionist:** Purchasing is a part of SCM, as SCM completely subsumes purchasing, as well as logistics, marketing, and operations management.

4. **Intersectionist:** SCM is not the union of logistics, operations, and purchasing. Rather, it includes elements from all of these disciplines as well as marketing, organizational behavior, and strategic management.[34]

The PRTM Group describes the transformation from a functionally focused supply chain to a cross-enterprise collaboration arrangement. The stages are:

- **Stage 1: Functional focus.** Functional departments within an organization focus on improving their own process steps and use of resources. Managers typically focus on their individual department's costs and functional performance. Processes that span across multiple functions or divisions are not well understood, resulting in limited effectiveness of complex supply chain processes.
- **Stage 2: Internal integration.** Division- or companywide processes are now defined, allowing individual functions to understand their roles in complex supply chain processes. Cross-functional performance measures are clearly defined, and individual functions are held accountable for their contributions to overall operational performance. A well-defined demand/supply balancing process that combines forecasting and planning with sourcing and manufacturing is evident at this stage.
- **Stage 3: External integration.** Stage 2 practices are now extended to the points of interface with customers and suppliers. The company has identified strategic customers and suppliers, as well as the key information it needs from them to support its business processes. Joint service agreements and scorecard practices are used, and corrective actions are taken when performance falls below expectations.
- **Stage 4: Cross-enterprise collaboration.** Customers and suppliers work to define a mutually beneficial strategy and set real-time performance targets. Information technology now automates the integration of the business processes across these enterprises in support of an explicit supply chain strategy.[35]

An alternative approach is to describe the evolution of a supply chain along the following steps.

1. Enterprise integration (functional/process)
2. Corporate excellence (intra-enterprise)
3. Partner collaboration (inter-enterprise)
4. Value chain collaboration (external)
5. Full network connectivity (total business system).[36]

A Comprehensive Supply Chain Model

We have developed a model to describe the journey from batch flow to an integrated, agile supply chain (Figure 10.2). The general structure of the model is as follows: the horizontal axis, reading from left to right, shows the progression over time from batch flow to an integrated and agile supply chain. Although the progression

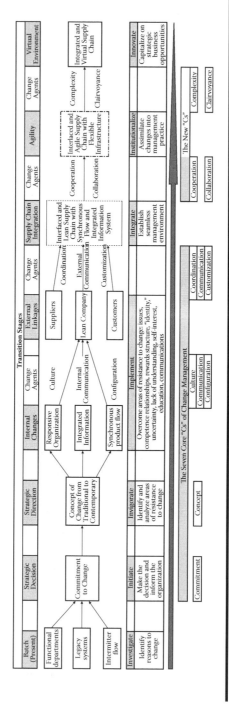

Figure 10.2 Transition from batch flow to an integrated and agile supply chain.

occurs over time, the axis does not represent a specific time scale. The journey varies for each supply chain but generally the time is in years, not months or weeks.

The vertical axis is composed of three major levels. The top level shows the stages that a company goes through: (1) present batch flow, (2) make the strategic decision to change, (3) develop a strategic direction for their program, (4) effect the internal changes needed, (5) establish linkages with key external entities, (6) attain an effective level of supply chain integration, (7) enhance the supply chain with agility, and (8) enter the virtual environment state. The change agents needed to move from stage to stage are interspersed between the stages.

The bottom area of the model shows the key change components, or "Cs," needed to make the change successful. In addition to the Cs, the diagram shows the initial starting conditions of functional departments, legacy systems, and intermittent process flow. Further to the right, it shows the changes to a responsive organization, integrated information flow and synchronous product flow. Even further to the right, it shows the beginnings of a supply chain and subsequent progression to integrated and agile supply chains.

At the middle of the diagram, we show the action steps necessary, or the "Is" of the program. These are the activities necessary to bring about the Cs described above.

As an overview, the left third of the diagram may be viewed as "starting the journey," the middle portion may be seen as "achieving internal synchronization," and the right third as "building the supply chain."

Figure 10.2 illustrates several salient points about the development of effective supply chains:

- Developing a supply chain is an evolutionary process. The movement from batch to a lean enterprise precedes the progression to a supply chain, then to an integrated supply chain, and, finally, to a lean and agile supply chain.
- The progression along the model shows movement from discrete, even fragmented, functions of an entity to the integration of multiple entities into a seamless vibrant entity.
- The movement to integration requires the blending of interdisciplinary functions within a company and between companies.
- As the evolution progresses, there is a shift in emphasis from transactions to strategies. Although attention to detail persists throughout, there is also a need to develop a holistic perspective of the program.
- Just as there is a need for interdisciplinary activity, there is a requirement for interactive movement from connectivity to coordination to cooperation to collaboration.

Each element of the model is explained more fully in the following sections with examples from reported case histories. First, we look at why many authorities advocate the shift from batch flow to lean production.

The World of Lean Production

In Chapter 5, we used lean production as an example of one of the major improvement programs that has become popular in the past few years. Although it arose in the manufacturing sector, its use is now advocated in service operations. Lean thinking has largely replaced Just-in-Time (JIT) in the number of articles written. The terms *lean production, lean manufacturing*, and *lean thinking* have become almost synonymous in the literature. The key values are overviewed below:

- **Create value for the customer.** Lean production creates value for the customer by reducing the amount of waste within companies and along the supply chain.
- **Create a smooth flow of goods.** One of lean production's objectives is to smooth the flow of goods along the supply chain by eliminating problems and other delays that interrupt flow.
- **Provide a fast response.** By eliminating waste and creating a smooth flow, lean production reduces the response times in product and service deliveries.
- **Detect quality problems quickly.** Lean production continues an emphasis on maintaining quality of products and services. The need for fast, even flow makes it possible to detect quality problems sooner.
- **Develop cross-trained employees.** Cross-trained employees are necessary to maintain the even flows and high quality required by lean production.
- **Illustrate a contemporary outlook.** Lean production reflects some of the latest thinking in operations management and fits nicely with the supply chain concepts.

Although lean production provides a number of benefits, it is not without some limitations or constraints in its implementation:

- **It takes time to implement lean production.** Companies that implement lean production should not view it as a quick fix. It takes time to implement it effectively. We will describe some of these steps in the following section.
- **Lean production requires significant change.** To implement a complete version of lean production, a company will probably have to make substantial changes in the way it operates. Changing the physical layout of a plant and moving from job specialization to job enlargement and enrichment are two areas that involve the greatest adjustment.
- **It is difficult to achieve and maintain.** The road to lean production is long and difficult. For those companies that achieve a level of competency, consistent effort is needed to maintain that level of competency. Just as it is difficult for world-class athletes to maintain their physical conditioning, it is difficult for companies to stay lean. To continue with that metaphor, they can become "lax and fat."

■ **It is difficult to extend throughout the supply chain.** Attaining a lean condition in one company is difficult; achieving that condition throughout the supply chain is even more difficult because there may be trade-offs that benefit one member of the supply chain while adversely affecting another. For example, reducing inventory at the retailer level may impose greater inventory levels upstream in the supply chain.

The potential benefits of a lean manufacturing world sound wonderful. So how does a company get there? To oversimplify, a company needs to reduce variation in its products or services, its purchasing process, its production process, and in its delivery process. These steps relate to the internal management of a business.

Product

Many companies attempt to increase revenues by expanding the variety of products or services they offer. Although this approach may generate sales in the short term or until a competitor offers the same products or services, it also increases complexity and variation in the product lines. This complexity inhibits the ability to go lean. To resolve this dilemma, a company needs to develop product or service modules that can be configured to suit the variety required. An example is kitchen cabinets that can be customized by assembling modules to fit the available space. A service example is a will prepared by a lawyer who adapts modules of features to prepare a customized will for specific clients.

Purchasing Process

The traditional purchasing process has involved placing a variety of potential suppliers in a competitive bidding situation to obtain the lowest cost. This introduced variation in the suppliers used, quality levels, response times, and confidence levels in the suppliers. To reduce these variances, companies need to develop long-term relationships with carefully selected suppliers. Contemporary lean purchasing practices dwell on the total cost of ownership and consider both tangible and intangible factors in their decision making.

Production Process

In production, variations can take the form of uneven work flow, smaller lot sizes, extra setups and changeovers, employee work assignments, more rush orders, and a host of other problems that make the production of goods and services a demanding task. To reduce the adverse effects of variation, the production process must become more flexible with shorter setup times, more versatile employees, and general-purpose equipment—to name a few. For example, automobile manufacturers are designing new plants, or redesigning old plants, capable of producing multiple models of cars on the same assembly lines. In addition, the planning horizon must be extended, with increased feedback from downstream customers, so that there is an increased opportunity to cope with the variation.

Delivery Process

Increased variation means more frequent but smaller deliveries. The increased frequency of deliveries has increased the scheduling difficulties, especially in large distribution centers. It is no longer possible for suppliers to deliver at their convenience; they must deliver at the time specified by the customer. If the supplier is early, they wait. If they are late, they pay both financial penalties and perhaps face the loss of future business. The answer is a focused approach to their delivery process that is dedicated to eliminating potential problems before they occur and in designing an integrated delivery system that considers customer needs, delivery equipment, employees, external factors, and potential acts of nature, such as hurricanes or snowstorms. Consequently, some manufacturers are turning to third-party logistics providers (3PL) to handle the design and operation of their transportation functions.

Demand Variation

Companies also need to reduce the variation in demand. There may be greater variation in demand than in any of the internal areas described above. Reducing demand variation may sound counterintuitive at first. Who wants to tell the customer when to place an order or request a service? However, there may be situations where customers benefit if they change their ordering habits to obtain faster service or avoid long waits. Collaboration with customers can sometimes pay dividends in reducing demand variation.

How are companies doing in their quest for a lean supply chain? Tables 10.1 and 10.2, introduced earlier, show results from a survey of several hundred companies.

Why is it so difficult to achieve the lean status? The literature suggests a gap in understanding the transition and change process. Managers understand the need, but they do not always understand how to make the necessary changes. Even if they understand, it is difficult to anticipate the problems and challenges in making the implementation. They need an integrative model of the common drivers (barriers and facilitators) of the change process. In the following section, we describe why achieving a lean and integrated supply chain is difficult.

Steps to Achieve a Lean and Agile Supply Chain

It is easy to be enthusiastic about the possibilities of change, but it is difficult to reap the benefits of those changes. We believe there are several important phases a company must go through to achieve success.

Commitment

The first requirement is a commitment to change. A decision to change should come only after a careful investigation reveals the necessity to change. A commitment to change requires top management support. Although this support does not necessarily require day-to-day participation in the details of the program, it does mean that top executives support the program and understand its importance in the company's future. Management should provide adequate funding for the project and commit to future allocations as needed. Top management should remember its role as the drivers of change in the structure of the organization and its practices. They monitor the progress of the program and recognize the achievement of milestones in the program's progress.

As part of the commitment to change, the entire organization should recognize and accept a role in the program. This realization means that employees must understand why the program is important and what their role is in its implementation. They should have a role in the design of the program to ensure their "buy-in" as a stakeholder in the process.

The commitment phase should involve enough of a plan to make sure that the initial presentations about the objectives, scope, and potential consequences of the program are accurately and objectively portrayed to employees at all levels of the organization. The intentions of the company should also be shared with other key stakeholders, whether it is customers, suppliers, financial institutions, or major investors. Planning should mean that the company is thinking several steps ahead and not just responding to random events.

Researchers have found that top management support is important, as evidenced by the following comments:

> Supply chain management continues to fall short of its great promise, according to a new assessment. Surveying nearly 200 senior executives from manufacturing and industrial companies in North America, Europe, Asia, and Latin America, a consulting firm cites reasons for the discipline's failure, including (1) top management's narrow view of SCM and superficial involvement in the design and guidance of the function, and (2) an unrealistic view of what technology can achieve.[37]

> Supply chain management requires strong top management leadership. Our survey found that in companies where responsibility for SCM resides below senior management, annual savings in the cost to serve customers are just 55 percent of what they are when SCM is a component of the overall business strategy.[38]

Commitment includes recognition that changes will be required in technology, infrastructure, and culture, as described in Chapters 7, 8, and 9.

Once a company has made the commitment to change, it has to act. However, action without a plan is premature. At this point, planning should focus on developing the concept that will carry the program through to a successful conclusion.

Concept

The next step is to develop a concept of the program—an understanding of what to do, what to accomplish, who will be responsible, and when it will be completed. The concept should be a "big picture" view of the program that provides a direction and scope for those who will participate in the design and implementation process.

Now is the time to involve the entire organization. This type of program needs the participation of the management team and nonmanagerial employees. Program managers should anticipate where they will find enthusiastic acceptance of the change plan and where they may encounter some resistance. Although it is desirable to gain acceptance throughout the organization, it is essential to have support during the initial change stages.

It is also important to adapt the program to fit the company. Although there are many elements of the program that may be generic to most companies, there are unique characteristics that will be positive factors if correctly incorporated into the improvement program. On the other hand, these same characteristics may become negative factors if ignored or disturbed.

As part of the program plan, project managers should decide what changes will be required in the structure of the company—the organization, policies, processes, and especially the effect on job contents and employee responsibilities and relationships. They can then decide how best to present these required changes to the rest of the organization.

As the planning progresses, it will become obvious that the program to adopt a lean and agile supply chain consists of a variety of activities and events. In other words, it is a project and needs a project planning approach. As with most improvement programs, the implementation efforts run parallel with the regular sustaining operations of the business. Unless there is a way of monitoring progress, the program will likely falter or even lose its identity as an improvement effort.

Configuration

This phase involves the physical changes necessary in the marketing, purchasing, manufacturing, distribution, finance, and top management processes. Although these changes may be relatively straightforward, they can cost significant amounts of money and often take substantial amounts of time. Many of the changes involve physical or process changes; however, in all cases, the employees involved will need to change their operating paradigms. We will discuss this aspect later in the Culture section. In the sections below, we list a number of the changes for each of the major functions of a business.

Marketing

Marketing will assume greater responsibility for retaining customers. The current thinking, represented by customer relationship management (CRM) programs, is that it is more beneficial for a company to cultivate lifelong relationships with customers than it is to keep trying to find new customers. They will have to change from a "make and sell" paradigm to a "sense and respond" paradigm.[39] The supply chain basics for the marketing function include the following:

■ Build databases to capture more data about existing customers.
■ Analyze existing customers to determine why they buy the products and services they do.
■ Develop programs designed to retain the customer.
■ Obtain feedback from customers for use in the design of products and services (sense and respond).
■ Participate in cross-functional teams to develop customer service programs.

To accomplish these and other changes, the marketing department will need to develop a structure that is flexible and operates with employees who endorse and participate in this new way of thinking.

Purchasing

Purchased goods and services represent a major portion of a company's expenditures. This cost category can range from 50 to 70 percent of sales in manufacturing industries and even higher in distribution and retailing industries. Consequently, the purchasing function will assume greater responsibilities in the future. The role of purchasing is changing from processing quotations from a wide variety of vendors to find the lowest price, to building relationships with a small group of suppliers. The supply chain basics for the purchasing function are as follows:

■ Reduce the number of suppliers and expand the expectations for the performance of every supplier.
■ Expand the areas of responsibility to include purchasing services as well as materials.
■ Become a facilitator to help other functional areas develop relationships with suppliers.
■ Participate in a cross-functional team to develop effective outsourcing programs.
■ Become involved in the strategic planning for the company.

This transition will require a change in buyer attitudes and an upgrade in the skills required of the purchasing staff. Purchasing will become a professional-level job, an even more responsible function in both strategic and operational decision making and implementation.

Manufacturing

Manufacturing will need to continue to make the transition from a progressive bundle (batch) system to a modular (flow) system.[40] A shift in focus from equipment and labor efficiencies to improving customer service will be needed. The supply chain basics for the manufacturing function are listed next:

- Rearrange plant layouts to improve the flow of materials.
- Increase worker versatility and empowerment.
- Increase equipment flexibility, such as in reducing setup times.
- Reduce process variation in output and quality.
- Change production performance measures to reflect the new environment.

To facilitate the transition from internal efficiencies to improving customer service, the manufacturing function will need to work more closely with purchasing and marketing to coordinate customer orders in reducing response times.

Distribution

The distribution function—which includes the movement of goods from the manufacturer to the distributor, then on to the retailer, and from there to the consumer—will also be increasingly concerned with the speed and timing of deliveries. Businesses will be especially concerned with having the right goods and services available to the customer. At the same time, they want to reduce inventories, especially excess and slow-moving inventories. The supply chain basics for the distribution function include the following:

- Retailers and e-commerce sellers will share point-of-sale information with their upstream suppliers.
- The distribution function will increase its capabilities to customize products and services.
- The distribution function will use a greater variety of transportation methods— truck, train, ship, and air—depending on the cost versus response-time balance.
- Distribution costs will receive closer attention in the total cost of ownership analyses, especially concerning offshore outsourcing.
- The distribution function will be increasingly outsourced to private carriers, such as UPS and FedEx.

Distribution costs have become a significant portion of the total costs of products and services. The distribution function is also becoming a major factor in customer service considerations. It will become an even more important part of the supply chain in the future as the timing and traceability of deliveries becomes more critical.

Finance and Accounting

The financial and accounting functions have an important role in improving flow in the supply chain. They must provide the funds to make the physical and structural changes and monitor the results to ensure effective utilization of the organization's resources. The supply chain basics for the finance and accounting functions are as follows:

- Develop a total cost of ownership analysis, especially for use in evaluating offshore outsourcing.
- Modify reporting systems to provide more meaningful information to operating functions.
- Change the performance measures to evaluate the entire supply chain outcomes, not just local outcomes.
- Work with supply chain partners to achieve an equitable distribution of costs and profits.
- Balance their efforts between external reporting and internal reporting requirements.

Finance and accounting must become participants in the company's efforts to build their supply chains, not just perform the functions of interested bystanders.

Top Management

Top management must become an active participant in building the supply chain. There are many decisions, both strategic and operational, that require their support. The supply chain basics for top management include the following:

- Propose the structural organizational changes that will be needed along with the accompanying responsibilities.
- Analyze the decisions needed to move actively into (or away from) offshore outsourcing.
- Initiate contacts with other top managers to facilitate developing supply chain partners.
- Propose agreements on changes in performance measures and compensation incentives.
- Serve in the role as the visible motivator of change.

In many companies, the transition to agile and integrated supply chains is a new and complex transition; top management must lead the way.

The configuration change is one of the most visible changes that a company makes in moving toward supply chains. There are many articles written about it because it is often one of the most analyzed and debated areas of change. Although configuration is a necessary step, it is not sufficient in itself. More is required. Many businesses work to make their chains faster or more cost effective, assuming that those steps are the keys to competitive advantage; however, companies must also build supply chains that are agile, adaptable, and aligned. Great companies create supply chains that adapt when markets or strategies change.[41]

Communication

Communication is the exchange of information between entities. It is facilitated by informal and formal (largely electronic) networks. It is technology-driven but people dependent; fraught with a lack of incompatible or local standards; involves various modes (e.g., telephone, fax, e-mail, intranet, EDI, Internet, social media); and, above all, is continually evolving.

This phase involves designing and implementing the data exchange systems needed to achieve the instantaneous transfer of data, both within the company and among the entities involved in the supply chain. This network can be expensive and time consuming; in addition, management must be willing to exchange confidential information.

Making the physical changes in the supply chain without upgrades in the communications systems, both within the company and between the company and its suppliers and customers, is not sufficient. New technology in the form of hardware and software may be required to enable rapid and reliable communication among employees in a department, among departments within a company, and among companies.

The implementation of new systems can be both expensive and time consuming. In addition to the technology, the employees who will use these new systems require time to accept and learn how to effectively utilize the new technology. Failure to get the employees successfully involved can lead to poor results, or even failure of the entire effort.

As difficult as it may be to implement the communications system within a company, it is even more difficult to implement inter-company communications. Although communications technology is readily available in the form of EDI or the Internet, the reluctance of companies to adapt between companies often prevents a high level of success. Darren Cooper and Michael Tracey believe that "the strategic aspects of their selection and implementation have for the most part been neglected until recently, and much of the literature fails to address the human and process factors involved in deploying this technology."[42]

IOS's are not a dream. They exist, although not all company relationships have reached the collaboration stage. That will emerge as the technology, infrastructures, and relationships of companies come into alignment.

Social media sites present challenges for many businesses. They offer a wealth of data about how individuals feel about a variety of subjects, some of which could be of value to companies once they learn how to collect, analyze, and interpret their findings.

Culture

Companies often ignore or de-emphasize the need to match the changes necessary with the existing culture of the organization, but this is often the key to successful implementation, and if ignored may limit success. Changing culture requires an understanding of human relations and a willingness and ability to effect major changes in the thinking of individuals and the actions of teams.

Making the configuration and communication changes requires a high level of acceptance on the part of employees. The change from batch to lean requires a major cultural shift in the company. For many employees, it may signal the end of an era. No more mass production or local optimization thinking—it is time to shift paradigms from the traditional to the contemporary.

The program to change or adapt the culture of the organization to the program of building supply chains often lacks a systematic approach that can be done in a series of sequential steps. How best to blend or adapt the program to the culture or modify the culture to the program takes insight that involves an understanding of human nature. Failing to account for the impact changes have on the employees can stall or even derail the most desirable of changes.

A decade ago when the *Harvard Business Review* convened a panel of leading thinkers in the field of supply chain management, technology was not the top concern. People and relationships were the dominant issues of the day. The opportunities and problems created by globalization, for example, are requiring companies to establish relationships with new types of suppliers. The ever-present pressure for speed and cost containment is making it even more important to break down stubbornly high internal barriers and establish more effective cross-functional relationships. The costs of failure have never been higher. The leading supply chain performers are applying new technology, new innovations, and process thinking to far greater advantage than the companies that are lagging. This roundtable gathered many of the leading thinkers and doers in the field of supply chain management. Together, they took a wide-ranging view of such topics as developing talent, the role of the chief executive, and the latest technologies, exploring both the tactical and the strategic in the current state of supply chain management.[43] Cultural differences make a difference in the "fit" among supply chain members. Although trade legislation may influence supply chain combinations, they must consider cultural differences as well.[44]

When successfully implemented, these changes will help a company approach a lean production status. However, is a lean company enough? Or will it be necessary to move to a lean and integrated supply chain?

Customization

Mass customization is an extension of agile manufacturing. Lean means being able to produce customized goods or services at optimal efficiency. Agile means being able to move from one product or service to another within the known product lines without noticeable interruption. Mass customization is the ability to make any of the known products at a high volume and to customize that product to a customer's individual specification, at a cost comparable to that of making a standard product.

Is the market-of-one imminent? Can businesses provide a unique product or service to each customer? Although it appears unlikely, a number of companies are working toward that objective. Where it is possible, companies are using modular components to facilitate postponement of the customizing process. They are moving from the make-to-stock (MTS) mode to the assemble-to-order mode. In time, they may be able to move back even further to the make-to-order mode if they can reach an acceptable compromise between level of customization and response time.

Another interim step is to provide greater variety when it is not possible to use the postponement strategy, such as in grocery items or small tools. Greater variety can be used to appeal to smaller market niches, but it is unlikely to approach the market of one. One of the negatives about greater variety is that it forces consumers to make more decisions about which item they want, a decision-making process that may add stress to their lives.[45] The answer lies in combining product modularity, process flexibility, customer relationships, and employee versatility into a dynamic system. All of these are necessary features for a business.

The Integrated Supply Chain

Everyone talks about an integrated supply chain, but how do you get there? In most cases, the transition passes through three distinct and often overlapping phases: interfacing, interlacing, and integrating. During the interfacing phase, two separate entities begin to pass information back and forth through dissimilar information processing systems. The movement of goods improves, but everyone realizes it could be better. The next move is interlacing, where changes connect the flow of information more closely and directly between the two entities. In the integrating phase, a level of connection is achieved where information flow is efficient and effective for both entities. Few companies have reached the final phase of integration. To achieve even the beginning levels of supply chain integration requires extending communications into cooperation and collaboration. Increasing competition stimulates independent firms to collaborate in a supply chain that allows them to gain mutual benefits. This requires the collective knowledge of the coordination mode, including the ability to synchronize interdependent processes, to integrate information systems, and to cope with distributed learning.[46]

Another study found that the degree of trust, power, continuity, and communication between supply chain partners would enhance commitment and, consequently, the integration of the SCM business process.[47]

Coordination

Coordination means working together to achieve a mutual objective. It suggests that companies begin to understand what their suppliers or customers want and need, and what each participant should do to help achieve the objectives. It can be as simple as sending an order acknowledgment to the customer, or as involved as meeting a one-hour delivery window imposed by the customer.

Coordination provides the infrastructure and opportunities to make it possible to work together. It includes working in an "open-system" environment and linking entities together both inside and outside the company. It involves policies, procedures, contracts, and other means necessary to work together and is among the early attempts to align the components of the supply chain.

Collaboration

Collaboration implies a greater level of interconnectivity between customer and supplier. Collaboration involves freely sharing information and working together to develop a more meaningful plan for the flow of both information and goods through the supply chain. Several programs enhance collaboration, most recently collaboration, planning, forecasting, and replenishment (CPFR), a program that attempts to integrate the entire stock replenishment process.

Collaboration includes the motivation to work together effectively for the welfare of the entire supply chain. It builds on the coordination infrastructure and requires less tangible qualities such as trust (see Managerial Comment 10.1). It maximizes the benefits to the supply chain, not the individual entity, and provides the ultimate service to the customer.

Managerial Comment 10.1   Trust

Robert Handfield and Christian Bechtel report that one of the most misunderstood concepts and a ripe area for research in SCM is in the area of trust. Trust (and its cousin, collaboration) seems to be the single most discussed element in making supply chains function effectively and efficiently. They found that four bodies of theory (transaction cost economics, organizational design, relational theory, and network theory) support the premise that effective communication of requirements with appropriate safeguards and approaches is critical to effective customer and supplier relationship management.

They conclude that, although trust is generally considered critical to successful SCM, it is one of the most difficult attributes to measure. There are different conceptual paradigms of trust: reliability, competence, goodwill (openness), goodwill (benevolence), vulnerability, loyalty, multiple forms of trust, combining trust with vulnerability, and nonpartisan proactive-based trust. When it is all said and done, integrity in action and thinking seem to be the foundations of trust between companies. Although much of the research to date has focused on analytical approaches to managing supply chains, the area that requires the greatest work is managing supply chain relationships.[48]

Although coordination and collaboration are necessary, they are difficult to achieve. Internal company objectives may inhibit the attainment of supply chain objectives. The barriers include economic, technological, and cultural matters. Some things change slowly, especially those that may have an unknown effect on the jobs of employees at all levels of the organization.

Once the framework of a lean supply chain has been established, companies must move from a program mode to an institutionalized version in which the key features of the program become part of good management. Specific programs can be sustained for a limited period; however, at some point the main elements of the program must be assimilated into the management practice of the company or become lost as the program declines into ineffectiveness or is terminated by management.

Steps in the Change Process

Investigate

Although it is easy to be carried away with enthusiasm about the possibilities of the change, the first need is to investigate why the change would be beneficial. A company should go through the process of deciding what it wants to accomplish from the change. If the company is satisfied that the change to lean production will provide the desired results, then a commitment can be made to going through with the change.

Initiate

Once management has made the commitment to change, they have to initiate action. But action without a plan is premature. At this point, they need to develop the concept that will carry the program through to a successful conclusion.

Invigorate

This is the time to get the rest of the organization involved. The participation of the management team and the nonmanagerial employees is required. At this stage, management should anticipate where it will find enthusiastic acceptance of the change plan and where it may encounter resistance. Although acceptance throughout the organization is necessary, it is vital to have support when the program is in the critical beginning stages.

Implement

The actual implementation stage should follow three main areas. First, the configuration of the physical elements of the production or service processes needs to be addressed. Second, communication within the organization and with suppliers and customers needs to be established in which the details of the program are outlined. Finally, the organizational culture of the firm must be changed.

Integrate

Integration comes slowly. It begins with communications and gradually progresses through coordination, cooperation, and collaboration. One of the primary requirements is trust among supply chain partners, and trust comes from a series of successful experiences in working together. It begins at the individual level and gradually builds to a collective trust relationship between entities. Integration comes as much from human relationships as from technology and structure changes.

Institutionalize

Specific programs can only be sustained for a limited period; at some point, the main elements of the program must be assimilated into the management practice of the company, or they will be lost as the program declines into ineffectiveness or is terminated by management. The goal, of course, is to make the program part of the normal operating life of the organization. In other words, as the novelty of the new supply chain configuration gives way to familiarity and effectiveness, it gradually becomes institutionalized into the business.

Innovate

Once a company builds a facsimile of an integrated supply chain, it is time to change it. The business world is constantly changing and there is a need to adjust the supply chains to keep pace. We will describe this further in the following section.

A Look Ahead

Suppose that lean production is achieved and then extended to a lean supply chain. Is that the final answer, or should the company move on to agile manufacturing?

Agile manufacturing is first lean and then agile. It is an extension of lean manufacturing and is sometimes viewed as a competitive asset for the future. The three interrelated support areas for agile manufacturing are technology, organization, and people. Each of these areas was described in Chapters 7, 8, and 9. The following quotes point out the need for collaboration among supply chain members:

> You can be lean by yourself, but you cannot be agile by yourself.[49]

> Virtual companies are "the epitome of the agile enterprise"—a vehicle to tap the knowledge of participating companies.[50]

> In agile manufacturing, the aim is to combine the organization, people, and technology into an integrated and coordinated whole.[51]

Is agile manufacturing the final answer? No, there may be a need to move even further, toward mass customization and complexity management.

Up to now, we have described actions that are difficult but reasonably straightforward. In the next two phases, we believe that companies will have to demonstrate significant innovation skills.

Complexity

Managing the complexity of a business in today's global competitive environment is challenging. For management, it may be an even greater challenge than achieving mass customization. It goes beyond customizing a known product for an individual customer. It involves addressing all the internal and external issues that may exist, while at the same time making the product at a competitive cost with minimal lead time. Operating in a complex or chaotic environment tests the capabilities of any company.

In Chapter 3, we described open systems and the need for organizations to deal with the complexities of the forces that are external to the organization's control. These forces, especially those resulting from global competition, require that businesses develop global strategies affecting product design (products must be designed to fit the market they serve), processes (to reflect the best blend of resource costs, work rules, and local preferences), and locations (how to blend a myriad of decision criteria in the face of potentially dramatic change).

Complexity is difficult, and introduces some paradoxes that increase the confusion:

- **The optimization paradox.** Despite the potentially huge economies from designing supply chains from a global view, most manufacturers optimize locally.

- **The customer collaboration paradox.** Despite the need to be much more responsive to customers, few manufacturers are collaborating closely with them.
- **The innovation paradox.** Product innovation is continuing to accelerate, yet few manufacturers are preparing their supply chains for faster new product introduction.
- **The flexibility paradox.** Flexibility is a key priority, but it is being sacrificed in the drive to cut unit cost.
- **The risk paradox.** Keeping supply chain quality high is critical, yet manufacturers' risk of supply chain failures keeps growing.[52]

Traditional decision making has relied on linear thinking—the assumption that the future will be largely like the past. Today, however, many businesses are finding that using the past as a basis for planning the future may not be very useful; in fact, it may be downright misleading. But how can organizations apply nonlinear thinking? It takes a completely new theory, appropriately called complexity theory. William Frederick describes complexity theory as a variant of chaos theory that explains the organizational and evolutionary dynamics that occur as complex living systems interact with each other and their environments. He suggests that both the corporation and the community are natural systems interacting and coevolving in response to environmental factors. The corporation is hypothesized as a complex system that must adapt itself to the environment in which it operates.[53]

Dealing with complexity will be a necessary core competency for any successful organization. Frederick poses the following question: "Is the corporation a self-organized complex adaptive system (CAS) housing an autocatalytic component, operating on a fitness landscape, and exposed to the risk of chaotic change while being held in its niche by a strange attractor?"[54]

Many companies will probably find complexity, or its variant, chaos theory, as too "far out" to be considered seriously. However, a few companies are already trying to decide how to incorporate this theory in their thinking. A full explanation is beyond the scope of this book; we can only suggest that complexity theory will be an important consideration for the future.

Clairvoyance

Clairvoyance is the power to perceive things that are out of the natural range of human senses. Few articles or books combine clairvoyance and business, presumably because businesses align with rationality and logic rather than with a discipline that has been associated with frauds and horoscopes. Up until now, businesses have been dealing with the known world, even though it may be complex and unexpected. In the future, managers will have to deal with greater uncertainty. The present forecasting techniques will be increasingly inadequate. Current methods rely heavily on extension, usually linear, from present known positions. At some point, organizations will have to build bridges between disconnected islands of products or services. In other words, they will have to be clairvoyant. Organizations will need people who have the "gift" to foretell the future, and that is a skill not taught in business schools.

Perhaps it would be more palatable to say that businesses need people who possess the wisdom to assimilate fragments of leading indicator information into a holistic and reasonably accurate view of the future. Wisdom is the distillation and organization of knowledge in such a way that it has relevance to the decision at hand. Some individuals seem to possess an uncanny ability to make correct decisions while others struggle in a hopeless maze of contradictory information. For the moment, we will call that first ability *clairvoyance*. It raises several questions. What does it take to see into the future? Is it a skill that is inherited—a sixth sense—or can it be learned? If so, how? Is clairvoyance something like "thinking outside the box"? Does it result from empowered cross-functional teams? Can it be created through the use of biotechnology? Can it be directed and controlled, or is it the result of random insights? We touched on knowledge management in earlier chapters and will address it again in Chapter 12.

Just as with chaos and complexity theory, the detailed discussion of clairvoyance is beyond the scope of this book. We can suggest books that we believe make what we have called clairvoyance a bit more tangible and realistic. In *Blink*, Malcolm Gladwell describes how some people have the knack of making decisions without the need for exhaustive analyses; they see the big picture and just intuitively know![55] In *A Whole New Mind*, Daniel Pink describes the emergence of "right-brain" thinking as coming of age in the world.[56] Henry Mintzberg also discussed left- and right-brain implications (see Managerial Comment 10.2). These ideas represent significantly different ways of making decisions from those approaches that have been practiced for a long time in the business world. Although practitioners of these techniques may not be clairvoyants, they will be advanced beyond most of today's managers.

Managerial Comment 10.2   Implications for the Left Hemisphere: Left-Brain Thinking

First, I would not like to suggest that planners and management scientists pack up their bags of techniques and leave organizations, or that they take up basket-weaving or meditation in their spare time. (I haven't—at least not yet!) It seems to me that the left hemisphere is alive and well; the analytic community is firmly established, and indispensable, at the operating and middle levels of most organizations. Its real problems occur at the senior levels. Here analysis must coexist with—perhaps even take its lead from—intuition, a fact that many analysts and planners have been slow to accept. To my mind, organizational effectiveness does not lie in that narrow-minded concept called "rationality"; it lies in a blend of clearheaded logic and powerful intuition.

A major thrust of development in our organizations, ever since Frederick Taylor began experimenting in factories late in the last century, has been to shift activities out of the realm of intuition, toward conscious analysis. That trend will continue. But managers, and those who work with them, need to be careful to distinguish that which is best handled analytically from that which must remain in the realm of intuition. That is where we shall have to continue looking for the lost keys to management.[57]

Conclusions

Earlier, we described the Cs of change as moving from batch flow to a lean and agile supply chain. A few progressive and venturesome companies are well along on their journey. Some have started while others are still thinking about it.

There are additional Cs to be traveled. Some have been discovered and many companies are well beyond the introductory stage, such as with mass customization and partner collaboration. Others, such as complexity management, hold promise but are not well defined or understood. Some, such as clairvoyance, are still in the untapped potential stage.

We close the chapter with these thoughts:

- Companies understand batch, but is it not time to change?
- From batch to lean is just the beginning.
- Agile and integrated supply chains may be better, but they are difficult to implement.
- Virtual configurations may be best, but it is still a goal of the future.
- Technology is an enabler, but it requires organization and people to make it happen.
- Not to change is an unrealistic alternative.
- Collaboration will be the most strategic capability in the extended supply chains.
- Service and support will become as important as the product itself.
- Companies will improve their service capabilities to adapt in turbulent environments.
- Assets and functions not at the core of value delivery are to be divested.
- The success in turbulent business environments depends on delivering new products and services at a faster rate and transforming value chains into customer-focused virtual networks.[58]

In this chapter, we have described the myriad of things that companies must accomplish to become participants in an agile and integrated supply chain. In Chapter 11, we describe the many services that will be required to support the supply chains.

References

1. Davis, E.W. and Spekman, R.E., *The Extended Enterprise: Gaining Competitive Advantage through Collaborative Supply Chains,* Prentice Hall, Upper Saddle River, NJ, 2004.
2. Poirier, C.C. and Quinn, F.J., Survey of supply chain progress: Still waiting for the breakthrough, *Supply Chain Management Review,* 10, 8, 18, 2006.
3. Poirier, C.C. and Quinn, F.J., How are we doing? A survey of supply chain progress, *Supply Chain Management Review,* 8, 8, 24, 2004.
4. Davis, E.W. and Spekman, R.E., *The Extended Enterprise: Gaining Competitive Advantage through Collaborative Supply Chains,* Prentice Hall, Upper Saddle River, NJ, 2004.
5. Manrodt, K.B., Abott, J., and Vitasek, K., *Understanding the Lean Supply Chain: Beginning the Journey,* 2005 Report on Lean Practices in the Supply Chain, APICS, p. 7.
6. Walker, W.T., *Supply Chain Architecture: A Blueprint for Networking the Flow of Material, Information, and Cash,* CRC Press, Boca Raton, FL, 2005.
7. Taylor, F.W., *The Principles of Scientific Management,* Harper & Row, New York, 1911.
8. Blackstone, J.H., *APICS Dictionary* (13th ed.), APICS—The Educational Society for Resource Management, Chicago, IL, 2005.
9. Manrodt, K.B., Abott, J., and Vitasek, K., *Understanding the Lean Supply Chain: Beginning the Journey,* 2005 Report on Lean Practices in the Supply Chain, APICS.
10. Lee, H.L., The Triple-A supply chain, *Harvard Business Review,* 82, 10, 102, 2004.
11. Heckmann, P., Shorten, D., and Engel, H., *Supply Chain Management at 21: The Hard Road to Adulthood,* Booz Allen Hamilton, New York, 2003.
12. Kotzab, H. and Otto, A., General process-oriented management principles to manage supply chains: Theoretical identification and discussion, *Business Process Management Journal,* 10, 3, 336, 2004.
13. Haeckel, S.H., *Adaptive Enterprise: Creating and Leading Sense-and-Respond Organizations,* Harvard Business School, Boston, 1999.
14. Kidd, J., Richter, F.J., and Li, X., Learning and trust in supply chain management, *Management Decision,* 41, 7, 603, 2003.
15. Manrodt, K.B., Abott, J., and Vitasek, K., *Understanding the Lean Supply Chain: Beginning the Journey,* 2005 Report on Lean Practices in the Supply Chain, APICS.
16. Poirier, C.C. and Quinn, F.J., Survey of supply chain progress: Still waiting for the breakthrough, *Supply Chain Management Review,* 10, 8, 18, 2004.
17. Krikke, H., LeBlanc, L., and van de Velde, S., Product modularity and the design of closed-loop supply chains, *California Management Review,* 46, 2, 23, 2004.
17a. Gyerey, T., Jochim, M., and Norton, S. *The Challenges Ahead for Supply Chains,* McKinsey & Company, New York, 2010.

17b. CSC (2012). Supply Chain Trends Survey: Secrets of Supply Chain Success, accessed 9/24/2012, http://assets1.csc.com/insights/downloads/Supply_Chain_2012FinalReport3.7a.pdf.

17c. APICS (2011). APICS 2011 Supply Chain Strategy Challenges and Practices, APICS—The Association for Operations Management.

17d. White, C. 2012 Supply Chain Survey Results, IDC Insights, accessed 9/19/2012, https://idc-insights-community.com/manufacturing/manufacturing-value-chain/2012-supply-chain-survey-results.

18. Anonymous, The challenge of complexity in global manufacturing, Special report, Deloitte Touche Tohmatsu, 2003.

19. Friedman, T.L., *The World Is Flat: A Brief History of the Twenty-First Century*, Farrar, Straus and Giroux, New York, 2005.

20. Beth, S., Burt, D.N., Copacino, W., and Gobal, C., Supply chain challenges: Building relationships, *Harvard Business Review*, 81, 7, 64, 2003.

21. PRTM, Supply Chain Maturity Model, http://www.prtm.com/services/supply_chain_maturity_model.asp, 2005.

22. Cigolini, R., Cozzi, M., and Perona, M., A new framework for supply chain management: Conceptual model and empirical test, *International Journal of Operations & Production Management*, 24, 1/2, 7, 2004.

23. Simatupang, T.M, Wright, A.C., and Sridharan, R., The knowledge of coordination for supply chain integration, *Business Process Management Journal*, 8, 3, 289, 2002.

24. Kempainen, K. and Vepsalainen, A.P.J., Trends in industrial supply chains and networks, *International Journal of Physical Distribution & Logistics Management*, 33, 8, 701, 2003.

25. Kannan, V.R. and Tan, K.C., Attitudes of US and European managers to supplier selection and assessment and implications for business performance, *Benchmarking*, 10, 5, 472, 2003.

26. Wong, C.Y., Hvolby, H.H., and Johanses, J., Why use loosely coupled supply chains? Report from the Center of Industrial Production, Aalborg University, Denmark, 2004.

27. Poirier, C.C. and Quinn, F.J., Survey of supply chain progress: Still waiting for the breakthrough, *Supply Chain Management Review*, 10, 8, 18, 2006.

28. Handfield, R.B. and Bechtel, C., Trust, power, dependence, and economics: Can SCM research borrow paradigms? *International Journal of Integrated Supply Management*, 1, 1, 3, 2004.

29. Crandall, R.E., Dream or reality? Achieving lean and agile integrated supply chains, *APICS Magazine*, 15, 10, 20, 2005.

30. Ghosh, S., Thorton, J., DeHondt, G., and Faley, R.H., The paradox of overinvestment in enterprise integration, *Align Journal*, 1, 2, 40, 2007.

31. Cigolini, R., Cozzi, M., and Perona, M., A new framework for supply chain management: Conceptual model and empirical test, *International Journal of Operations & Production Management*, 24, 1/2, 7, 2004.

32. Lambert, D.M., Garcia-Dastugue, S.J., and Croxton, K., An evaluation of process-oriented supply chain management, *Journal of Business Logistics*, 26, 1, 25, 2005.

33. Lambert, D.M., Garcia-Dastugue, S.J., and Croxton, K., An evaluation of process-oriented supply chain management, *Journal of Business Logistics*, 26, 1, 25, 2005.

34. Lambert, D.M., Garcia-Dastugue, S.J., and Croxton, K., An evaluation of process-oriented supply chain management, *Journal of Business Logistics*, 26, 1, 25, 2005.

35. PRTM, Supply Chain Maturity Model, http://www.prtm.com/services/supply_chain_maturity_model.asp, 2005.

36. Poirier, C.C. and Quinn, F.J., How are we doing? A survey of supply chain progress, *Supply Chain Management Review*, 8, 8, 24, 2004.

37. Booz Allen Hamilton, Senior execs: SCM hasn't delivered, *Industrial Engineer*, 35, 9, 12, 2003.

38. Heckmann, P., Shorten, D., and Engel, H., *Supply Chain Management at 21: The Hard Road to Adulthood*, Booz Allen Hamilton Inc., New York, 2003.

39. Haeckel, S.H., *Adaptive Enterprise: Creating and Leading Sense-and-Respond Organizations*, Harvard Business School, Boston, 1999.

40. Abernathy, F.H., Dunlop, J.T., Hammond, J.H., and Weil, D., *A Stitch in Time: Lean Retailing and the Transformation of Manufacturing—Lessons from the Apparel and Textile Industries*, Oxford University Press, New York, 1999.

41. Lee, H.L., The Triple-A supply chain, *Harvard Business Review*, 82, 10, 102, 2004.

42. Cooper, D.P. and Tracey, M., Supply chain integration via information technology: Strategic implications and future trends, *International Journal of Integrated Supply Management*, 1, 3, 237, 2005.

43. Beth, S., Burt, D.N., Copacino, W., and Gobal, C., Supply chain challenges: Building relationships, *Harvard Business Review*, 81, 7, 64, 2003.

44. Kidd, J., Richter, F.J., and Li, X., Learning and trust in supply chain management, *Management Decision*, 41, 7, 603, 2003.

45. Schwartz, B., *The Paradox of Choice, Why More Is Less*, Harper Collins, New York, 2004.

46. Simatupang, T.M, Wright, A.C., and Sridharan, R., The knowledge of coordination for supply chain integration, *Business Process Management Journal*, 8, 3, 289, 2002.

47. Wu, W.Y., Chaing, C.Y., Wu, Y.J., and Tu, H.J., The influencing factors of commitment and business integration on supply chain management, *Industrial Management + Data Systems*, 104, 3/4, 322, 2004.

48. Handfield, R.B. and Bechtel, C., Trust, power, dependence, and economics: Can SCM research borrow paradigms? *International Journal of Integrated Supply Management*, 1, 1, 3, 2004.

49. Sheridan, J.H., Agile manufacturing: Stepping beyond lean production, *Industry Week*, 242, 8, 30, 1993.

50. Sheridan, J.H., Agile manufacturing: Stepping beyond lean production, *Industry Week*, 242, 8, 30, 1993.

51. Kidd, P.T., *Agile Manufacturing: Forging New Frontiers*, Addison-Wesley, Reading, MA, 1994

52. Anonymous, The challenge of complexity in global manufacturing, Special report, Deloitte Touche Tohmatsu, 2003.

53. Frederick, W.C., Creatures, corporations, communities, chaos, complexity, *Business and Society*, 37, 3, 358, 1998.

54. Frederick, W.C., Creatures, corporations, communities, chaos, complexity, *Business and Society*, 37, 3, 358, 1998.

55. Gladwell, M., *Blink, The Power of Thinking without Thinking*, Little, Brown and Company, Time Warner Book Group, New York, NY, 2005.

56. Pink, D. H., *A Whole New Mind, Moving from the Information Age to the Conceptual Age*, Riverhead Books, a member of Penguin Group (USA), New York, NY, 2005.

57. Mintzberg, H., *Mintzberg on Management, Inside Our Strange World of Organizations,* The Free Press, 1989.
58. Kempainen, K. and Vepsalainen, A.P.J., Trends in industrial supply chains and networks, *International Journal of Physical Distribution & Logistics Management,* 33, 8, 701, 2003.

Chapter 11

The Role of Services to Complement the Supply Chain

Introduction

In Chapter 10 we discussed the evolution of localized processes into integrated and agile supply chains. We showed that the final version of the supply chain is a combination of all types of industries. Figure 11.1 shows the relationships along the supply chain. In this chapter, we describe two additional complementary types of industries—producer services and social services. Although these industries are not a direct part of the supply chain, they are complementary and essential to the success of the supply chain. Figure 11.2 shows their relationship to the basic supply chain.

We also describe the area designated in Figure 11.2 as consumer, or personal, services. They fit in this chapter because they are complementary to the supply chain, and although some occur during the supply chain operation, many occur after the direct sale of the product or service to the consumer. Examples include the service or repair of automobiles, computer help centers, and the processing of extended credit transactions.

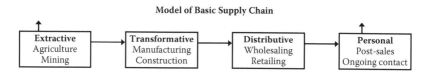

Figure 11.1   Model of basic supply chain.

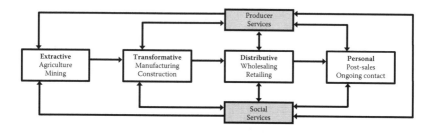

Figure 11.2   Model of services provided to businesses.

Producer Services versus Consumer Services

Although the North American Industry Classification System (NAICS) is useful, it concentrates primarily on the function of the business and does not distinguish among the types of customers served or the type of institution involved in the distribution of the service. Customers fall into two categories: other producers of goods or services (businesses) and individual consumers. Gershuny and Miles offer an extended classification, which they credit to Joachim Singelmann (Table 11.1). It shows agriculture and mining as extractive industries, and construction and manu-facturing as transformative industries. The service industries are grouped into four sectors: distributive, producer, social, and personal.[1]

This distinction helps to identify the strong relationship that exists between producer service businesses and the goods-producing industries. If a manufactur-ing firm goes out of business, that firm will no longer need such services as group insurance, commercial loans, and consulting services. On the other hand, if a service company has been buying computers, office equipment and supplies, and delivery trucks to conduct their business, the goods-producing side of the economy has apparently benefited due to the added sales.

Figure 11.2 shows that the producer and social services provide services to all four of the industry groups along the supply chain—extractive, transformative, dis-tributive, and personal. It also shows that both producer and social service sectors receive outputs from the transformative, distributive services, and personal services sectors. Interdependencies exist among all the sectors. Table 11.1 lists the major industries found in each of the sectors shown in Figure 11.2.

Table 11.1 An Extended Sectoral Classification

I. Extractive

(1) Agriculture, fishing, and forestry

(2) Mining

II. Transformative

(3) Construction

(4) Food

(5) Textile

(6) Metal

(7) Machinery

(8) Chemical

(9) Miscellaneous manufacturing

(10) Utilities

III. Distributive Services

(11) Transportation and storage

(12) Communication

(13) Wholesale trade

(14) Retail trade (except eating and drinking places)

IV. Producer Services

(15) Banking, credit, and other financial services

(16) Insurance

(17) Real estate

(18) Engineering and architectural services

(19) Accounting and bookkeeping

(20) Miscellaneous business services

(21) Legal services

V. Social Services

(22) Medical and health services

(23) Hospitals

(24) Education

(25) Welfare and religious services

(26) Nonprofit organizations

(27) Postal services

(28) Government

(29) Miscellaneous professional and social services

continued

Table 11.1 (continued)   An Extended Sectoral Classification

> **VI. Personal Services**
>
> (30) Domestic services
> (31) Hotels and lodging places
> (32) Eating and drinking places
> (33) Repair services
> (34) Laundry and dry cleaning
> (35) Barber and beauty shops
> (36) Entertainment and recreational services
> (37) Miscellaneous personal services

Source: Adapted from Gershuny, J.I. and Miles, I.D., *The New Service Economy: The Transformation of Employment in Industrial Societies*, Praeger Publishers, 1983.

Figure 11.3 shows the relationships among the different classifications provided in Table 11.1. Extractive industries provide output primarily to the transformative industries; however, they receive participation from both the producer services group and the social services group. A copper mining company supplies copper ore to a refining company and, in return, is a customer of an insurance company or bank (producer services) and the postal service (social services). The transformative industries receive inputs from the extractive industries, the producer service group, and the social services group; they provide output to all the service groups primarily through the distributive services group. A computer manufacturer (transformative) receives sheet metal from extractive sources, manufactures computers, and sells them to an engineering firm (producer services); to universities (social services); and indirectly, through a distribution network, for example, to hotels that service the ultimate consumer.

The distributive services receive inputs from the transformative industries and both producer services and social services; in turn, they provide outputs to all three service groups. A wholesale distributor leases its delivery fleet from a truck rental company (producer services) and obtains employee hiring and training services from the local government (social services). The distributor, in turn, distributes office supplies from manufacturers (transformative) to all types of service companies. Producer service industries provide their output to the value-adding industries of agriculture, mining, and manufacturing; in turn, they use inputs from the distributive services group. Banks (producer services) provide a variety of services to all other industries and receive output from all groups except extractive and personal services.

Social services also provide their output to all of the other groups and, in turn, use outputs from the distributive and producer service groups. A hospital (social services) receives services from the transformative and distributive groups and

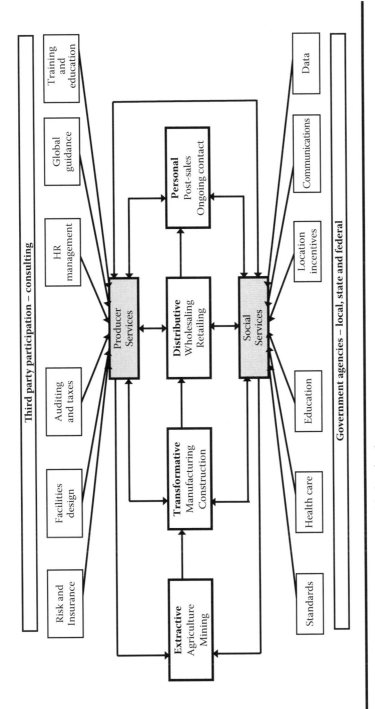

Figure 11.3 Model of services provided to businesses.

provides services to all other groups. The personal services group receives inputs from distributive services, producer services, and social services; they provide their outputs to the individual consumers. A restaurant (personal services) is primarily a receiver of goods from distributive services as well as services from both producer and social service groups to provide output to the consumer.

What Are Producer Services?

Much of the early research in service industries was concerned with consumer, or personal, services. Harry I. Greenfield prepared a classic monograph in 1966, "Manpower and the Growth of Producer Services."[2] This was the first attempt to segregate producer services from consumer services and to identify the economic impact of producer services in the general economy. Greenfield defined producer services as "those services which business firms, nonprofit institutions, and governments provide and usually sell to the producer rather than to the consumer" (p. 1). The term *business services* also is used to denote "services provided by businesses for businesses."[3] Managerial Comment 11.1 describes the demand for producer services.

Managerial Comment 11.1   Determinants of Demand for Producer Services

The first determinant to be considered is the desire to produce the existing level of output at a lower cost. The usual explanation of the source of the saving is the greater productivity of the external labor force, which is at once more specialized and more closely supervised. Examples: General maintenance—custodial, laundry service, window cleaning, food service.

A second determinant of the demand for external services, closely related to the first, is the desire to increase the quantity or quality of the output produced by currently employed resources. Example: Use of a market research firm to supplement the internal staff.

A third motive for using external business services is the desire of a firm to eliminate what might be termed troublesome functions, such as staffing low-status jobs, dealing with complex jurisdictional or other problems in union-management relations, or coping with irregular work schedules. Example: Use of construction or subcontractor firms to handle non-routine maintenance.

A fourth reason for the use of external services is the periodic need for specialized personnel that many small and medium-sized firms cannot afford to employ on a full-time basis. Example: Legal and tax services.

A fifth source of demand for external services arises from the need of companies to adjust to erratic labor requirements. Examples: Hiring part-time or temporary employees for peak loads, such as in taking physical inventory.

A sixth motive for using external services is the desire on the part of many firms to maintain a small, compact, relatively homogeneous labor force. Examples: Concentrate on the core competencies and outsource noncore functions, such as IT maintenance.

Another reason that can be identified as contributory to the increased use of external services is management's desire to enhance the firm's image by being in the forefront of technological or organizational innovations. Example: Management consultants, especially IT.

Finally, external services are often called upon to provide guidance with respect to the firm's growth. Examples: Economic and financial consultants, management consulting firms, market research firms, investment counselors, and other type of firms specializing in long-range planning.[4]

Greenfield also provides a functional classification of producer services in Table 11.2.

Why Are Producer Services Important?

Producer services are important because they are essential to the operation of any business; yet most businesses are not able to perform all of the services for themselves. They have to depend on other businesses to provide the needed services.

A study by William Beyers summarizes the reasons businesses need outside producer services. The firms surveyed ranked the lack of technical expertise as the top reason for seeking outside producer services. The second most important reason, for all but the largest firms, was the lack of size of the business. The third most important reason was the perception that outside firms can perform better. Other reasons listed as a top-four choice by at least two companies included avoidance of risks for in-house service and increased government regulation. Finally, reasons given by at least one group included lack of financial resources to perform services and infrequent need for services. Of the sixteen reasons to choose from, only seven were considered among the top four reasons to seek outside producer services.[5]

Table 11.2 A Functional Classification of Producer Services

Policy Making	
Current Problems	**Longer-Range Planning**
Banking, insurance, and real estate advisors	Business management consulting services
Legal problems	Economic and market research
Human resources and labor relations consultants	Research development laboratories
System analysts	Testing laboratories

Administrative	
Production-Related	**Nonproduction**
Armored car services	Advertising
Detectives, protective services	Bookkeeping, accounting
Equipment rental, freight forwarding	Coin-operated machine rental
Photofinishing laboratories	Credit bureaus and collection agencies
Safety inspectors	Direct mail advertising
Services to dwellings and other buildings	Duplicating, copying, and stenographic services
Testing laboratories	News syndication
	Private employment agencies
	Routine data processing
	Telephone answering

Source: Adapted from Greenfield, H.I., *Manpower and the Growth of Producer Services*, Columbia University Press, New York, 1966.

The Role of Outsourcing

The outsourcing of producer services is common; however, the motivation for outsourcing may vary, depending on the type of producer service. Some producer services, such as institutional food service or custodial service, can usually be outsourced for a lower cost, reducing the complexity of the workforce and investment requirements. Producer services, such as legal or architectural services, are outsourced because it would be difficult to maintain the expertise for these services in-house. Other services, such as operating the information technology function, may be outsourced to acquire a higher level of quality and knowledge in a rapidly changing environment. Cloud computing, or outsourcing a variety of IT functions to third-party providers, is rapidly gaining a significant position in today's IT environment. Whatever the reason, the outsourcing of producer services appears to be increasing.

The Role of Producer Services

The primary role of producer services is to support the supply chain. It should perform a supportive role but should not be so aggressive or dominant that it influences the strategic direction or effectiveness of the supply chain.

Some producer services involve continuous contact with the supplier. Call centers and banks provide this type of service. Some producer services require only occasional participation in a business. Executive search firms or public accounting audit teams represent this type of service. Even when the service is ongoing, producer services have many of the characteristics of a project and probably should be managed as a project to recognize that it is not a part of the internal operation of a business.

Classes of Producer Services

Producer services can fall into the same categories of functions described in Chapter 4—sustaining operations, problem-solving activities, or continuous improvement programs. An example of sustaining operations is custodial or food services. A problem-solving type of producer service is the use of a crisis management consultant to help during a toy recall triggered by an infant death. A continuous improvement program can use management consultants to provide program direction and training of employees.

Reasons to Acquire Producer Services

In this section we look more closely at specific types of producer services and some of the reasons companies elect to acquire these services. We provide representative examples of each type of service.

Gain a Cost Advantage

One of the most obvious reasons to outsource producer services is to gain a cost advantage. If this can be accomplished without a loss of quality and without an increase in risk of causing disruption in normal business operations, it is a popular choice among most businesses. Examples include the following:

- Call centers. One of the most publicized forms of producer services is call centers. This option is popular but controversial. Businesses can have the call center handled for less cost, but they also run the risk of irritating customers who do not find the service at the call center satisfactory.
- IT programming. IT services, especially in the programming area, are an increasingly active area of outsourcing. Companies are taking advantage of electronic communications capabilities by using businesses in lower-labor-cost areas, such as India, to perform routine IT services. As previously mentioned, cloud computing is another way to outsource computing services.
- In addition to reducing costs, outsourcing reduces the need for capital investment in systems design and the resources to operate those systems in-house.

Remove Noncore Types of Activities

These activities are necessary but are not usually similar to the primary functions of a business.

- Custodial services. Most office locations use a cleaning service to perform custodial services. This service can often be done at a lower cost, and it removes a dissimilar operation from the company, one that usually has differences in pay grades, employee turnover, special training requirements, and different inventory requirements. This service is usually carried out at night when the regular staff is not there. Even manufacturing operations find it is desirable to outsource custodial services.
- Food service. There is an abundance of food service operations near most manufacturing and service facilities, so many businesses have abandoned their food service operations. Those that have retained them consider it an added benefit for employees rather than as a cost-saving measure. Even where the food service is on-site, it is usually an externally contracted service rather than an internal operation.

Supplement Internal Staffs with Added Expertise

Many businesses want to have representation in a particular activity but do not want to provide capacity in those functions to handle all the situations that may arise. In many cases, they want to depend on the suppliers of these services to possess the latest knowledge and external contacts necessary to do the complete job. The internal staff will have the general knowledge and the capability to know how best to utilize the external suppliers to provide the most appropriate service to the company.

- **Advertising.** Most small- to medium-size businesses do not have full-time staff devoted strictly to advertising. Although some may be building their own web pages, advertising through newspapers, magazines, and electronic media is usually handled through agencies that specialize in that form of service. Although large companies may have an internal advertising staff, they usually go through agencies to place their ads.
- **Market research.** Firms that specialize in market research have two main advantages over in-house research departments. First, they have a vast amount of information already available from their previous work. Second, they have the knowledge and skills to collect additional information through surveys or other research. They can usually perform a better job, and faster, than in-house research departments.
- **Employee screening.** Selection of qualified candidates is an increasingly difficult task. It takes personnel who can verify the information on applications, to screen out individuals who do not fit the position requirements. It takes a combination of hard facts and judgment to eliminate applicants with undesirable personal characteristics, without violating the rights of those individuals. Although HR departments may be able to do this, they often call on employment agencies or search firms to handle the preliminary stage of the employee selection process.
- **Employee training.** All employees, new and experienced, need some level of training. Some of this is conducted before the individual is hired. Learning to drive a tractor-trailer rig, learning to weld, or learning computer literacy are examples of these skills. Community colleges usually provide this type of training. Once the selection has been made, some additional training may be necessary. Where there are sufficient numbers, or where the training is generic, a business will likely outsource this function.
- **Banking.** Most businesses, even small ones, have employees who are knowledgeable about the company's finances. However, all businesses have the need to develop a relationship with their bank(s). Businesses need to be able to borrow money, to finance working capital or capital equipment requirements. They cannot accomplish this on the spur of the moment; therefore, they need to establish their relationship with a bank before they need financial resources, and then they must continue that relationship to ensure that they receive those resources and the services the bank can provide.
- **Engineering testing.** Most engineering departments have some testing capability. However, where specialized testing is required, they usually prefer to outsource this service. They may also use service companies to verify their own testing so that the product can be marketed with an approval rating from the testing company.
- **Employee benefits.** All HR departments have some information about employee benefits. Sometimes they want to supplement their information

with other data available from companies specializing in that area. Similar to market research consultants, employee benefits consultants can provide up-to-date information and overall knowledge in this area that is usually superior to the internal HR department.

■ **Wellness programs.** Health care costs are increasing rapidly. This trend is of great concern to businesses because providing health care coverage is a significant cost of conducting business and the potential of it becoming a mandatory obligation is even more threatening. Consequently, many businesses are moving to a preventive strategy of promoting wellness programs. The basic concept is that not only will the costs be lower, but employees will be more productive and less prone to absenteeism.[6]

Provide Flexible Capacity or Avoid Overload on Key Departments

This reason is similar to the previous one when an internal staff supplements its own capacity with external suppliers. In this case, the reason is not to gain added expertise but to provide capacity to handle demand variability. The variability may result in predictable demand patterns, such as end-of-fiscal-year activities, or in seasonal buildups or contractions in the number of employees required.

■ **Inventory-taking service.** Internal employees could take the physical inventory and, in many businesses, do. However, a growing number of businesses are outsourcing the inventory-taking activity to external suppliers. In most cases, the external supplier has employees who are more highly trained and efficient in taking inventory. It also avoids the interruptions to the normal operation of a business when employees do not have to take time to do a job for which they lack skills and generally dislike.

■ **Reservation systems.** For a long time, airlines have used travel agencies to handle a part of the ticket reservation workload. However, online reservations have become more popular, to the point where some travel agencies have gone out of business. Hotels and other lodging businesses also use this online capability. This trend is beginning to extend to making reservations with restaurants, doctors, and any place where reservations or appointments are used.

■ **Registration at colleges and universities.** One of the most effective systems to relieve the internal staff of peak loads is the use of computerized registration systems at colleges and universities. Although it may not involve an outside business offering the service, the computerized system acts in much the same way a service provider would act. It handles the registration without requiring the direct attention of the internal staff. The system also relieves the students of the burden of standing in line to register and pay their fees.

Acquire Expertise Not Available Internally

It is not always prudent to develop internal resources and specialized skills, especially when the requirement for those capabilities may be intermittent or occasional. Outside service providers can usually fill this void quite well.

- **Facilities location.** Selecting a site for a new manufacturing plant, distribution center, or retail outlet is a daunting task for businesses with little experience in this area. It starts with selecting a country (if considering a foreign location), then a region, and finally a specific site. Most businesses do not have the data or the knowledge to use the data in considering demographics, incomes, traffic patterns, local ordinances, and a host of other variables. Facilities location providers can meet this service need.
- **Architectural or facilities design.** It is rare that companies have their own architectural or facilities design staffs; they must usually rely on outside service providers. Exceptions could be retail companies that have hundreds of retail outlets. Even some of these firms may use outside architects and designers to gain experience and expertise with working with external providers.
- **Construction.** Construction work is also a very specialized service. Even though it is considered as part of the transformative industries, it is a service to most businesses. It involves major project planning capabilities, a capability not available in most other types of businesses.
- **Security services.** Businesses have purchased security services for a long time. When the concerns were primarily with early identification of fires or screening vendors for purchasing agents, the business could adequately handle this with their own employees. As the needs increased—industrial espionage, computer piracy, or possible terrorism threats—it has become obvious that specialized services, supplemented with specialized equipment, are needed.
- **Investment counseling.** In addition to the banking relationships described earlier, larger businesses, especially public companies, may need the services of companies that can advise on investment strategies. Public companies need help in developing strategies to issue stock or to buy back stock, to use debt or equity to finance future expansion plans, or to develop strategies to present their best face forward to the investing public.
- **Risk and insurance.** Historically, business insurance was for fires, floods, thefts, and other relatively tangible occurrences. Over time, the risks have increased and now businesses must be protected against less tangible threats such as lawsuits from product recalls (children's toys), accidents on the business premises (falls on slippery floors), consequences of inappropriate use of the product (hot coffee spills), and failure to prevent employee misappropriation of funds (Enron). Insurance protection requires ever-increasing specialized knowledge that is difficult to maintain within a business.

- **Management consulting.** There are management consultants available to meet almost any need. At some point, even small companies can benefit from the use of consultants. Businesses may need help in planning and implementing new software or some management program to reduce costs or improve quality. We described a number of these programs in earlier chapters. It is unwise for a business to attempt these programs without inadequate knowledge—knowledge that can be available from professional consulting firms.

Assist in Strategic Planning

In the preceding examples, most of the supplier services were in the form of a hands-on arrangement. Suppliers actually did some or all of the work that would have been otherwise performed by internal employees. Assistance in strategic planning is different. It is more of an advisory or training role and the internal staff does most of the actual work.

- **Global guidance consultants.** Almost all businesses, even small ones, are lured by the attraction of global markets. Most need a great deal of help in finding their way through the maze of legal and cultural differences they may encounter in their entry into just one new market. They need help in selecting the market with the most potential and then developing a plan to enter that market. Global guidance consultants are available with these types of services.
- **Merger and acquisition specialists.** Some industries, such as financial institutions and department stores, are experiencing a wave of mergers as companies move toward economies of scale. There is little doubt that electronic communications capability makes it easier for companies to run larger companies by facilitating information flow. Deciding who and how to buy other companies, or sell a business, is not for novices. Specialists are available for these types of situations.
- **Education providers.** Businesses are finding the need for employees with skills beyond those in the past, especially in the use of computers and automated equipment. To acquire these employees, it is necessary to work closely with community colleges and universities who can provide on-site training programs, certification review courses, and even advanced management education necessary to satisfy a company's need for trained and educated employees.

Comply with Regulatory or Otherwise Mandatory Requirements

Sometimes a business has no choice as to whether to use the external service providers. Either the service is required by law or mandated in some form, such as by a major customer.

- **Auditing.** Audited financial statements are a requirement for public companies. Even private companies may need audited statements if required by their creditors or other stakeholders. Auditing is not only a specialized activity;

it implies independence, an independence that requires an outside agency. As companies become more complex in product and service lines, and more global in markets, the skills required to perform a financial audit are challenging to even the most knowledgeable of public accounting firms. As these requirements mount, there is a growing potential for business-auditor conflict in the interpretation and application of various regulatory agency rules and guidelines.

- **Environmental compliance.** Businesses are experiencing only the tip of the iceberg in environmental compliance issues. What was once a local issue—the dumping of toxic materials into the river—is now a part of the global warming concern. Even when there are knowledgeable employees within the company, many businesses can benefit from outside services to ensure they are complying with local, national, and other environmental regulations. This task will become a more demanding requirement in the future.
- **Technology compliance.** Technology is a major driver of change. An example is the requirement by Walmart that its suppliers use RFID tags on products they purchase in the future. Although this program will be implemented over a period of years, it represents only one of many technology requirements that will change the way businesses operate. Another similar program is the increasing demand for businesses to reduce environmentally or ethically harmful ingredients or processes—for example, hazardous materials in computers or sweat shop conditions in clothing manufacturers.
- **Import/export services.** These services classify products for import, secure export licenses, and check shipments against denial lists.
- **Traceability of product flow.** This is especially important in pharmaceutical and food industries, where regulators have a responsibility for human life.

Manage Major Programs

Most of the preceding reasons for outsourcing producer services are to sustain the normal operations of a business. A final reason is to ensure that the business has expertise available to handle programs—both of the problem-solving and continuous improvement kind.

- **Crisis management.** Even in the most well-managed business, a crisis can be catastrophic. Arthur Andersen was one of the premier public accounting firms in the world, yet the Enron debacle caused its demise. On the other hand, Johnson and Johnson handled the sabotage of its product, Extra Strength Tylenol, so well that it actually gained market share after the crisis was over. Whether it is toys that choke children, canned food that kills pets, slippery floors that cause falls, or computer systems that fail to protect individual identities, crises happen. Crisis management consulting firms are available to assist businesses in setting up crisis management plans. These firms can also coach management on the best ways to respond to specific crisis events.

■ **Continuous improvement programs.** In Chapters 4, 5, and 6 we described a variety of programs that companies use to make improvements in their operations. Although some companies have the capability to implement these programs on their own, most do not. They need help from outside service providers, whether it is software vendors or some other form of management consultant.

Steps to Interfacing Business with Producer Services

We have described a number of producer services that businesses may need. In this section we outline a general approach to obtaining these services. Buying services is often more demanding than buying products because although it is possible to see, feel, and even try out a product, this is not possible with services. Businesses should take the following steps in planning and implementing the use of producer services in the company.

Recognize the Role of Producer Services in the Company

Most companies do a good job of deciding which products to make and which to buy. They are less diligent in deciding which services to perform internally and which to outsource. They have less formal procedures for evaluating this decision area, because although the cost of doing the service internally vs. outsourcing may be determined, the benefits are often less tangible and more difficult to assign a value to. It is important to recognize the need to make the evaluation and to determine the reason for having each of the service activities. Management should understand that the activity area described in Chapter 10—the integrated supply chain—is the main reason businesses exist. The services described in this chapter supplement the supply chain. Although they are essential and can often provide a competitive advantage, they augment but do not replace the activities along the supply chain.

Develop a Strategy for Each Producer Services Area

Each producer service area is different and must be planned as a unique part of a business. The actions taken in the accounting area are not the same as in the engineering area. In accounting, the emphasis may be on the implementation of an enterprise resources planning (ERP) system; in engineering the focus may be on designing a new quality performance program that involves the purchase of custom test equipment and rigorous training of quality technicians. Once a strategy has been developed, the business must decide whether to perform that service internally or to outsource it.

Select Suppliers to Provide a Continuing Relationship

If the decision is to perform the service internally, management must decide if the internal staff is adequate to handle the service. Are more people required? Is the present staff adequately trained? Are more professionally trained employees required? Do they need more equipment? If so, what kind? Do they have adequate IT resources? If the decision is to outsource the service, good supplier selection is critical. Many of these questions must be asked about the supplier. In addition, a total cost comparison should be made. What is the likelihood that the supplier will meet the service level requirements? Can the supplier handle future volume and variety requirements?

Prepare a Plan That Interfaces with the Supply Chain Plan

We reported earlier that the producer services areas are to supplement the integrated supply chain. It takes skill and perseverance to match the capabilities of internal and external producer services suppliers with the needs of their customer—the supply chain part of the business. The service suppliers must also remember that although they are directly responsible to their company's part of the supply chain, they may also have a responsibility to other members of the supply chain. For example, the IT department may provide sales information for all products and the customers to whom those products were sold. The IT department may also need to provide sales information to suppliers for those products that each supplier sells to the company. Resolving reporting relationships and personality conflicts is no small part of this matching process.

Implement the Service—Fit It to the Situation

Planning for the services is difficult, and implementation can also be difficult and time-consuming. One of the major obstacles is fitting the right people to the processes, or vice versa. Should policies and procedures be the best that can be conceived or should they be modified to fit the people involved? The practical answer is that there is a high likelihood that some adjustments will need to be made to reflect the situation. People can eventually compensate for system deficiencies, but systems can seldom be designed to compensate for people deficiencies.

Measure the Progress

Developing the producer services package to fit with the integrated supply chain is a major project. Even after the services have been designed and put into place, it is a continuing project to maintain, adjust, and improve the producer services package. The composition of the producer services package is dynamic—there are continuing changes in internal employees, supplier relationships, economic conditions, competitor innovations, and governmental regulations. In addition to the changes in the producer services package, there will be changes in the integrated supply chain served. This changes the relationships of the package elements. The changes must somehow be identified and their effect on the overall performance of the supply chain evaluated. Performance measurement is a critical aspect of any business.

Revise and Assimilate

No matter how well designed and implemented, changes force revisions to the producer services package. Expect this and adjust accordingly. It is a normal part of business. Just as a business must continually update its product and service lines, it must also update its producer services package. It may help to view this as a separate part of the business to be planned in much the same way as the integrated supply chain part of the business.

The Future of Producer Services

What changes can we expect in producer services in the future? Several ideas stand out as likely. There will be an increasing use of IT in producer services. Intra- and inter-organizational systems will become standard in most companies. The need to communicate will drive the use of technology. With it comes a reduction in the personal contact that some employees cherish. The depersonalization of communications will be more acceptable in business-to-business transactions than in business-to-consumer transactions. Even with what appears to be widespread use of IT in business today, it is only the beginning.

With the increased ability to communicate across distances, there will be a continued increase in outsourcing producer services. Most companies do not believe that producer services are part of their core competencies or offer a competitive advantage. Therefore, they will prefer to have these services provided by outside suppliers. At some point, this trend will slow and perhaps reverse itself as companies assign greater value to their information; however, that time still seems to be in the future.

Companies have concentrated their improvement programs primarily along the supply chain. At some point, they will decide that improvement in producer service areas will offer equal, or greater, benefits. In some cases, they will be able to apply the same improvement programs to producer services that they used in supply chain applications. More likely, they will find that other types of programs will be needed to achieve the improvement results they desire.

One of the promising areas for improvement comes in the services sciences. This is the application of quantitative techniques to services. Although the core idea is not new, the applications are being packaged into a composite of techniques and methodologies to facilitate the transfer of knowledge from organization to organization.

Producer Services Firms Buy Producer Services

Manufacturing companies are not the only industry group to buy producer services. Producer service firms also need to buy producer services from other firms.[7] In other words, there are no firms that supply a full range of producer services, at least for now. Accounting firms buy HR services; architectural firms buy custodial services; custodial firms buy tax services; and so on. Although it appears likely that producer services firms will continue to be specialists, that may change over time.

Producer services have been fragmented parts of a business operation in the past. In the future, they will be part of an integrated package that will play a more important role in the operation of any type of business or organization.

What Are Social Services?

Social services are similar to producer services in that they are services provided to a business. They differ in that they are usually provided by nonprofit organizations, or government agencies, as opposed to the profit-oriented organizations that provide producer services. The bottom half of Figure 11.3 shows the categories of social services.

Why Are Social Services Important?

We saw that producer services are usually necessary in running any type of business. Some social services may be necessary, such as the U.S. Postal Service or intellectual property protection. Some may be optional, such as the use of census data or other business-related information. Some may be welcome as an indirect benefit, such as education systems or protection of assets. Finally, some may be mandatory, such as waste disposal regulations or financial reporting requirements. Whatever the reason, social services play an important role in the lives of most organizations.

The Role of Social Services

In theory, businesses could operate without social services; in practice, it would be difficult. In some cases, social services provide benefits that might not otherwise be available. In some cases, social services act as a constraint on the operations of a business, although they may provide a benefit to society in general.

A primary role of social services is to provide direct services to businesses that would be beyond the capability of individual businesses to provide, regardless of size. Although some of these services are mandated by federal law (Bureau of Labor statistics and interstate highways), many are provided within the jurisdiction of state (university school system) and local governments (landfill operations). Although it is conceivable that private companies could operate some of the government agencies, the privatization of government agencies is a slow, but not impossible, process.

Another role of social services is to provide indirect services to both businesses and individuals. Police departments and fire departments represent this type of social services. Although private businesses may maintain their own staff to provide security protection and have designated employees available for fire protection, such measures are usually minimal and rely heavily on the local agencies to provide the much-needed support in case of a major occurrence.

Sometimes social services promote social movements that motivate businesses and other organizations to begin practices that provide for the general welfare of the population. Recycling programs are an example of this. Although private businesses are participants in the recycling process, the driver is government at all levels. Where necessary, the government can move this activity from a voluntary recommendation to a mandatory requirement by enacting legislation that prescribes what businesses must do or face the penalties.

One of the more valuable social services is the work done by the federal government to provide a vast repository of information. Agencies, such as the Census Bureau or the Bureau of Labor Statistics, compile and make available a vast amount of data and information, usually free of charge, to everyone. Individual firms would not have access to the information in these sources, much less have the financial or physical resources needed to set up and maintain these information sources.

Classes of Social Services

Three major classes of social service organizations—governments, nonprofit organizations, and quasi-private organizations are examined here. Each group includes a variety of entities that can provide social services to businesses.

Government

Federal, state, and local governments operate throughout the business world. They provide a variety of services that are described in detail in a later section. In some cases, businesses welcome the service; in other cases, businesses consider the services provided by government as unnecessary or detrimental to the business.

Nonprofit Organizations

These include professional organizations that provide education courses, reference materials, and conferences where employees of a business can enhance their individual and collective knowledge. This class includes organizations that focus their efforts on improving some situation or group of subjects—endangered species, sweatshops, or global warming. In some cases, the business community may endorse these programs; in others, they may resist.

Quasi-Private Firms

These organizations may have government sponsorship or operate as a nonprofit entity; however, they compete with private businesses. A hospital financially supported by the local government may compete directly with private hospitals. A state-supported university competes with privately funded universities. The U.S. Postal Service competes with Federal Express and United Parcel Service.

Business organizations may have a mixed view of these classes of organizations that provide social services. Do they really help? Or are their services more negative than positive? Some of these issues will be sorted out in the next section.

Specific Social Services

A few of the specific social services provided to businesses are described here, concentrating on those that most businesses view as positive.

Information

Government agencies, especially at the federal level, such as the Bureau of Economic Analysis (BEA), collect, organize, and make available a vast amount of information, especially economic news. It is largely free and easily available with today's electronic communications systems.

Communications Systems

In addition to providing information, the federal government provides a variety of communication networks that make it possible for businesses to communicate with one another. Most notably is the Internet, a tool that most businesses would be lost without in today's global trade environment.

Standards

The business community could not operate without standards. It may be units of measure such as for weights or distances. It could be standards for radio frequencies or bar code symbology. It could even be standards of conduct in boardrooms. Without standards, businesses would be in chaos. Standards are also necessary to protect society and businesses from harmful forces such as environmental pollution, potential automobile safety issues, and workplace discrimination problems. Usually, these standards are brought forth by the government after a considerable amount of public outcry that "something should be done."[8]

Education and Training

There is little question that we are entering the information age. Companies increasingly depend on knowledge to provide them with a competitive advantage. One way to attain knowledge is to combine information and education. Educational systems—public schools and universities—have been a major factor in the evolution of the business world to its present state. In addition to the broader vista of education, social service providers also provide employee training to meet specific job requirements. One of the problems today is getting agreement among the interested parties regarding just how this should be accomplished.

Location Incentives

One of the newer services that appeals to many businesses is that of location incentives. Governments, often a combination of state and local agencies, work together to attract new businesses to communities. This service is often a subject of debate among rival politicians; nevertheless, it is a social service to businesses.

Support Infrastructure

Businesses are dependent on the infrastructure provided by governments. Businesses need roads with signs and stoplights, water purification systems, waste disposal systems, and zoning regulations to ensure a comfortable relationship between businesses and the local community. This is an area where business and government must work together to achieve realistic solutions to the needs of a variety of groups.

Protection

Businesses have large investments in resources—facilities, equipment, and most important, people. These assets need protection, and police and fire departments provide at least some of the protection. Businesses also need protection from other impediments to their success—predatory pricing, intellectual property theft, trivial lawsuits, and unfair competitive practices.[9] Although the individual organization may have to solicit help, governments at all levels have an obligation to protect a business.

Health Care

Health care is another area of social services that is looming larger in each generation. Some of the services described above, such as information, education, and protection, contribute to the physical well-being of the population. However, businesses have a special interest in the health of their employees. Healthy employees, both physically and mentally, are of great value to the organization. Businesses are reorienting their focus to wellness, a more proactive perspective than simply taking care of the ill employee.[10] This requires a collaborative effort between the business community and the social service providers.

Steps to Interfacing Business with Social Services

The approach outlined above to integrate producer services into a business also applies to social services, but with some qualifications or cautions:

- **Businesses may have less choice in selecting the supplier.** Although a business will be able to select the location of a new facility, the business will have to accept the service of the local government once the new location has been settled. This will be especially important in considering foreign locations. In many cases, the business will have a "sole source" situation and must maximize the benefits from that situation, rather than having the flexibility to move to another supplier.
- **The relationships with the supplier may be more permanent.** As a corollary to the "sole source" situation, relationships with social service providers will last longer. It will be up to the business to work out the best arrangement with its provider. The form of government may not change, the site of most universities does not change, and the campaigners for purer water may span generations.
- **Political climates affect social supplier actions.** Although the form of the government agency may not change, it is likely that the political climates may change, sometimes frequently or abruptly. When this happens, the relationships with businesses may move in a more positive or negative direction. For example, taxes may go up or regulations may increase. Businesses have to adjust to this change and make the best of it, or move, which is often a less desirable choice.
- **Consequences may be more than economic.** When businesses work collaboratively with social service providers, there are usually positive economic results. On the other hand, if the relationship deteriorates, the consequences to the business could be more than just economic. They could face fines, for example, for failure to follow waste disposal regulations and thereby hurt their public image. If the violations are more serious—for example, criminal acts are committed such as falsifying financial results—then the managers involved could face imprisonment. Sometimes the effect of these acts causes the demise of the business, such as for Enron.
- **Social services may not be as good as private enterprises.** Government and nonprofit agencies are thought of as not being as effective or efficient as private organizations. Although this may not always be true, it is something to consider. These agencies exist to provide services to a mass group of cus-

tomers; they cannot choose their markets. An individual business will gain the most benefit by building a long-term relationship with its providers and working with them through the disruptions or transitions.

The Future of Social Services

How will social services change in the future? Some of the major issues that will have an effect on business are discussed here.

Level of Regulation

In recent years, several industries, such as transportation, have seen a reduction in the amount of regulation. On the other hand, there is a growing concern about the level of regulation in personal health-related issues, such as the food supply or the quality of medicines. There is also likely to be more emphasis on environmental issues. Increased regulations may pose additional constraints on businesses that require a transition to a new operating environment. Through making these changes in processes, businesses may find ways to make improvements that will increase their competitive advantage in the global marketplace. At least one industry has seen both a relaxing of some regulations and a tightening of others. The airline industry was deregulated in terms of route selections during the 1980s, but after 9/11, safety and passenger regulations have been understandably increased. Such events make this industry one of the most difficult ones in which to operate.

Level of Privatization

Governments and academic institutions are slow to change. However, there are indications that the outsourcing movement will eventually reach even the most staid of organizations. Local governments are finding that privatization works in a variety of public service areas. During his tenure as mayor of Indianapolis, Stephen Goldsmith spearheaded a number of successful privatizing projects in such varied areas of city government as highway litter pickup, printing and copying, courier service, wastewater treatment, construction permits issue, construction and operation of correctional facilities, and welfare administration.[11] More recently, Emily Thornton describes a number of cases where governments have sold major assets such as bridges, airports, parking garages, and shipping ports to private investors to recover cash to "solve short-term fiscal problems." She believes that "the conditions are ripe for an unprecedented burst of buying and selling. All told, some $100 billion worth of public property could change hands in the next two years."[12]

Universities are facing more competition from distance learning institutions and corporate colleges. It will be interesting to see if state and private universities develop their own in-house distance learning capabilities or outsource this to private institutions. It appears likely that there will be more privatization of social services, not less, in the future.

Level of Effectiveness and Efficiency

The cost and quality of social services is attracting increased attention from the business community, as well as the individual citizen. It is not only logical but also politically perceptive for government and government-supported institutions to begin a program to improve the effectiveness of their services (provide the right services) and efficiency (do the services right). This will provide an opportunity for business and nonprofit organizations to work together in optimizing their mutual offerings.

Emphasis on Business-Related Topics

If business and nonprofit organizations work together more closely, it follows that government agencies will become more business savvy. If they do, they will be better able to identify the right issues to work on and come up with a better balance of solutions. Although this is rational, it may never reach a satisfactory resolution because of continuing diversions into international conflicts and needs. The subject of sustainability, and its relationship to business interests, is one that is of concern to a number of entities. There is no lack of issues or problems to work on. Most of these will benefit from the collaboration of business and nonprofit organizations. Some of the most obvious issues include health care, global warming, ethical behavior, terrorism, and secular vs. religious disputes.

What Are Consumer Services?

Consumer services are those services that provide outputs directly to the consumer, or business-to-consumer (B2C). Personal services differ from the distributive services shown in Table 11.1. Distributive services include retail and other selling outlets that deliver products and services to the consumer, although their emphasis is usually on the product complemented with services. Personal services tend to be service businesses that may also deliver products as part of the service. This is the ultimate purpose of business—to service the customer. Most businesses tend to concentrate on how to supply services to the consumer; only a few are consciously considering the services that the consumer provides to the business.

Roles of Consumer Services

Consumer services have two major functions. First, they are the way that many products and services are delivered to the ultimate customer—the individual consumer. The second function is to accumulate feedback from consumers about their impressions, both good and bad, about the products and services they have received. The first role is obvious and is the culmination of the efforts described in Chapter 10 to develop an integrated and agile supply chain to make and deliver the products and services.

The second role is not so obvious. Retail organizations, both those that sell through storefronts and those that sell through websites, are developing very sophisticated ways of collecting data about what the customer buys and, in some cases, what the customer would like to buy. See Chapter 10 for a fuller discussion of the work being done in this area. The process for collecting data in the consumer services businesses is less formal. A hair stylist may remember how an individual customer wants her hair styled and a local restaurant may remember which table you prefer when you come to dine. However, much of this knowledge is in the minds of individuals, known as tacit knowledge, and is not compiled as collective information available for the organization—explicit knowledge. There is some evidence that this is changing. An automobile dealer may keep a file of individual customers and a record of the services performed on their vehicles. A hotel or motel chain may retain a record of a customer's stays, regardless of location. In most cases, the information is inadequate to develop decisions that will influence product or service design. However, some companies, such as Amazon, are developing more comprehensive profiles of their customers and are able to offer suggestions about future purchases that may prove desirable from a customer's viewpoint and profitable from Amazon's perspective. The use of data analytics is rapidly emerging as a way to find meaning in heretofore unstructured and unanalyzed data.

Future of Consumer Services

Consumer services will continue to grow as the overall economic prosperity of a country increases. Many personal services might be considered luxuries, not necessities. Because most businesses providing personal services are small, they have not developed the management information systems that larger businesses have. The exceptions are those in the hospitality industries, particularly resorts and higher-end hotels, where their IT capabilities are equal to or better than those of other industries.

Another trend already evident is for larger companies, especially those in the retail industry, to add select personal services to their portfolio of services. Department stores provide beauty stylists, many malls house movie theaters, and building supply retailers sponsor pet adoption Saturdays.

Integrated Service Package

We have described a number of services that a business, whether for-profit or nonprofit, needs to have. In the past, most organizations viewed these services as individual functional areas that were a necessary cost to conducting business. In the future, more organizations will view these services as complementary services that make the supply chain part of the business better. Many companies will begin to view this set of services as an integrated service package, not as simply additional business obligations.

The Need for an Integrated Service Package

Why does a company need an integrated service package? Why can it not continue to operate with fragmented islands of services throughout the company? Do the benefits outweigh the costs?

We do not know the answers to these questions, because the premise that an integrated service package is better remains to be tested. We can only speculate that it seems logical. Integration is defined as "to make into a harmonious whole by bringing all parts together."[13] That not only sounds logical; it sounds desirable and financially feasible. Based on the results from other types of integrating efforts, including the one described in Chapter 10, it may not be easy. In the next section we outline how we think organizations can achieve an integrated service package.

Steps in Developing the Integrated Service Package

How does a company go about developing its integrated service package? The steps outlined below follow closely the steps outlined in the strategic planning process in Chapter 2.

Recognize the Need to Change

Is the present combination of services adequate? Could they be improved? Just as management considers its need to improve its supply chain arrangement, it must also recognize the need to make improvements in the company's service package. Management must also recognize that there are no quick and easy fixes. The transition program will take time and perseverance; unfortunately, most companies have competitors who are also just beginning to recognize the need to make this transition.

Identify the Service Package Components

Once a company has decided that it needs to develop a service package, the next step is to decide which services should be included in the package. Combining services that work in synchronization with each other is especially important to consider. Any services unnecessary for future operation should be eliminated. The company should look particularly for duplication among different functional areas. Is there redundancy in data collection or reporting? If so, the organization should decide where and how best to perform the work. If the service cannot be eliminated, there may be a way to simplify it. This is a good time to convert a marginally running process into an excellent process. After simplification, it may be possible to combine some of the services into a new process that spans previously separated functional areas. Companies should not be bound by the old functional arrangements; the new combination of services may offer a way to streamline the organization.

Develop Objectives for the Service Package

In this phase, management should decide what they want the service package to accomplish. Is the objective to increase the effectiveness of the supply chain part of the business? Is it to reduce the cost of providing the services? Is it to increase the effectiveness and efficiency of the service package components? Specific objectives for the supply chain part of the business should be developed. Such traditional measures as sales volume, profit, return on investment, and inventory turns are a few of the useful tangible measures. It is more difficult to measure the effect of the service package on the effectiveness of the supply chain, or even the efficiency of the service package. Nevertheless, it is important that there be some way to measure the performance of the service package. For example, a university that is seeking to add an online program to its academic offerings may set a goal to attain a certain level of enrollment while staying within a specified budget. Developing performance measures in this area will be an interesting challenge for both practitioners and academics.

Develop Strategies for the Service Package

Strategies are the action plans to achieve the organization's goals. Companies should first develop systemwide or cross-functional strategies that involve integration, collaboration, and all the other steps necessary for success. Initially, this step will be difficult because it has not been done in most companies. Rather than frustrate all the individual functions with an overabundance of business strategies, it is a good idea to go slowly in developing overall service package strategies. Let them come naturally, one at a time. Implementing one good strategy that gains acceptance throughout the organization is better than failing to implement multiple strategies. One of the first strategies will need to be the organization structure for the service package. This may require some changes in the basic functions of the business. The traditional structure may no longer be suitable for the integrated service package approach. For example, the university seeking to develop its online and nontraditional education programs may set up a new department to oversee this endeavor.

Develop Objectives for Each Component

Once management decides on a strategy or strategies for the total integrated service package, each component, or functional area, should develop objectives for their part of the total service package. These objectives should support the overall strategies. Managers of the functional areas need to recognize that their objectives must fit with the total organizational strategies and not just for their functional areas—a global optimum, not a local optimum.

Develop Strategies for Each Component

Once the objectives are identified, strategies to meet these objectives should be developed. These action plans will usually be of a project nature and should have schedule dates, costs, and outcomes well defined, along with the responsibilities assigned for each part of the plan. It may not be realistic to plan every strategy at the same level of detail. The most critical should be carefully planned; other plans of less importance may depend on more "hands-on" management.

Implement the Strategic Plan

After the planning phase, the plans must be implemented. Although implementing an integrated service package may not carry the excitement of implementing the supply chain part of the business, it is important and should be done with the professionalism deserved. It will be easier to see the costs of this program than its benefits, so top management must have faith that it will eventually pay off in the long run.

Evaluate the Results

Just as with any type of management program, the progress of the implementation and the value of the outcomes need to be carefully assessed. This does not mean that the monitoring has to be in real-time; however, it should be timely, accurate, and with the full awareness of not only the participants but also the other stakeholders in the company.

Revise as Needed

Any plan should be dynamic, and be revised to reflect changing conditions. The service package program is supplementary; therefore, it will also be reactive to the supply chain part of the business, not proactive in its own right. Revision to reflect changing needs is appropriate; this type of planning is, after all, a fluid endeavor.

Summary

In Chapter 10, we described how a company could move toward an integrated and agile supply chain. Regardless of the company's role in the supply chain—fabricator, assembler, distributor, or retailer—it is an essential part of the total supply chain. The power of the supply chain is that it integrates a number of activities that were previously performed independently or, at best, with localized interfacing in a limited number of companies.

In this chapter, we have described the services that a business must provide, either internally or through outside suppliers, to be successful. There can be no disputing the fact that services are essential to complement the supply chain functions. We believe that organizations will be served best if they view these services as being a part of an integrated package of services that integrate smoothly into the supply chain part of a company's business.

References

1. Gershuny, J.I. and Miles, I.D., *The New Service Economy, The Transformation of Employment in Industrial Societies,* Praeger, New York, 1983.
2. Greenfield, H.I., *Manpower and the Growth of Producer Services,* Columbia University Press, New York, 1966.
3. Sako, M., Outsourcing and offshoring: Implications for productivity of business services, *Oxford Review of Economic Policy,* 22, 4, 499, 2006.
4. Greenfield, H.I., *Manpower and the Growth of Producer Services,* Columbia University Press, New York, 1966.
5. Beyers, W.B., Impacts of IT advances and e-commerce on transportation in producer services, *Growth and Change,* 34, 4, 433, 2003.
6. Collins, J., Workplace wellness, *Business and Economic Review,* 51, 1, 3, 2004.
7. Juleff-Tranter, L.E., Advanced producer services: Just a service to manufacturing? *The Service Industries Journal,* 16, 3, 389, 1996.
8. Hartley, R., *Business Ethics: Violations of the Public Trust,* John Wiley & Sons, New York, 1993.
9. Steiner, G. and Steiner, J., *Business, Government, and Society: A Managerial Perspective* (11th ed.), McGraw-Hill Irwin, New York, 2005.
10. Robbins, S. and Coulter, M., *Management* (9th ed.), Pearson-Prentice Hall, Upper Saddle River, NJ, 2007.
11. Goldsmith, S., Can business really do business with government? *Harvard Business Review,* 75, 3, 110, 1997.
12. Thornton, E., Roads to riches, *Business Week,* May 7, 2007, p. 50.
13. *The American College Dictionary,* 2nd college ed., Houghton-Mifflin, Boston, 1982.

Chapter 12

The Future of Improvement Programs

Introduction

In Chapter 1, we presented a traditional approach to improvement programs—logical and sequential. In this chapter we look ahead to describe how improvement programs in the future will enter a murkier environment—more complex and chaotic.[1] Although improvement programs will still be essential to the success of a business, they will become more closely intertwined with the sustaining and problem-solving parts of the business so that it will be more difficult to separate one mode of operating from another.

The Background to Improvement Programs

To understand the future of improvement programs, we will take a brief look at where we have been in this book. Each chapter presented a topic that served as a building block to initiating improvement programs. We showed the vanishing boundary between manufacturing and service industries in that improvements from manufacturing can be used in services and vice versa.

The Vanishing Manufacturing/Services Boundary

Chapter 1, The Vanishing Manufacturing/Services Boundary, described the ways in which the manufacturing/services boundary is being eliminated, or bridged over. The material was organized to tie in with the rest of the book.

449

The Foundation Topics

Chapters 2, 3, and 4 described the key topics that relate to improvement programs.

In Chapter 2, Critical Success Factors and Strategic Planning, the increasing array of critical success factors (CSFs) was presented. CSFs were identified as those items that determine what it takes to compete; things a company "must" do well to compete effectively. This chapter explained that the manufacturing/services boundaries are eroding. Rapidly increasing global competition is forcing companies to compete more effectively to survive. They do this by identifying the CSFs needed and building them into their strategic and tactical planning.

Chapter 3 discussed The ITO Model—the basic component of the supply chain. It was explained that the manufacturing/service boundaries are disappearing because we "can"; that is, we have the capability to build relationships, both physical and personal, that are meaningful and productive. We are learning how to connect between entities in a supply chain and build the product–service bundle to be both effective and efficient.

Chapter 4, The Role of Management Programs in Continuous Improvement, showed that business innovation and improvement involves the blending and packaging of management theories into programs. Businesses have found that it is often convenient to use "programs" to focus attention on one, or a few, CSFs. To reduce costs, we have a program of continuous improvement. To improve quality, we have a total quality management program. Programs involve packaging basic management ideas into a package that addresses a specific need—from management theory to a systems approach using contingency theory to customize it.

Knowledge Transfer across the Manufacturing/ Services Boundary

In Chapters 5 and 6, some of the ways in which knowledge is being transferred across the manufacturing-services boundary was shown. Chapter 5 described how improvement programs that originated in manufacturing have been extended to the services sector. Chapter 6 described how improvement programs that originated in the services sector have been extended to the manufacturing sector. These chapters showed that the lines between manufacturing and services are blurring so much that it is becoming more difficult to identify where programs originate and how they move from one sector to the other. Some programs, such as supply chain management, share in the knowledge from both sectors.

Chapter 5, Adapting Manufacturing Techniques to Services, described the major programs that have been used to move manufacturing concepts and techniques into service applications. Cost reduction and quality improvement have been primary objectives, such as in JIT, which has morphed into lean production, and TQM, which has been largely superseded by Six Sigma, although there have been a number of variations in a program's appearance. Two companies—United Parcel Service (UPS) and Amazon—were described as companies that have adopted improvement programs developed in manufacturing to further their success.

Chapter 6, Extending Service Techniques to Manufacturing, described the concepts and techniques that originated in service industries and extended into the manufacturing areas. These efforts have centered largely on customer service programs, such as CRM and Quick Response programs. In services, much of the attention is on the proper use of employees, as well as equipment and facilities. Two companies were described—General Electric (GE) and Hewlett-Packard (HP)—that have moved into services, but with mixed results.

Agents of Change

In Chapters 7, 8, and 9, three change agents that affect the success of any improvement program—technology, structure, and culture—were described. These disparate elements must blend as a synchronized force, making it possible for improvement programs to reach their intended goals.

Chapter 7, The Role of Technology in Continuous Improvement, described the progression in using technology to develop more sophisticated, and complex, processes that facilitate the application of resources, such as people, equipment, and facilities. Technologies that drive the processes, such as CNC and FMS, were described as well as the major increases in the application of information processing technology, such as ERP systems, RFID, SOA, and cloud computing.

Chapter 8, The Role of Infrastructure in Continuous Improvement, showed that structure is how companies organize and how they support the input–transformation–output (ITO) process. It was shown that in many cases, companies have stressed the application of technology but have not changed their structures, and that one of the most significant needs is to move from a functional orientation to a process or horizontal orientation. The need for accurate, complete, and rapid communication necessitates a different way of organizing. Key issues mentioned included outsourcing, teams, and project management.

Chapter 9, Understanding Organizational Culture—The Elusive Key to Change, discussed that just as with structure, companies have often forgotten to proactively consider the implications of change for employees. The chapter described how employees are expected to understand intuitively, accept, and participate willingly in the change process, even though in many cases the change may be detrimental to their personal welfare—loss of earnings or even their job. It was shown that as a result, it is important for managers to understand the fundamentals of organizational culture. This chapter presented those fundamentals, and outlined how to manage the culture change process.

Integration of Related Entities

In Chapters 10 and 11, the segments described in earlier chapters were blended. Chapter 10 showed that the main stream of a business flows along the supply chain and builds from batch operations to an agile and integrated system. Chapter 11 showed how the support services necessary for a business to be successful could be organized into an integrated services package to support the supply chain described in Chapter 10.

Chapter 10, Integrated Supply Chains—From Dream to Reality, described the core supply chain from raw materials to end consumer. An overview was given of the phases that a company goes through in developing its supply chain and the way the pieces fit together. Building an agile and integrated supply chain is a complex process that requires a blending of technology, structure, and cultures. This is not as sequential and well-orchestrated as companies would like and often involves false starts and the need to reorient as circumstances change, but some companies persevere and eventually achieve a workable version of a supply chain. Some of the current issues in supply chains were also described, including how to achieve equity among participants, how to use the ITO process to move from a gathering of independent entities into a closely linked unit, and how to make the supply chain relevant to today's needs but flexible enough to change to meet tomorrow's needs.

Chapter 11, The Role of Services to Complement the Supply Chain, addressed the links from support services to business, such as providing group health care plans, equipment financing, product liability insurance, and management consulting. This chapter also presented support services provided to the individual consumer, such as hospital services, home mortgages, automobile insurance, road maintenance, and preparation of tax returns.

In the first eleven chapters, a rapidly changing business world is described. This will change even more in the future. In this final chapter we offer some ideas about changes that could face the business community in the years ahead. The need to discover how to move from functional silos to integrated entities, from left-brain to right-brain thinking, from local to global optimization, and from repetitive perfection to dynamic "satisficing"[2] is discussed.

Future Areas of Emphasis

The future will most likely not be a linear extension of the past; it will have twists and turns that may not be apparent at first, but will appear logical after they occur. This follows the reasoning offered in chaos theory, a subject we describe briefly in a later section. We offer the following as our opinion about future areas of emphasis for organizations.

Services Will Continue to Increase as a Critical Success Factor in Business

Several factors will force companies to increase their dependence on services as CSFs in the future. To compete successfully, they will need to keep increasing the level of the product/service package they offer to consumers. In many cases it will be easier to add services than it will be to add product features.

Continuing Need to Integrate the Product and Service Bundle

As businesses add services to their product offerings, there will be a greater need to add these services as part of an integrated product/service package. A piecemeal approach will not have the appeal of one that has a holistic appearance.

Continuing Increase in Globalization

Globalization will continue to expand, for two main reasons: the result of a search for lower costs, and, perhaps even more important, the desire to locate operations closer to the market. This will necessitate the ability to manage a diverse and widespread organization structure effectively. Of particular interest is the movement of nearly every major automaker to China. Partnerships with Chinese automakers have allowed automakers to market their cars in China and take advantage of low labor costs as well. The future configuration of the auto industry will be intriguing to watch. Is it possible that U.S. automakers will manufacture vehicles in China, only to export them back to the United States? This scenario may seem worrisome, and only time will tell what the U.S. auto industry may look like in the future.

Outsourcing Will Become a More Focused Activity

As part of the globalization expansion, outsourcing will continue to be a major consideration. Although the past motivation to outsource has been largely the quest for lower costs, this will begin to change in the future. When businesses analyze outsourcing from a total cost of ownership perspective and consider such indirect costs as longer lead times, increased inventories, transportation costs of products back to the home country, and flexibility constraints, they will become more selective in their outsourcing efforts. Although they will continue to outsource, it will be viewed from a strategic objective.

Increased Need for Project Management Competencies

The requirements for good project management will grow. The dynamic nature of business will increase the need to view product development, outsourcing, and improvement programs as projects that need careful management not only to achieve desired results but also to ensure that projects are completed and do not interfere with the normal operation of the business.

Decision Making Will Deal with a Blend of Hard and Soft Variables

Managers strive to make good decisions. They try to accomplish this task by analyzing facts—hard numbers about costs, sales, productivity, response times—or a variety of other bits of data. Scholars have built an impressive array of methodologies to help in the decision-making process—time value of money, linear programming, expected value analysis, and economic value added, just to name a few. These analyses have had difficulty in dealing with soft numbers and bits of information that are more difficult to quantify—customer satisfaction, consumer optimism, propensity to buy, and job satisfaction. Several noted authors warn that decision making in the future will require a blend of hard and soft variables,[3–5] so we will have to rethink our dependence on numbers and rely more heavily on the judgment or wisdom of the decision makers.

Decisions Will Become More Complex

Future decisions will be more complex because they will have further-reaching consequences. They will require more advanced decision-making methods. Will the answer be in more quantitative solution techniques? Will they come from greater involvement of consumers—through surveys, focus groups, and blogs? Data analytics, or the analysis of "big data," is being advocated by several of the major consulting firms as an answer to the value of using all the data being collected at POS terminals, online transactions, and even social media sites. Will companies have to be satisfied with less than optimum decisions because of the insurmountable complexity or, to use Simon's term, "satisficing"?[6]

Future of Improvement Programs

Some of the areas of emphasis we see ahead for organizations have been outlined, especially business organizations. In this section the changes to expect in this area are described.

Improvement Programs Will Become Increasingly Important. If businesses do not continue to improve, they will ultimately fail.

Improvement Programs Will Be Customized to Specific Applications. Companies will become more proficient in selecting and adapting the improvement program to their specific need and situation.

Improvement Programs Will Be Better Managed. Businesses will plan, introduce, implement, monitor, and measure each step of the program's life so that it will be accepted and used for the benefit of the company, the employees, the customers, and other stakeholders.

Improvement Programs Will Become a Regular Part of Business Operations. It will be important that the business assimilate the positive features of the program into its sustaining operations. Bolton and Heap call this the "lock-in" phase, in which the business takes time to consolidate its gains before going on to the next improvement program.[7] Juran calls this the control phase of a company's life before moving on to a "breakthrough" or the next improvement program.[8]

Improvement Programs Will Lose Their Manufacturing or Service Heritage. In the future, improvement programs will be a blend of manufacturing and services joint efforts, not a product of one or the other.

The Drivers of Change

As we described in Chapters 7, 8, and 9, the three major drivers of change are technology, structure, and culture. In the past, these have been separate entities, sometimes at odds with one another as businesses attempt to introduce improvement programs. These conflicts will have to change in the future, but how will it happen? An analogy that may seem out of place in a business setting—children, families, and communities—illustrates the relationships.

Technology

Technology is like children. It is spontaneous, sometimes radical, often disruptive, and seldom traditional. It is a driver of change. It wants to go places, see things, pose questions, and rebel against the status quo. Technology wants to deviate from the traditional ways of doing things but is often constrained by the structure of the organization, just as children are constrained by the structure of the family.

In the past, children were more influenced by their parents than by their peers or society in general. This is changing as children have greater exposure to the outside world. In a sense, technology is going through the same transition. At one time, technology grew from focused efforts within an organization. Today, technology advances come from knowledge transfer across multiple disciplines, fostered by electronic communication systems and more open information accessibility.

Structure

The structure of an organization is like the family structure, providing vision, mission, strategies, programs, and processes to carry on the daily routines of operation.

Although it may sometimes be a barrier to change introduced by technology, eventually it accepts good ideas and squashes bad ideas.

Structures of today's organizations are changing, some rapidly. The technology of the Internet, the openness of outsourcing, and the kinetic energy of employee empowerment are examples of the need for technology and structure to accommodate each other if there is to be harmony in the operation of the business.

Culture

The culture of an organization is like the community. It responds to the antics of technology, often with disapproval or rejection. Only over time will it adjust to, or accept, the change that technology offers. Organizations with deeply embedded cultures will find it difficult to implement any kind of improvement or change program. To improve, they must find a way to change the culture of the organization, a formidable task.

Sometimes communities take the initiative in change, perhaps as the result of a crisis or a reaction to some anticipated action with which they can mount organized opposition. When this happens, the culture can become a driver of change, although the momentum often is short-lived and somewhat unpredictable. It is not likely that cultures will change dramatically in the future. Companies that implement improvement programs will need to manage their organizational culture change as well.

Most Likely Future Methodologies

What are some of the methodologies, or programs, that will become popular in the future? Some are already well along the program life cycle, while others are in the early stages of discovery. They are described below, in no particular order of importance or timing.

Integrated Supply Chains

In Chapter 10, we described how to build an agile and integrated supply chain. To our knowledge, no company is there yet. Although some have moved well along the path described, most businesses are still in the early stages. This journey will consume a large part of a company's effort in the future. They will probably move quickly in setting up the physical parts of the supply chain—the customers, the products, and the suppliers. It will take much longer to develop the collaborative relationships that will make the supply chains most effective. Unfortunately, once a specific supply chain is achieved, it will become obsolete and a new supply chain will be required. Companies will have to be agile in creating new supply chains, perhaps using the virtual organization idea described in a later section.

Outsourcing

Outsourcing, especially offshore outsourcing, is currently in the growth part of its life cycle. From its humble beginnings as a "make or buy" decision area in manufacturing, outsourcing has grown to become a strategic weapon to gain a competitive advantage in both manufacturing and service activities. As it matures as an improvement program, outsourcing will be refined and adapted to fit specific business needs best. Some of the expectations for the future include the following:

- The outsourcing trend will continue in the United States; however, many companies are being more cautious and some are recalling outsourced work because of unsatisfactory results.[9] *Nearsourcing* and *reshoring* are terms currently discussed in the literature and have been applied in a limited number of companies.
- Extensive outsourcing will require a major restructuring of the purchasing, or procurement, function. Purchased services will increase as a percentage of the total costs and their composition will increase in complexity. It will take a multifunctional team to manage the outsourcing programs effectively.[10]
- Project managers and project management skills will become increasingly important. Although some in-house projects can be nursed along informally, outsourcing requires the formal coordination of functions and tasks, both internal and external.
- Although politicians lament the trend, it appears unlikely that the federal government will do anything substantial to stop the outsourcing movement.[11] Although there may be short-term hurdles, there will not be permanent barriers.
- U.S. companies will continue to outsource, but at a more deliberate pace as they move up the learning curve. Companies will view outsourcing as a management function and will learn to analyze, plan, manage, evaluate, and control the process. When appropriate, outsourcing will become a core function of a business.

Developing a successful outsourcing program is a little like making a soufflé. When done correctly, it is a masterpiece to enjoy. When done incorrectly, it is a mess to clean up.[12]

Total Cost of Ownership

One of the outgrowths of the outsourcing movement will be a more comprehensive method of determining the total cost of ownership of products and services. At the present time, the direct savings in labor costs is often enough to justify an outsourcing program. The major negative is increased lead time; other problems include variances in quality and communication barriers. Of course, any time domestic jobs are lost to offshore outsourcing, the local community suffers. At some point in the future, these direct savings will decrease and indirect benefits and costs will become a more important consideration in the decision-making process. Such factors as value will assume greater importance. What is the value of more or less customer service? What is the value of proprietary knowledge that has been transferred to a supplier? What is the value of being a global market participant? Answering such questions will require a blend of known quantitative techniques coupled with new approaches to be discovered and refined.

Performance Measurement

Just as decision making will involve a blend of quantitative and qualitative variables, so will performance measurement. The traditional performance measurement systems have been focused on overall company financial performance, and sometimes the specific measures used at lower levels of management were difficult to relate to overall performance. Consequently, depending on the use of the measures, they may or may not have been meaningful. To make performance measurement systems more effective in the future, several things have to happen:

- Performance measures should be selected to meet specific needs. It may not be possible to use the same measure to assess multiple tasks. For example, inventory turns may be a useful measure for inventory management purposes but not necessarily for measuring the performance of a supervisor of a parts fabrication shop. Some measures work better for groups than for individuals.
- Performance measures should reflect the strategic and operational plans throughout the organization. If the plan is to minimize the level of work-in-process inventory, then machine utilization is probably not a fair measure to evaluate the effectiveness of a department.
- Local optimization should be discouraged and global objectives encouraged. Most businesses are trying to optimize the performance of their supply chains, which may not be conducive to optimizing the performance of local functions or departments. It is difficult to encourage and facilitate interdepartmental or intercompany collaboration; using performance measures that introduce conflicts only makes it more difficult.
- Both quantitative and qualitative measures should be included. As discussed in earlier sections, both quantitative and qualitative measures should be considered in decision-making analyses. It follows that businesses should learn to use qualitative variables as performance measures. This will encourage the acceptance of such concepts as "satisficing"[13] and "it is better to be roughly right than precisely wrong."[14]
- Performance measures should be accepted by everyone who uses them. Although sometimes difficult and always time-consuming, it is a good idea to

get acceptance of the performance measures by all of those being measured. Most employees do not like to have their performance measured but find it tolerable if they believe the measure used is fair and appropriate.

■ Today, the concept of data analytics, or the analysis of huge amounts of unorganized data, is becoming a topic of great interest. Technology is making it easier to collect the data; however, there is a need to find ways to analyze the data and convert the results into meaningful measures of business performance.

Project Management

Project management is an underutilized methodology. It has been around for several decades and is used in the military and in major construction projects. However, it can be useful within organizations to plan and monitor new product development, improvement programs, outsourcing projects, and other unique activities. As company operations become more volatile and complex, there will be greater opportunities to use project management techniques. Businesses are changing from a repetitive transaction orientation to a customizing process orientation.

Mass Customization

Like outsourcing, mass customization is in the growth part of its life cycle. The objective of mass customization is to provide a customized product or service to an individual customer—a market of one. Examples include Dell's customizing a computer, Subway's customizing a sandwich, and a physical therapist designing a treatment program for a patient. The objective is to provide high-volume customized products or services at prices competitive with standard products and services. Mass customization is closely allied with the concept of postponement where the customizing is completed on standard modules.

One approach may be mistakenly thought of as mass customization. Having thirty-six varieties of canned tomatoes at a grocery store is not mass customization, but rather is niche marketing. Introducing greater variety may be more an attempt by a company to coax out extra sales than a response to consumer needs or wants. It has been suggested that having so many choices causes stress to consumers because of the time involved and anxiety caused by making the decisions.[15] Increased variety in manufacturing or service operations introduces another form of variance that is difficult to manage.

An interesting conflict arises between mass customization and offshore outsourcing. Mass customization needs short response times; offshore outsourcing requires longer response times. It will take a careful balancing of the two programs to arrive at an acceptable blend of the two concepts. One approach is to outsource the standard modules and perform the customizing as a domestic operation.

Virtual Organizations

Virtual organizations make it possible to adapt the organization to the specific need, whether it is to launch a new product—the Boeing 787—or to assemble assistance for responding to a natural disaster—a hurricane. Virtual organizations may exist for varying lengths of time—enough to get the job done. It is no longer feasible, and often does not make economic sense, for a company to design a product, build facilities to produce it, design and launch a marketing program, arrange financing, and do all the other things necessary to introduce a new product. This made sense when the product life expectancy was years or even decades and changes in the succeeding products were minor. Today product life cycles are usually in months or years at most, and changes in succeeding products may be substantial. Today, the answer is to fit together a group of participants who can work together to accomplish the overall objectives—the virtual organization. Although the concept is still in the early stages, it makes sense and will become a more prominent factor in the future.

Information Technology

Information technology (IT) has enabled organizations to do things that previous generations only dreamed of when reading Jules Verne or H.G. Wells. IT is only in the early stages of its potential. To accomplish supply chain integration, offshore outsourcing, mass customization, and virtual organizations, IT is necessary. Inter-organizational systems (IOS) to facilitate greater cooperation and coordination among businesses on projects will become standard. Advancement in such technologies as service-oriented architecture (SOA), cloud computing, and virtualization will make it possible for "my computer to talk with your computer." As John Teresko, veteran columnist with *Industry Week,* puts it: "Remember when the factory floor and enterprise IT were two separate, unconnected worlds? Unfortunately, integrating these two disparate departments still remains a challenge for much of manufacturing, according to Cisco Systems and Rockwell Automation. Their advice: Think of integrating them as a way of future-proofing your business."[16] IT will be essential for companies of all sizes, both manufacturing and service.

Environmental Design

Business, society, and government do not always work well together. One exception in the future may be in a program to protect the global environment. Society is becoming increasingly aware of and alarmed at the effects of climate change. Governments are noticing that society is concerned and are taking steps to announce their intentions to act. Businesses are looking for ways to accommodate the need to change their ways and, if possible, to make money in the process. This mixture of concerns will likely lead to some positive actions in the future. Right now, the efforts are piecemeal and cautious; however, there seems to be enough momentum to mount a significant effort. We may be nearing the tipping point in the sustainability movement.[16a] This will necessitate the use of supply chains, knowledge transfer, cross-functional teams, virtual organizations, IOSs, and a host of other tools and techniques to make such an effort successful. In a sense, it is a massive improvement program that incorporates the ITO Model introduced in Chapter 3, with the objective to transform an input of an unprotected environment into an output of a protected environment. Incorporated in this transformation is the concept of reverse logistics.

Service Sciences

Chapter 6 described some of the improvement programs that originated in services and how they could be extended into manufacturing and other service organizations. Chapter 10 described how distributive services could be combined with extractive and transformative industries to form integrated and agile supply chains. Chapter 11 described how producer and social services form an integrated package to supplement the supply chains. Services are a vital and essential part of the world's economy.

However, most of the positive results from services have been achieved without a formal science or theory of services. Today there is a growing interest in the field of services sciences. There is a need to enrich the contributions of services by adding a scientific approach—in the form of quantitative analysis theories and increased IT applications—to the already significant intuitive and ad hoc approach to service applications. Services sciences is a field that is gaining recognition among scholars and practitioners.

Chaos and Complexity

There are numerous variations of the "butterfly" illustration used to describe events in chaos theory, one of which is that a butterfly flapping its wings in Brazil can create tiny air currents that eventually lead to a hurricane in the Gulf of Mexico. The butterfly example illustrates a fundamental concept in chaos theory called "sensitive dependence to initial conditions."[17] This concept maintains that small differences in initial conditions can have vastly different outcomes. This theory is useful to business researchers who attempt to understand why similar companies in the same industry may have very different success outcomes. Chaos theory has also had applications in the area of understanding crisis events in the life of an organization.[18]

Perhaps the most promising aspect of chaos theory is its use in understanding complex phenomena, or the type of environment in which most organizations find themselves. Although technically chaos theory is rooted in mathematics and science, its use in understanding complex business phenomena is mainly metaphorical. While this area may seem somewhat abstract, applications of chaos theory have been made in a number of areas, including the fields of psychology,[19] social work,[20] strategic management,[21] health care management,[22] public relations,[23] and public management.[24] Its future use looks promising. We need to develop a quantitative methodology for use in business applications.

Most Likely Improvement Programs

What kinds of improvement programs will we see in the future? What will be their goals? Probably much the same as today. Here we look at the types of improvement programs that will continue to receive attention, and in a later section we look at the industries in which these improvement programs will have the greatest potential.

Cost Reduction

The pressure to reduce costs is almost a given in considering improvement program objectives. No matter how much businesses do to reduce costs, customers expect them to do more. Although there will be continued interest in reducing product and service costs, the area of emphasis may shift in the future, to include greater efforts on reducing capital costs.

Product or Service Costs

Historically, companies have focused on reducing the costs of making products or providing services. With the help of engineers and empowered employees, companies have reduced direct labor costs to a very low percentage of the total cost. Many companies then switched their emphasis to the direct purchase side of the costs and worked to reduce direct material and other direct purchases through a variety of improvement programs and, more recently, outsourcing. In addition, some companies are working to reduce overhead costs. This is more difficult because while it is easy to identify the components of direct materials and direct labor costs, it is more difficult to identify the components of overhead costs because most accounting systems conceal their true identity. However, overhead costs, as a group, are now usually two to four times as large as direct labor costs in manufacturing plants, and cost reduction programs must find ways to address these areas. Cost reduction programs of the future will continue to address direct labor costs, but the emphasis will also shift to direct purchases and overhead expense areas.

Working Capital Costs

Working capital consists of accounts receivable, inventories, and accounts payable. Although it may at first seem that businesses do not want to reduce accounts receivable, there may be another reason. Customers may withhold payment for products or services because of an unresolved problem. This may be partial delivery, defective units, or errors in the invoice. Whatever the reason, it means that the business has excess accounts receivable because of unresolved issues. This could go unnoticed if the management information system does not report it to someone who can take action. As a result, there is a cost associated with having excess accounts receivables in that cash flow is decreased.

Management pays a lot of attention to inventory, but often the focus is on having enough of individual items versus having too much in total. Inventory has two primary costs—having inventory (carrying costs) and the cost of being out of inventory (stockout costs). Any program to reduce costs must recognize the need to find the right level of inventory not just in total but also for individual items. That is a huge task because many businesses, especially those in wholesale and retail, have thousands of items to manage. Although computer systems can do a lot of the work on staple items, considerable human input is required for seasonal and style items. Inventory will remain a key area for cost reduction; however, this area requires finesse, not brute force.

Accounts payable has also been an area where many companies have taken the approach that they just want concessions from their suppliers. Large companies sometimes use their clout to force smaller suppliers to grant extended terms, or sometimes they do not pay their suppliers on time. As building collaborative relationships becomes more important in future supply chains, greater consideration will be given to the total purchase package so that accounts payable will become part of a holistic solution.

In addition to lean production programs, another program that stresses the need for throughput improvement is the Theory of Constraints (TOC). It is defined as a "holistic management philosophy developed by Dr. Eliyahu M. Goldratt that is based on the principle that complex systems exhibit inherent simplicity."[24a]

Capital Equipment and Facilities

Companies spend time and effort to justify their acquisition of capital equipment and facilities; less time is spent considering the costs incurred in operating these investments. Even small companies perform time value of money analyses and consider buy versus lease alternatives. However, just as with overhead costs, it is more difficult to determine the cost of operating capital investments and the cost reduction potential in that area. There are isolated instances such as when the cost of fuel attracts attention for a large truck fleet, or when seasonal surges overload the capacity of a distribution center. However, the availability of such programs as outsourcing makes it possible to consider their impact on reducing the costs of capital equipment and facilities. Such analyses require reliable operating cost information.

Response Time Reduction

As described in Chapter 2, response times are rapidly becoming a key CSF for many businesses, both manufacturing and service. James Womack and Daniel

Jones reported that it takes an average of 309 days to move from aluminum oxide to an aluminum soft drink can. Although that seems like a long time, the amazing part is that during these 309 days, only about three hours' worth of work is performed.[25] That suggests that the total response time could be reduced by reducing inventories and move times. If the business community is to move to mass customization, it must find ways to reduce response times.

- ■ Existing products or services. There are several variations of response time. One is the time to respond to a customer's order for an existing product or service—a can of peas at a grocery store or a book from an online seller. A variation of that is the time to deliver a product or service that requires some slight modification—a sub at a deli counter or a computer with selected accessories added. An even longer response time results when the supplier must order the raw materials or assemble the service package—a wedding cake or public stock offering. Finally, the longest time is when the product or service must be designed—a house or an entire wedding. These examples follow a hierarchy of approaches known in manufacturing as "make to stock" (MTS), "assemble to order" (ATO), "make to order" (MTO), and "engineer to order" (ETO). MTS has the shortest lead time and the least customization. ETO has the greatest customization and the longest lead times. Although the concepts are not new, the emphasis is now on how to increase customization while reducing lead times.

At some point in these examples, the product or service may cease to be an existing model but becomes a new model. This moves the discussion to response times to design new products or services. The automotive industry has moved rapidly in reducing the time it takes to design and move new models into production. What used to take five to ten years now takes twelve to twenty-four months. Techniques such as concurrent engineering are used to achieve these reductions. The reductions in time have come largely because of reducing wait times caused by sequential processing, where one functional area waits for upstream participants to finish their work. Another reduction results from less recycling of designs as participants resolve questions. These reductions are similar to the reduction of wait and move times described above. The actual design may be reduced somewhat, but the greatest reductions have been in reducing non-value-added activities.

Response times receive little attention in the area of response time to implement improvement programs. This book is filled with ideas about improvement programs. If these programs can provide a benefit to companies, the sooner the programs are implemented, the sooner the company can begin receiving the benefits. In the future, companies will need to pay attention to faster planning and implementing of improvement programs.

Quality

The quality of products and services is a basic CSF, as are cost reduction and response time reduction. Both TQM and Six Sigma are improvement programs aimed directly at improving quality. Although they both originated in manufacturing, they are being successfully implemented in a number of service applications, as described in Chapter 5. Unfortunately, there are many examples of only partial successes or failures.

Product quality lends itself to measurement of tangible attributes. Does the product conform to its specifications? It is usually easy to know when a product is defective, a sure measure of quality. Product failure can often be corrected, hopefully before there is major damage or injury. Companies will continue to emphasize their efforts to improve product quality.

Another aspect of product quality is reliability. Historically, most consumers were more concerned with quality at the time of their purchase; today they are also concerned with the expected reliability of the product over the time they will own it. This affects their estimate of value, measured as the lifetime cost of a product. A growing number of consumers will pay more initially for a product if they believe it will last longer and have lower operating costs. Automobiles and appliances fall into this category. As with overhead and stockout costs, it is still difficult to determine the lifetime costs of a product.

The quality of services is even more difficult to evaluate because a service may conform to specifications but be unsatisfactory to a consumer. In evaluating service quality, a consumer looks at both the service and the process of how the service was provided. Did the cashier smile when she scanned your groceries? Was the waiter pleasant when he brought you the menus? Did the doctor spend enough time explaining the surgical process he was going to perform? Did the technician ask if you were comfortable as she started the ultrasound to clean your teeth? Did the help desk attendant adequately understand your problem and offer an answer that not only solved your problem but also matched your level of computer literacy? As the world moves to a greater dependence on services to provide competitive advantage, it becomes more important to understand service quality and how to improve it.

Customer Service Level

Customer service has moved into the forefront in recent years, as the movement to be responsive to customer needs becomes the foundation on which companies build their strategies. It could be considered part of service quality, but it is probably better to view it as a separate CSF. We discussed the role of the customer extensively in Chapter 6. To paraphrase Drucker, businesses only exist to serve the customer.[26] Customer service has many facets. It requires that the order taker quote a realistic delivery date, the inventory manager has the item in stock, the distribution center sends the correct item, the delivery person delivers the item to the correct address, the invoice is correct, and the check from the customer is credited to the correct account. These actions and many more are part of customer service. The service side of customer service is as important as the product side of the transaction.

Some customer service items can be measured, such as the fill rate on orders or the number of invoice errors. The number of buttons telephone callers have to push before being connected with a person, and how many transfers they must wait through before they connect with someone who can help them can be traced. No matter how proficient companies become at measuring the measurables, they will still have to deal with the intangibles of individual customer differences. When we can identify and understand the DNA of humans and what makes them different, maybe we will be able to anticipate and prevent a customer from becoming dissatisfied.

Flexibility

Emphasis on becoming lean and reducing inventories has required companies to become more flexible. Flexibility in this case means handling volume and mix fluctuations within the present system. Manufacturing and distributive companies have previously handled volume fluctuation by building inventories in the slow periods and selling them down during the busy periods. This worked, but at the expense of higher average inventories. It also worked best with stable product lines in which new products were introduced at a pace that enabled companies to phase out the older products in an orderly fashion. To operate with lower inventories and with greater turnover in products from year to year, companies have to formulate better forecasts. To accomplish this, companies need to share information. Flexibility is as much about information sharing as it is about processes that can adjust to volume and mix variances.

Companies have worked to design flexible capacities to match flexible demands. This is especially true in scheduling employees. Service companies have become very good at scheduling employees to match demand in such businesses as restaurants, grocery stores, and clothing stores. They have volume fluctuations daily, weekly, and seasonally. In the future, they will need to become even better at this type of scheduling. One of the constraints will be moving from the traditional five-day, eight-hour workweek that has been the norm. The business community and employees will have to find a way to accommodate this transition.

Flexibility also implies the capability to handle product and service mix variations. The emphasis has been to reduce the lot size so that the existing process can respond to changes in mix by making a wider variety of products during a single day. To do this, manufacturers have had to reduce setup times so that there is a less severe time penalty in switching from one product to another. This will continue to be the emphasis, and in addition, manufacturing processes will be designed to make setups easier.

A number of service businesses already deal with customer mix variation. Employees have to have the versatility to handle what amounts to a lot size of one. A call center attendant can expect that the next call will be different from the last call. An order taker at L.L. Bean or Land's End can expect that no two callers will want the same product. The process provides information that, in combination with the ability of the employee, makes it possible for the employee to handle a variety of calls seamlessly. Where the process is automated, such as with online bill payments, the computer is programmed to handle the variety. Flexibility will probably join cost, response time, and quality as a primary CSF for most businesses.

Agility

Agility is an extension of flexibility. It is the ability to change from one system of manufacturing or service to a new system to remain competitive. In manufacturing, it is the ability of an automobile manufacturer to move a substantial part of its production capacity from SUVs to hybrids. In retailing, it could be for a retail bookstore chain to launch an online bookselling service (Barnes and Noble), the conversion of a "mom and pop" local chain to a "big box" national chain (Lowe's), or the conversion of a clothing manufacturer to a chain of retail clothing stores (Jos. A. Bank). In banking, it is the introduction of investment advisory services. In health care, it is the addition of a wellness center to help prevent health problems.

The key to being agile is in designing processes that can be converted or added to without being detrimental to the existing process. An even greater requirement is the ability of management to recognize the need and to introduce a program that will successfully achieve the transition. Agility is a strategic and long-term consideration. Only a few companies are doing it, some are thinking about it, and many have not seriously considered it.

Compatibility

As supply chains mature, there will be an increasing emphasis on the compatibility of supply chain members. They will have to work together, and this means increased communication, cooperation, coordination, and collaboration. This will require new technologies in communication as well as a great deal of human interaction, a challenge that will take years to achieve. One of the essential ingredients is trust, and trust is not well understood in the business community. A great deal of research is being done to understand trust and how it is developed not only between individuals but also between organizations.[27] Although this area holds great promise, its realization will take time and patience.

Integration

Compatibility implies separate and distinct organizations working together. The next stage is integration of organizations until they become almost inseparable. This arrangement could mean the integration of individual entities into the supply chain until they are so completely coordinated that they become an integrated chain of separate units. In addition to supply chains, virtual organizations also hold promise as a way to develop integrated entities, if only for a short period of time. This concept could be extended beyond business organizations to blend the interests and capabilities of social and political organizations. The effort in global environmental initiatives would be a worthy objective for an integrated system of organizations.

Integration requires increased information flow. As information is more widely disseminated, there is a greater need for increased cyber security. While supply chains thrive on integrated information, they can also be seriously disrupted by the failure of information flow.

Sustainability

For supply chains to be successful, they must incorporate sustainability practices— they must not only become "green" in their forward direction (from farm or mine to consumer) but must also include the reverse flow of goods (reverse logistics) at the end of their useful life. More sustainable practices can reduce costs and increase the marketability of products and services to an enlightened marketplace.

Risk Management

The world is becoming a more complex environment for businesses and other organizations. Globalization of markets and extended supply chains increase the likelihood of significant reductions in the flow of goods and services. Companies must develop strategies and plans to deal with a wide range of risks, from late deliveries to weather-related disasters.

Industries Most Likely to Stress Continuous Improvement

Which industries will be the most active in implementing improvement programs? We believe that the emphasis will continue to shift to the services areas of business. Some of the industries that should be the most involved in continuous improvement programs are described below.

Health Care: Hospitals and Wellness Centers

One industry that presents tremendous opportunities for improvement is health care. There are governmental and social pressures, as well as business pressures, to effect change. Interested parties are concerned with reducing costs through prevention or mitigation of chronic illnesses, improving quality by reducing errors in patient medication and treatments, reducing response times or lengths of stay, improving flexibility in both volume and mix, increasing agility in handling a greater variety of illnesses and surgical procedures, becoming more compatible (patient friendly), and providing smooth transitions involving a numerous independent operators (including primary care physicians, specialist physicians, physical therapists, lifestyle counselors). The Patient-Centered Medical Home (PCMH), a provision in the Patient Protection and Affordable Care Act of 2010, is being promoted as a way to accomplish some of these objectives. Care providers are expected to become active in all types of improvement programs while at the same time continuing the sustaining and problem-solving parts of their business. They will have a host of new technologies, archaic organization structures (most have to outsource their primary transformation process of surgery), and a culture that is under attack from a number of influential outside sources (including insurance companies, governments, privacy advocates, and trial lawyers).

The traditional role of the health care profession has been to treat the sick. That is no longer good enough. The goal now is to extend life expectancies and to improve the quality of life during that extended period. *Prevent* is the watchword, and there has been an explosion of wellness centers, personal trainers, weight reduction systems and procedures, physical and psychological trainers, assisted living centers, extended care centers, and life extension centers.

In addition, there is the financial consideration of medical insurance that is taxing the best minds of government and industry. What a wonderful opportunity for a few good improvement programs!

Pharmaceuticals

Intimately connected with the health care industries is the pharmaceutical industry. Advertisements for medications abound. Do you watch the six o'clock news? It is difficult to imagine a single person who is not taking medicine for some perceived or real ailment. It begins at infancy and continues throughout life. If there is an industry that views people as lifelong customers, it is the pharmaceutical industry. There is a medication for every part of the body, and for every conceivable ailment. Perhaps in part because of this intense marketing of medications, humans are living longer and, in some cases, such as infantile paralysis or measles, diseases have been essentially eradicated where vaccines are available. How many people do you know that don't take some kind of medication?

There has been vast improvement in developing medications for a variety of ailments. However, the approach has been similar to that in manufacturing—mass production or group solutions. Just as with clothing, one size may not fit all. There may be side effects from medications, a patient may be given the wrong medication, or there may be problems with combinations of medications. The pharmaceutical profession struggles with trying to achieve a Six Sigma level of effectiveness.

Some research, such as in DNA, offers exciting promise. If the medication could be adapted to an individual's DNA, then the right medication could be prescribed for each individual—that is, if we could link up the supply chain that includes the manufacturer, the physician, the pharmacy, and a foolproof system to get a willing and totally compliant patient to take the correct medication. Of course, this means that the pharmaceutical manufacturer would have to enter the mass customization arena to match their medication with the patient. They might also have to make their medication compatible with a competitor's medication and share some of their proprietary information.

Local Government

Government agencies are expected to be effective—do the right things. They are also expected to be efficient. This challenge means they also need improvement programs. It may mean computerized processes to reduce the response time to getting automobile license renewals, or it could mean scheduling nighttime operating hours to accommodate the working public, or simplified income tax legislation to make it easier for individuals to file tax returns. There are cost, quality, response time, and flexibility opportunities in government, just as in industry.

One area that has had some success is privatization. Perhaps some governments, especially at the local level, are too complex and they need to outsource some of their functions to private industry, such as janitorial services or transportation systems. What are the core competencies of local governments?

Retail

Although retail is one of the oldest industries, it continues to evolve as it attempts to keep pace with the changing demands of customers. The advent of the Internet has forced retailers to seek continuous improvement. The Internet has made retailing a 24/7/365 option for the consumer. Even when their physical stores are closed, retailers' websites can continue to generate sales in "virtually" any part of the world where there is Internet access. The Internet has also forced retailers to compete on the basis of time. Deliveries are expected to be prompt and at a competitive price. Customers are usually given a choice of delivery options, depending on how fast they want their product and what they are willing to pay.

Retailers are also candidates for improvement programs related to quality and inventory control practices. Quality of service is expected from all retailers today, as consumers have many options to choose from and have high expectations on how they want to be treated. Inventory control is still important for all the traditional reasons; the product must be available to the customer, and yet holding costs must not be prohibitive. Online selling is also changing the logistics. More packages are being delivered directly from the manufacturer or distribution center to the customer, and bypassing the retail store.

Education

Improvement in our current educational system is necessary, especially in science and technology. Managerial Comment 12.1 cites a joint report from the National

Academy of Sciences, the National Academy of Engineering, and the Institute of Medicine.

Managerial Comment 12.1   The Future of Education in the United States

CHARGE TO THE COMMITTEE

The National Academies was asked by Senator Lamar Alexander and Senator Jeff Bingaman of the Committee on Energy and Natural Resources, with endorsement by Representative Sherwood Boehlert and Representative Bart Gordon of the House Committee on Science, to respond to the following questions:

> What are the top ten actions, in priority order, that federal policymakers could take to enhance the science and technology enterprise so that the United States can successfully compete, prosper, and be secure in the global community of the twenty-first century?
>
> What strategy, with several concrete steps, could be used to implement each of those actions?

FINDINGS

Although the U.S. economy is doing well today, current trends in each of those criteria indicate that the United States may not fare as well in the future without government intervention. The nation must prepare with great urgency to preserve its strategic and economic security. Because other nations have, and probably will continue to have, the competitive advantage of a low-wage structure, the United States must compete by optimizing its knowledge-based resources, particularly in science and technology, and by sustaining the most fertile environment for new and revitalized industries and the well-paying jobs they bring. We have already seen that capital, factories, and laboratories readily move wherever they are thought to have the greatest promise of return to investors.

RECOMMENDATIONS

The committee identified two key challenges that are tightly coupled to scientific and engineering prowess: creating high-quality jobs for Americans, and responding to the nation's need for clean, affordable, and reliable energy. To address those challenges, the committee structured its ideas according to four basic recommendations that focus on the human, financial, and knowledge capital necessary for U.S. prosperity.

The four recommendations focus on actions in K-12 education (10,000 Teachers, 10 Million Minds), research (Sowing the Seeds), higher education (Best and Brightest), and economic policy (Incentives for Innovation) that are set forth in the following sections.

Some actions involve changes in the law. Others require financial support that would come from reallocation of existing funds or, if necessary, from new funds. Overall, the committee believes that the investments are modest relative to the magnitude of the return the nation can expect in the creation of new high-quality jobs and in responding to its energy needs.

The committee notes that the nation is unlikely to receive some sudden "wake-up" call: rather, the problem is one that is likely to evidence itself gradually over a surprisingly short period.

> Recommendation A: Increase America's talent pool by vastly improving K-12 science and mathematics education.
>
> Recommendation B: Sustain and strengthen the nation's traditional commitment to long-term basic research that has the potential to be transformational to maintain the flow of new ideas that fuel the economy, provide security, and enhance the quality of life.
>
> Recommendation C: Make the United States the most attractive setting in which to study and perform research so that we can develop, recruit, and retain the best and brightest students, scientists, and engineers from within the United States and throughout the world.
>
> Recommendation D: Ensure that the United States is the premier place in the world to innovate; invest in downstream activities such as manufacturing and marketing; and create high-paying jobs based on innovation by such actions as modernizing the patent system, realigning tax policies to encourage innovation, and ensuring affordable broadband access.

CONCLUSION

We have led the world for decades, and we continue to do so in many research fields today, but the world is changing rapidly, and our advantages are no longer unique…. Without a renewed effort to bolster the foundations of our competitiveness, we can expect to lose our privileged position…. We owe our current prosperity, security, and good health to the investments of past generations, and we are obliged to renew those commitments in education, research, and innovation policies to ensure that the American people continue to benefit from the remarkable opportunities provided by the rapid development of the global economy and its not inconsiderable underpinnings in science and technology.[28]

The improvement program outlined above is comprehensive in its scope. It implies the need for improvements in practically all the areas that we have described in earlier sections in this chapter.

Knowledge Management: Where Does It Fit?

As organizations consider their need to improve, it is clear that one of the keys to success is in their ability to manage knowledge. The emphasis on improvement programs will shift from the physical orientation toward products and services to information flow and knowledge management. We discuss the following questions here.

- How does knowledge flow along the knowledge corridor from data to information to knowledge to wisdom?
- Can we use the supply chain analogy to show the flow of information?
- Will increased knowledge replace some of the need for "things"?
- How does knowledge management relate to the educational system?

Figure 12.1 shows the flow along what we describe as "the knowledge corridor."

From Data to Information

Figure 12.1 is a series of transformation processes moving from left to right. The first transformation is in converting data to information. In the top section, along the horizontal axis, we begin with data. It is transformed into information, largely by organizing bits of data into meaningful clusters of information. An example is to take individual daily sales by item, by customer, and by store and summarize this into stock replenishment orders by item, buying trends by customer, and revenue performance by store. It is usually collected transaction by transaction, and in today's environment, computers, in the form of point-of-sale terminals, do most of the data collection. Computers are also programmed to do most of the

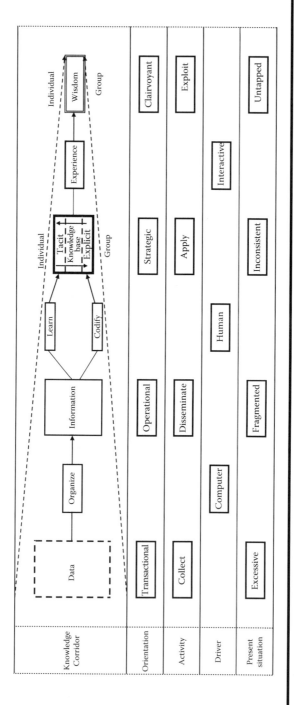

Figure 12.1   The knowledge chain.

data-organizing activities. At present, most organizations are faced with the problem of having too much data, not too little. Their problem is how to make this data more useful; for example, convert it to information. Data analytics techniques are expected to be a major player in this effort to convert raw data to meaningful information.

Most organizations are still learning to effectively use the information that has accumulated. The intent is to use it to make routine decisions in inventory management, customer relationship management (CRM), and resource utilization. To do this, they need an effective system for disseminating the relevant information to the right people to make the best decisions. They want to send enough information to users so that they can make better decisions, but they do not want to overburden users with excess information. It is a complex and never-ending process to tailor the collection and organization processes so that there is an effective flow from the data collection point to the decision-making point. The effectiveness of information systems today ranges from leaders with smooth and comprehensive flows to those who have undirected data collection and erratic information flow. We have labeled the overall situation as fragmented.

Figure 12.1 also shows converging lines that move from a widely separated state at the data stage of the knowledge chain to a narrowly separated state on the right-hand side of the diagram. This is to show that there must be a selection process that separates and preserves the most important data into more concise elements of information.

From Information to Knowledge

The next major transformation process is in converting information into knowledge. Although the conversion of data to information can be handled largely by computers, people are still needed to convert information to knowledge. This conversion has two major paths—learning and codifying. Learning is a process in which individuals convert information to tacit knowledge, or knowledge that is lodged within their own minds. Tacit knowledge remains with an individual until it is shared with another individual or codified to make it available to groups. From an organization's perspective, tacit knowledge is only valuable as long as that individual stays with the organization. It is to the organization's benefit to convert tacit knowledge to explicit knowledge.

Codifying information involves documenting in some formal process the rules, policies, and procedures of an organization; the result is known as explicit knowledge. Explicit knowledge is often the result of a group effort and is available to a wide cross-section of the organization. It extends beyond the tenure of any individual or group of individuals; it is part of the organization's knowledge base.

Knowledge is used in strategic planning and making decisions. The goal is to use the knowledge that has been derived from the collecting and disseminating processes described earlier. At present, it appears that there is even greater dispersion among organizations on how to use knowledge than there is on how to use information. Information comes from systems, and there is great consistency among information systems because of the efforts of many stakeholders to standardize information systems to facilitate intercompany communications. Knowledge, however, comes from individuals and there is great variation among people. Companies are often willing to share information; they are less willing to share knowledge. At present, it appears that the handling of knowledge is inconsistent within and among organizations.

As with the movement from data to information, Figure 12.1 continues the converging lines to convey the concept that knowledge is extracted from information into more concentrated and focused resources.

From Knowledge to Wisdom

The final stage shown in Figure 12.1 is the transformation of knowledge into wisdom. This is unexplored territory for most companies and for the individuals within those companies. Wisdom implies a level of understanding beyond that shown for knowledge. Wisdom is largely individually oriented, rather than emanating from an organization. Individuals gain wisdom in a variety of ways, largely through their own experiences. Some individuals gain wisdom while others with similar experiences do not. The expression "twenty years of experience" versus "one year of experience twenty times" seems to capture this distinction. Individuals with wisdom may seem to have an uncanny knack of predicting or spotting problems or opportunities before others do. They may even appear clairvoyant, a characteristic that we suggested in Chapter 10 would be a desirable capability for businesses to have in the future. When an organization discovers wisdom in an individual, they should exploit it for the good of both the organization and the individual.

Gaining wisdom is a learning process; how that process works is still largely unknown. Wisdom may be an untapped resource in organizations because it often goes unrecognized. It takes a rare combination of circumstances and individuals to recognize wisdom. The converging lines previously described continue to narrow as the diagram reaches wisdom.

The knowledge chain is a useful concept for continuous improvement. It implies that improvement programs can help in the development along the knowledge chain. The concept also suggests that the selection and implementation of improvement programs will gain effectiveness and efficiency through the application of information, knowledge, and wisdom.

We described the ITO Model in Chapter 3 as the DNA of the supply chain. It is also a model for other kinds of transformations. As described previously, the flow of knowledge can also be viewed as a form of supply chain, with the ultimate consumer being the individual with wisdom, and the fragments of data that enter the information chain being the inputs. We hope that this book will aid those who read it in their own quest for knowledge and wisdom.

Notes

1. Emblemsvag, J. and Bras, B., Process thinking—A new paradigm for science and engineering, *Futures,* 32, 7, 635, 2000.
2. Simon, H.A., *Administrative Behavior: A Study of Decision Making Processes in Administrative Organizations* (4th ed.), The Free Press, New York, 1997.
3. Emblemsvag, J. and Bras, B., Process thinking—A new paradigm for science and engineering, *Futures,* 32, 7, 635, 2000.
4. Mintzberg, H., *Mintzberg on Management, The Free Press, Inside Our Strange World of Organizations,* The Free Press, New York, 1989.
5. Pink, D.H., *A Whole New Mind: Moving from the Information Age to the Conceptual Age,* Riverhead Books, New York, 2005.
6. Simon, H.A., *Administrative Behavior: A Study of Decision Making Processes in Administrative Organizations* (4th ed.), The Free Press, New York, 1997.
7. Bolton, M. and Heap, J., The myth of continuous improvement, *Work Study,* 51, 6/7, 309, 2002.
8. Juran, J.M., *Managerial Breakthrough, A New Concept of the Manager's Job,* McGraw-Hill, Inc., New York, 1964.
9. Landis, K.M., Mishra, S., and Porrello, K., Calling a change in the outsourcing market: The realities for the world's largest organizations, Deloitte Consulting, e-mail: callingachange@deloitte.com, April 2005.
10. Venkatesan, R., Strategic Sourcing: To make or not to make, *Harvard Business Review,* 70, 2, 98, 1992.
11. Gibson, S., Outsourcing Is Growing Up, http://www.eweek.com, March 21, 2005.
12. Crandall, R.E., Device or strategy? Exploring which role outsourcing should play, *APICS Magazine,* 15, 7, 21, 2005.
13. Simon, H.A., *Administrative Behavior: A Study of Decision Making Processes in Administrative Organizations* (4th ed.), The Free Press, New York, 1997.
14. This quote has been attributed to several economists—Greenspan and Keynes, to name two. We expect that the quote goes back a long way, but we haven't found the sources yet.
15. Schwartz, B., *The Paradox of Choice, Why More Is Less,* Harper Collins, New York, 2004.
16. Teresko, J., Integrating IT with manufacturing, http://www.industryweek.com/PrintArticle.aspx?ArticleID=14151, June 1, 2007.
16a. Crandall, R.E. (2010). Putting together a global sustainability movement, *APICS Magazine,* 20(6), 26–29.
17. "The butterfly effect" was coined and discovered by Edward Lorenz. See Lorenz, E., *The Essence of Chaos,* The University of Washington Press, Seattle, 1993. Interested readers should also consult James Gleick's popular book for a less technical explanation of chaos theory. See Gleick, J., *Chaos: Making a New Science,* Penguin Books, New York, 1987.
18. Crandall, W., Crisis, chaos, and creative destruction: Getting better from bad, in *Re-Discovering Schumpeter Four Score Years Later: Creative Destruction Evolving into "Mode 3,"* Carayannis, E.G. and Ziemnowicz, C., Eds., MacMillan Palgrave Press, 2007.
19. See Barton, S., Chaos, self-organization, and psychology, *American Psychologist,* 49(1), 5–14, 1994; Carver, C., Dynamical social psychology: Chaos and catastrophe for all, *Psychological Inquiry,* 8(2), 110–119, 1997.

20. See Bolland, K. and Atherton, C., Chaos theory: An alternative approach to social work practice and research, *Families in Society: The Journal of Contemporary Human Services*, 80(4), 367–373, 1999; Hudson, C., At the edge of chaos: A new paradigm for social work? *Journal of Social Work Education*, 36(2), 215–230, 2000.

21. See Dervitsiotis, K., Navigating in turbulent environmental conditions for sustainable business excellence, *Total Quality Management*, 15(5–6), 807–827, 2004; Edgar, D. and Nisbet, L., A matter of chaos—Some issues for hospitality businesses, *International Journal of Contemporary Hospitality Management*, 8(2), 6–9, 1996.

22. McDaniel, R., Jordan, M., and Fleeman, B., Surprise, surprise, surprise! A complexity view of the unexpected, *Health Care Management Review*, 28(3), 266–278, 2003.

23. Murphy, P., Chaos theory as a model for managing issues and crises, *Public Relations Review*, 22(2), 95–113, 1996.

24. Farazmand, A., Chaos and transformation theories: A theoretical analysis with implications for organization theory and public management, *Public Organization Review*, 3, 339–372, 2003.

24a. Blackstone, J.H., Jr.. *APICS Dictionary* (13th ed.), APICS—The Organization for Operations Management, Chicago, Illinois, 2010, p. 151.

25. Womack, J.P. and Jones, D.T., *Lean Thinking: Banish Waste and Create Wealth in Your Corporation*, Simon & Schuster, New York, 1996.

26. Drucker, P.F., *Management: Task, Responsibilities, Practices*, Harper & Row, New York, 1974.

27. Magsood, T., Walker, D., and Finegan, A., Extending the "knowledge advantage": Creating learning chains, *The Learning Organization*, 14, 2, 123, 2007.

28. Committee on Prospering in the Global Economy of the 21st Century: An Agenda for American Science and Technology, National Academy of Sciences, National Academy of Engineering, Institute of Medicine, Rising Above the Gathering Storm: Energizing and Employing America for a Brighter Economic Future, The National Academies Press, 2006; also see Fahey, L. and Prusak, L., The eleven deadliest sins of knowledge management, *California Management Review*, 40(3), 265–276, 1998; Girad, J., Where is the knowledge we have lost in managers? *Journal of Knowledge Management*, 10(6), 22–38, 2006; Lester, D. and Parnell, J., *Organizational Theory: A Strategic Perspective*, Atomic Dog Publishing, Mason, OH, 2007.

Index

A

ABC, *see* Activity-based costing
Activity-based costing (ABC), 131
Adhocracy culture, 367
Advertising, 429
Agilent Technologies, 256, 258
AGV, *see* Automated guided vehicles
Amazon, 6, 109, 133, 135, 138–149, 172, 451
   agency pricing, 143
   Amazon Cloud, 145
   Amazon Web Services, 139, 143
   antitrust lawsuit, 143
   cloud computing services, 148
   as competitor, 149
   distribution (fulfillment centers), 143–145
   e-books, 145, 148
   employees, 144
   expansion into new businesses, 138
   fulfillment centers, 139
   incorporation, 138
   inventory management, 145
   major business segments, 148
   manufacturing (Kindle), 145–146
   origination (book authors), 146
   overview, 138–140
   primary source of revenue, 138
   retailing (electronic products), 142–143
   retailing (physical products), 140142
   sales, income, and cash flow, 140
   self-publishing, 146
   services, 138
   shipping companies, 144
   Simple Storage Service, 140
   summary, 146–149
   supply chain, 146, 147
   variable costs, 139
   virtual manufacturers, 145

American Express, 336
American Society of Mechanical Engineers
   (ASME), 91
Annual plans, 33
Antitrust lawsuit, 143
Apple, 142, 143, 255, 387
ARAMARK, 322, 333
Arena, 290
Arthur Andersen, 81
ASME, *see* American Society of Mechanical
   Engineers
Assemble-to-order (ATO) strategy, 43, 113
ATO strategy, *see* Assemble-to-order strategy
AT&T, 172, 362
Augmented product, 208
Automated guided vehicles (AGV), 297
Avery-Dennison, 336

B

Banking, 101, 429
Barnes & Noble, 142
BCG, *see* Boston Consulting Group
BEA, *see* Bureau of Economic Analysis
Best Buy, 6
Big data, 243
Boeing, 290, 332, 337
BookSurge, 146
Bootlegging, 361
Boston Consulting Group (BCG), 355
Bounded rationality, making decisions in, 66
BPR, *see* Business Process Re-engineering
Break-even analysis, 308
Buckman Laboratories, 346
Bullwhip effect, 191
Bureaucratic management, 94
Bureau of Economic Analysis (BEA), 439

Business intelligence, 206–207
Business Process Re-engineering (BPR), 82

C

Canon, 267
CAS, *see* Complex adaptive system
Chaos theory, 13, 411, 461
Chase model, 114
CIM, *see* Computer-integrated manufacturing
Cisco Systems, 267, 460
Clairvoyance, 411–412
Clan culture, 368
Cleveland Clinic, 135
Closed-loop supply chain (CLSC), 71
Cloud computing services (Amazon), 148
CLSC, *see* Closed-loop supply chain
CNG vehicles, *see* Compressed natural gas
	vehicles
Collaborative Planning Forecasting and
	Replenishment (CPFR), 191,
	193–194
Comcast, 245
Compaq, 267
Complex adaptive system (CAS), 411
Complexity theory, 411
Compressed natural gas (CNG) vehicles, 151
Computer-integrated manufacturing (CIM),
	297
Computer Sciences Corporation, 382, 387
Conceptual models, 51
Concurrent engineering, 121
Consumer services, 443–444
	data analytics, 444
	explicit knowledge, 444
	future of, 444
	hospitality industries, 444
	roles of, 443–444
	tacit knowledge, 444
	ways of collecting customer data, 444
Container Store, 356
Contingency theory, 95–97, 450
Continuous replenishment programs (CRP), 192
Core competencies, 67, 173, 331
Corning, 337
Corporate culture, 283
Costco, 207
Cost leadership strategy, 323
CPFR, *see* Collaborative Planning Forecasting
	and Replenishment
CPM, *see* Critical path method
Creditworthiness, 8

Critical path method (CPM), 311
Critical success chains (CSCs), 42
Critical success factors (CSFs)
	business identification of, 20
	need to identify, 4
	outputs reflecting, 59
Critical success factors (CSFs) and strategic
		planning, 19–48
	corporate life, 19
	description of critical success factors, 19–21
		definition of CSFs, 20
		main thing, 20
		technology perspective, 20
	evolution of CSFs in the United States,
		21–25
		agriculture, 21
		"American system" of manufacturing, 22
		competitive situation, 21
		demand for new CSFs, 25
		globalization, 24
		improved customer service, 21
		Industrial Revolution, 23
		mass production, 23–24
		order qualifiers, 21
	hierarchy of critical success factors, 34–40
		background of CSFs in strategic
			planning, 35–36
		by-product technique, 36
		characteristics of strategic CSFs, 37–38
		churning out the CSFs, 37
		decline, 40
		examples of interrelated CSFs, 35
		growth, 39
		key indicator system, 36
		maturity, 40
		null approach, 36
		personal CSFs, 37
		procedures for providing executive
			management information needs, 36
		product life cycle, 40
		startup, 39
		strategic business unit, 34
		temporal nature of CSFs, 38–40
		top-down analysis, 38
		total study process, 36
	hierarchy of the planning process, 29–34
		adjustment of business plan, 33
		annual plans, 33
		business operations, dynamic, 29
		business planning process, 33
		company objectives, development of, 32
		considerations, 32

contemporary view of strategic planning, 33
CSFs plan, 34
distinctive competencies, 31
essence of strategic planning, 30
kaizen blitz sessions, 34
knowledge of external environment, 30
links among plans, 31
opportunities, 30
project implementation, stalled, 34
project and program plans, 34
statistical process control, 34
strategic planning process, 31–32
strengths, 31
threats, 30
unforeseen circumstances, 33
weaknesses, 31
what strategic planning is, 30–31
need to be effective, 27–29
effectiveness, 28
efficiency, measurement of, 28
manager's job, 28
status quo, 28
vital few, 28
operational planning, role of CSFs in, 40–43
accounting, 41–42
assemble-to-order strategy, 43
critical success chains, 42
human resources, 41
make-to-order strategy, 43
make-to-stock strategy, 43
management information systems, 42
manufacturing flows, 43
manufacturing strategies, 43
operations, 42
organizational effectiveness, 41
positioning strategy, 42
product focus, 42
purchasing, 41
selection of processing strategies, 43
other changes during a country's economic life cycle, 25–27
increase in complexity, 25
Industrial Revolution, 25
job specialization, 26
satisficing, 27
Scientific Management philosophy (Taylor), 26
service-oriented economy, move to, 26
time horizon for decisions, 26
triple bottom line, 27

performance measurement and CSFs, 45–46
accounting systems, 46
key performance indicators, 46
quantitative goals, 45–46
selecting management programs, role of CSFs in, 44–45
business-critical factors, 45
ERP, 44
field study, 45
information center, 44
inter-organizational information system, 44
lean manufacturing, 44
technology absorptive capacity, 45
web-based supply chain management systems, 44
summary, 46
CRM, *see* Customer relationship management
CRP, *see* Continuous replenishment programs
CSCs, *see* Critical success chains
CSFs, *see* Critical success factors
Culture, *see* Organizational culture
Customer(s)
categories, 420
collaboration paradox, 411
contact, 316, 317
electronically empowered, 313
-engaged organization, 176–177
ITO Model, 58
maturity of, 173
profile, 444
Customer relationship management (CRM), 185–191, 475
background, 188
benefits, 189–190
definitions, 187–188
future of, 191
management fads, 188
marketing, 189
problems, 190
relation to supply chain, 190–191
straying acronym syndrome, 187
supply chain management, 188
what it does, 188–189
Y2K, 188

D

Data
analytics, 444, 475
quality, 207
Decision support system (DSS), 130

Delivery Information Acquisition Device
      (DIAD), 151
Dell, 17, 179, 207, 267, 303, 387, 459
Demand
      forecasting, 120
      management, 200–201
DIAD, *see* Delivery Information Acquisition
      Device
Digital document systems, 296
Discrimination lawsuits, 359
Disney Productions, 253
Distinctive competencies, 31
Door-to-balloon time, 126
DSS, *see* Decision support system

E

Eastern Airlines, 323
eBay, 207
E-books, 145, 148
E-business, 293–295
      characteristics of successful
            implementations, 295
      customer-focused processes, 294
      definitions and concepts, 293–295
      infrastructure, 294
      management-focused processes, 294
      processes, 294
      production-focused processes, 294
      transactions, 295
E-commerce, supply chains and, 389
ECR, *see* Efficient Consumer Response
EDI, *see* Electronic data interchange
Efficient Consumer Response (ECR), 191,
      192–193
EIR, *see* Enterprise Information Roadmap
Electronic data interchange (EDI), 192
Employees
      acceptance of new technology, 317
      benefits, 429–430
      commitment, 356–357
      empowerment of, 120, 335
      ethics training, 371
      formal recognition of, 376–377
      frontline, 130
      knowledgeable, 2
      management theories, 90–91
      manager communication to, 106
      morale, 359
      multiple tasks performed by, 125
      rejection of company mission, 356
      screening, 429

Southwest Airlines, 353
      training, 429
      turnover, 342, 343
Engineering testing, 429
Engineer-to-order (ETO) producers, 113
Enron, 81
Enterprise Information Roadmap (EIR), 187
Enterprise resource planning (ERP), 132, 278,
      284, 304
Entropy, 63
Equipment utilization, 118
Ernst & Young, 346
ERP, *see* Enterprise resource planning
Ethics, 370
ETO producers, *see* Engineer-to-order
            producers
Excel, 290
Explicit knowledge, 286, 444, 475
Extensible markup language (XML), 304, 390
Extractive industries, 422

F

Father of Modern Management, 94
Father of scientific management, 91
FedEx, 207, 402
Financial factors, 32
Flexibility paradox, 411
Flexible manufacturing systems (FMS), 287,
      297
FMS, *see* Flexible manufacturing systems
Focused factory approach (manufacturing), 282
Ford Motor Company, 106, 338, 362
"Functional foreman" recommendations, 384
Functional silos, 384
Functions of culture, 353
Future of improvement programs, 449–478
      background to improvement programs,
            449–452
         agents of change, 451
         contingency theory, 450
         employees, 451
         foundation topics, 450
         input–transformation–output process,
               451
         integration of related entities, 452
         ITO model, 450
         knowledge transfer across the
               manufacturing/services boundary,
               450–451
         satisficing, 452
         Six Sigma, 451

total quality management program, 450
vanishing manufacturing/services
    boundary, 449
drivers of change, 455–456
    communities, 456
    culture, 456
    structure, 455–456
    technology, 455
from knowledge to wisdom, 476
future areas of emphasis, 452–454
    continuing increase in globalization, 453
    continuing need to integrate the product
        and service bundle, 453
    decision making will deal with a blend
        of hard and soft variables, 454
    decisions will become more complex,
        454
    increased need for project management
        competencies, 453–454
    outsourcing will become a more focused
        activity, 453
    satisficing, 454
    services will continue to increase as a
        critical success factor in business, 453
future of improvement programs, 454–455
industries most likely to stress continuous
        improvement, 468–473
    education, 470–473
    health care (hospitals and wellness
        centers), 468–469
    local government, 470
    pharmaceuticals, 469–470
    retail, 470
knowledge management, 473–476
    codifying information, 475
    customer relationship management, 475
    data analytics techniques, 475
    explicit knowledge, 475
    from data to information, 473–475
    from information to knowledge,
        475–476
    goal, 475
    knowledge chain, 474
    tacit knowledge, 475
most likely future methodologies, 456–462
    chaos and complexity, 461–462
    environmental design, 460–461
    information technology, 460
    integrated supply chains, 456
    mass customization, 459
    outsourcing, 457
    performance measurement, 458–459

    project management, 459
    service sciences, 461
    total cost of ownership, 457–458
    virtual organizations, 460
most likely improvement programs,
        462–468
    agility, 467
    capital equipment and facilities, 463
    compatibility, 467
    cost reduction, 462–463
    customer service level, 465–466
    flexibility, 466–467
    integration, 467–468
    product or service costs, 462
    quality, 465
    response time reduction, 463–464
    risk management, 468
    sustainability, 468
    working capital costs, 462–463
F.W. Woolworth's, 323

G

Gap, The, 201
GE, *see* General Electric
General Electric (GE), 17, 207, 217–250, 346,
    451
    background, 217–218
    big data, 243
    early years, 217–218
    GE progress over time, 221–227
        business computers, 224
        commercial jet engines, 224
        community development and housing,
            226
        conclusions about Growth Council
            ventures, 226–227
        education, 226
        electronics, 221
        entertainment, 225
        evolution of products and services over
            stages of GE history, 222–223
        Growth Council product proposals
            (1970s), 224–225
        Growth Council service proposals
            (1970s), 225–226
        growth during World War II, 221–224
        industrial automation and productivity
            systems, 224
        materials, 224
        medicine, 226
        nuclear energy, 224

personal and financial services, 226
polymer chemicals, 224–225
propulsion systems, 221
model illustrating GE's strategy, 244–249
    adaptability, 249
    conclusions, 249–250
    early financing, 244
    evolution of services, 248
    influence, 249
    LATIN, 249
    leadership, 249
    networks, 249
    pattern, 244, 245, 246
    product life cycle, 245, 246
    product-to-service progression, 247
    strategies in developing services
        businesses, 246
    talent, 249
more recent product profiles, 227–244
    examples of segment products and
        services, 241–243
    GE revenues as percent of total, 234
    GE sales and income, 229–230,
        231–232, 235–236
    Immelt years, 234–237
    portfolio strategy, 240
    product line (2006), 238
    recent changes, 243–244
    segment revenues and profits, 239
    total revenues by products and services,
        233
    Welch years, 227–234
seven dwarfs, 224
ways for manufacturers to add services,
    218–221
    aggressive advertising and promotion,
        219
    benign cycle strategy, 219
    consignment selling, 219
    develop a market for the product,
        218–219
    enter a more profitable business,
        220–221
    extend known technology into a new
        product area, 219–220
    GE Credit, 219
    General Electric Supply Company, 219
    help sell products by financing sales, 220
    provide infrastructure in emerging
        countries, 220
    provide post-sales support, 220
    relationship selling (supporting the
        retailers), 219
    retailer franchising, 219
    why services are important to a
        manufacturing company, 218
Global Supply Chain Forum (GSCF), 391
Gobi radio technology, 153
Golden Rule, 360
Google, 172, 354
Green Supply Network, 72
GSCF, *see* Global Supply Chain Forum

H

Hachette SA, 143
Hallmark, 336
Hard technologies, 280
HarperCollins, 143
Heinz, 323
Hewlett-Packard, 157, 251–275, 337, 346, 451
    Adaptive Enterprise strategy, 263
    Apotheker as CEO (September 30, 2010, to
        September 22, 2011), 265
    distribution of revenues between products
        and services, 272
    financial results, 267, 270–271
    Fiorina as CEO (July 19, 1999, to February
        9, 2005), 263–264
    hierarchy of services provided to businesses,
        275
    Hurd as CEO (April 1, 2005, to August 6,
        2010), 264–265
    lawsuit, 265
    product groups, transformation in, 256
    product innovations, 251
    product segments revenues and income, 257
    revenues and income, 259, 262
    segment results before Agilent divestiture,
        260
    services since 2000, 267–273
    stages of evolution, 251
    summary of revenues and earnings by major
        product segment, 268–269
    timeline of product line evolution at HP,
        252
    transformation one (1939–1959), 253–254
    transformation two (1959–1968), 254
    transformation three (1968–1976), 254–255
    transformation four (1976–1986), 255–256

transformation five (1986–1999), 256–260
transformation six (1999–present), 260–267
Whitman as CEO (September 22, 2011, to present), 265–267
Hierarchy culture, 368
HomeBanc Mortgage, 360
Home Depot, 6, 172
Horizontal organization, 330
Hospitality industries, 444
Hot Topic, 355
Howard Johnson's restaurants, 323
HR, *see* Human resources
Human–machine interface, 11
Humanness, 277, 318
Human relations management, 91–92
Human resources (HR), 32, 41
Human resources (technology and), 296–300
    as aid to employee, 296–297
    automated guided vehicles, 297
    automated materials-handling systems, 296
    balance, 299
    computer-integrated manufacturing, 297
    demand for customization, 299
    digital document systems, 296
    displacement of people, 298
    examples of helpful technology, 297
    flexibility, 299
    flexible manufacturing systems, 297
    as integral part of process, 298–299
    interdisciplinary design methodology, 300
    lights-out factory, 298
    as substitute for employee, 297–298
Human Sigma, 128
Hype cycle, 85

I

IBM, 16, 178, 265, 273, 323
ICT, *see* Information and communications technology
IKEA, 291
Implicit knowledge, 285
Industrial Revolution, 23, 25
Information and communications technology (ICT), 73
Information technology (IT), 303–304, 305
    description of, 303
    function, 303
    issues matching IT to business, 305
Infrastructure, role of in continuous improvement, 321–350

alternate organizational structures, 329–333
    core competencies, 331
    cross-functional process teams, 331
    flatter organization, 331
    horizontal organization, 330–331
    inter-organizational virtual organization, 331
    matrix organization, 330
    re-engineering, 330
    virtual organization, 331–333
business strategy, 323–324
    cost leadership strategy, 323
    differentiation strategy, 324
    focus strategy, 324
    niche, 324
    premier guide, 323
    "stuck in the middle" strategy, 324
classical management functions, 324–327
    controlling, 327
    delays in product introduction, 327
    directing, 326–327
    goals, 325
    organizing, 325–326
    planning, 325
    theory of management, 325
    timetables, 325
corporate strategy, 322–323
    company growth, 323
    declining industry, 323
    growth, 323
    retrenchment, 323
description of infrastructure, 321–322
    components, 322
    meanings of term, 322
    terms, 321
knowledge management, integration of into organizational structure, 342–346
    culture, 344
    data, information, knowledge, wisdom, 342–345
    employee turnover, 342, 343
    forward thinking, 344
    intranets, 346
    medicine, 346
    new knowledge, 345
    "old guard" mentality, 344
    reasons why knowledge management may be impeded, 345–346
    why knowledge is not transferred, 345–346
    wisdom, 345

need for change in infrastructure, 347–348
  awareness of the need to change, 347
  difficulty of change, 347
  energy to carry out the plan, 348
  manufacturing companies, 347
  middle managers, 347
  a plan to make the change, including the
    project plan, 347
  resources, or monetary support, to
    complete the change, 348
  top management endorsement and
    continuing support, 347
organization structure, 327–329
  functional, 327–328
  geographic, 329
  product, 328
  visual representation, 327
role of the Internet in changing
    organizational structure, 340–342
  competition, 341
  information symmetry and asymmetry,
    342
  "make or buy" decision, 341
  outsourcing, 341
  transaction costs, 341
strategies, 322–324
  business strategy, 323–324
  corporate strategy, 322–323
trends in organizational structures,
    333–340
  employee empowerment, 335
  job specialization, 335
  lateral communication, 338
  mechanistic versus organic
    organizations, 340
  micro-management, 334
  moving from autocratic managers
    to more empowered employees,
    334–335
  moving from centralization to
    decentralization, 333
  moving from job specialization to higher
    skill variety, 335–336
  moving from line managers to self-
    directed work teams, 336–337
  moving from mechanistic structures to
    organic structures, 339–340
  moving from rigid policies and
    procedures to more flexibility,
    338–339
  moving from specialized departments to
    cross-functional teams, 337

  moving from top-down to
    multidirectional communications,
    337–338
  moving from vertical structures to
    horizontal structures, 334
  multi-skilling, 337
  policies, 339
  practices, 340
  processes, 339
  scientific management principles, 336
  self-managed teams, 337
  skunk works, 337
  "top-down" approach, 337
  trends, 333
Innovation paradox, 411
Input–Transformation–Output (ITO) Model,
    49–76, 386
  basic ITO Model (inputs, transformation,
    and output), 53–61
    addition of customers and suppliers, 56
    addition of resources to ITO Model, 55
    automation, 59
    components of the model, 57–61
    customers, 58
    examples, 61
    extension of basic model, 56–57
    general systems model, 54–56
    inputs, 60
    knowledge gained, 59
    modules, examples of, 54
    order processing links, 56
    outputs, 58–59
    transformation process, 59–60
  introduction to models, 50–53
    benefits of using models, 52–53
    conceptual models, 51
    conditions specified, 52
    graph, 51
    mathematical models, 51
    physical models, 50, 51
    problem components, 52
    types of models, 50–52
    "what if" questions, 53
    why models are used, 52
    wind tunnels, 50
  ITO Model, 49
  linkage of modules, 49
  nature of models, 50
  reverse logistics, 70–75
    barriers to implementation, 73
    benefits of, 72
    closed-loop supply chain, 71

forward and reverse supply chains, 74
green supply chains and, 70
information and communications
  technology support, 73
intent, 71
interest in, 71–72
laws, 72
public awareness of environmental
  concerns, 71
recommendations, 74
system design and implementation,
  73–74
update, 74–75
summary, 75
supply chain configurations, extension of
  ITO Model into, 62–70
building relationships, 68–70
classical management thinkers, 65–66
closed and open systems model, 62–68
closed systems, 62–63
closed-system strategy, 65
conflicts of interest, 69
core competency, 67
determinate system, 65
effect of variation on process, 67
entropy, 63
feedback, 67–68
making decisions in bounded rationality,
  66
marketing, 69
multiple supply chains, 70
open systems, 63–64
open-system strategy, 66–67
original conditions in relationship
  building, 68
progression to seamless flow, 69
radiofrequency identification tags, 64
realignment, 70
relationship building, 68
satisficing, 66
supply chain, 62
topics, 49
Integrated service package, 444–447
development steps, 445–447
develop objectives for each component,
  446–447
develop objectives for the service
  package, 446
develop strategies for each component,
  447
develop strategies for the service
  package, 446

evaluate the results, 447
identify the service package components,
  445–446
implement the strategic plan, 447
recognize the need to change, 445
revise as needed, 447
need for, 445
Integrated supply chains, *see* Supply chains,
  integrated
Internet
airline reservation systems, 59
banking, 303
bookseller (Amazon), 138
communication, 404, 439
company identity, 354
comparison shopping on, 173
competition, 470
EDI, 204, 205
examples of transactions, 295
goods and services purchased on, 302
impact of, 342
manufacturer websites, 8
opportunities for progressive companies, 74
organizational structure and, 340, 456
re-engineering initiatives and, 287
role of in changing organizational structure,
  340–342
  competition, 341
  information symmetry and asymmetry,
    342
  "make or buy" decision, 341
  outsourcing, 341
  transaction costs, 341
technology company, 362
wireless, 375
Inter-organizational communications (IOC),
  202–205
company culture, 204
Internet EDI, 203–205
order processing, 203
traditional EDI, 202–203
trust between participants, 204
value-added networks, 203
virtual private network, 204
Inter-organizational system (IOS), 203, 304,
  386
Inter-organizational virtual organization, 331
Intranets, 346
Inventory-taking service, 430
IOC, *see* Inter-organizational communications
IOS, *see* Inter-organizational system
ISO certification, 129

IT, *see* Information technology
ITO Model, *see* Input–Transformation–Output
   Model

J

"Jerk-free" atmosphere, 360
JIT, *see* Just-in-Time
Job
   content, 307
   division of into tasks, 116
   rotation, 92
   specialization, 26, 335
John Deere Equipment, 362
Johnson & Johnson, 81, 433
Just-in-Time (JIT), 396
   lean and, 124–127
      association with Six Sigma, 12
      bid quotation process, 125
      customer service, 126
      door-to-balloon time, 126
      kanban, 125
      Lean Sigma, 126
      local government, 125–126
      mass services, 125
      need for integrated system, 124
      service characteristics, 124–125
      state government, 125
   lean thinking replacing, 396
   as process-oriented waste eliminator, 110
   program, 78

K

Kaizen blitz sessions, 34
Kanban, 125
Key performance indicators (KPI), 46
Kindle, 145–146, 148
KMS, *see* Knowledge management systems
Knowledge
   demand forecasting, 120
   explicit, 286, 444, 475
   of external environment, 30
   graphing of, 51
   implicit, 285
   management
      organizational culture and, 284
      strategies, 98
      systems (KMS), 98
   management (future of improvement
      programs), 473–476
      codifying information, 475
      customer relationship management, 475
      data analytics techniques, 475
      explicit knowledge, 475
      from data to information, 473–475
      from information to knowledge,
         475–476
      goal, 475
      knowledge chain, 474
      tacit knowledge, 475
   management (integration of into
      organizational structure), 342–346
      culture, 344
      data, information, knowledge, wisdom,
         342–345
      employee turnover, 342, 343
      forward thinking, 344
      intranets, 346
      medicine, 346
      new knowledge, 345
      reasons why knowledge management
         may be impeded, 345–346
      why knowledge is not transferred,
         345–346
      wisdom, 345
   tacit, 286, 444, 475
KPI, *see* Key performance indicators

L

Labor, specialization of, 116
Lean, JIT and, 124–127
   association with Six Sigma, 12
   bid quotation process, 125
   customer service, 126
   door-to-balloon time, 126
   kanban, 125
   Lean Sigma, 126
   local government, 125–126
   mass services, 125
   need for integrated system, 124
   service characteristics, 124–125
   state government, 125
Lean manufacturing, 78
Lean production, 396–398
   collaboration with customers, 398
   delivery process, 398
   demand variation, 398
   goods producers, 287
   key values, 396
   limitations, 396–397
   product, 397

production process, 397–398
purchasing process, 397
third-party logistics providers, 398
Lean Sigma, 5, 126
Left-brain thinking, 412–413
Lenovo, 267
Lexmark, 157
Line flow management, 2
Logistics factors, 32
Lowe's, 6

M

Macmillan, 143
Make-to-order (MTO) strategy, 43
Make-and-sell strategy, 16
Make-to-stock (MTS) strategy, 43, 113
Management programs, role of in continuous
    improvement, 77–108
  description of management programs,
      78–85
    Business Process Re-engineering, 82
    continuous transformation process, 84
    crisis management, 84
    cycles, 83
    discontinuous versus continuous
        incremental change, 82–83
    examples of management programs, 79
    improvement programs, 82–85
    Just-in-Time program, 78
    lean manufacturing, 78
    normal or sustaining day-to-day
        operations, 81
    problem-solving activities, 81
    problem-solving process, 83
    questions, 77
    Total Quality Management program, 78
    Toyota Production System, 78
    types of business activities, 80
  future of management programs, 105–107
    future program emphases, 107
    management for the twenty-first century,
        105–107
    organizing around the needs of people,
        106
    Six Sigma, 106
    statistical process control, 106
    tasks, 105
  importance of management programs,
      89–90
    breakthrough, 89
    control, 89

crisis, 90
  scenario, 90
management program life cycles, 85–89
  academic researchers, 86
  end of the life cycle, 87–88
  hype cycle, 85
  implications of program life cycles for
      management, 88–89
  life-cycle stages, 85–87
  number of JIT and lean production
      articles, 87
  number of TQM and Six Sigma articles,
      88
  program implementation, requirement
      of, 89
  publication and management program
      life cycles, 86
origins of management programs, 90–97
  administrative management, 92–95
  bureaucratic management, 94
  characteristics for ideal bureaucracy, 94
  comparison of JIT and TQM, 97
  contingency theory, 95–97
  employee management theories, 90–91
  Father of Modern Management, 94
  father of scientific management, 91
  human relations management, 91–92
  job rotation, 92
  management principles (Fayol), 93–94
  program concepts derived from
      management theories, 97
  scientific management, 91
  systems theory, 95
program success, 98–105
  adaptation of programs to new
      conditions, 100
  assimilation, 105
  centralization versus decentralization,
      101
  create a receptive environment, 104
  cultures, 101
  evaluate and refine, 105
  evolution of management programs over
      time, 103
  execution and evaluation, 104–105
  failure to match program with need,
      98–102
  identify the need for improvement, 102
  implementing the program correctly,
      102–105
  industry traditions, 101–102
  knowledge management strategies, 98

organize for successful implementation, 104

outline the steps to be accomplished in the implementation process, 104

planning and preparation, 102–104

prepare for the next program, 105

program extensions and their chances of success, 99

select, or design, the right program, 102–104

strategic objectives, 99

tangible items, 105

top management support, 101

types of processes, 101

types of products or services, 101

summary, 107

Manufacturing

processes, types of, 111

resources planning (MRP II), 304

/services boundary, *see* Vanishing manufacturing/services boundary

Manufacturing, extending service techniques to, 171–276

adding services to product portfolios, 171

business intelligence, 206–207

customer relationship management, 185–191

background, 188

benefits, 189–190

definitions, 187–188

future of, 191

management fads, 188

marketing, 189

problems, 190

relation to supply chain, 190–191

straying acronym syndrome, 187

supply chain management, 188

what it does, 188–189

Y2K, 188

description of services, 179–183

attributes of services, 181

buyer involvement, 182

client-based relationships, 182

demand fluctuations, 182

intangibility, 181

ITO Model, 179

labor intensity, 182

lack of homogeneity, 181–182

lack of transportability, 181

model of service relationships, 180

perishability, 181

service, 181–182

distributive services, 182–183

economic growth and need for added services, 178

examples of programs developed in services, 185–207

business intelligence, 206–207

customer relationship management, 185–191

flexibility, 196–202

inter-organizational communications, 202–205

managing the customer encounter, 206

nonquantitative performance measurement, 206

product development as a result of customer inputs, 205

quality as customers' perceptions, not just conformance to specification, 206

response time reduction, 191–196

working in open-system environment, 205

flexibility, 196–202

absorbing strategies, 201

containing strategies, 201

demand management, 200–201

evolution from job specialization to self-directed teams, 202

flexible organization forms, 199–200

horizontal communication and organization structure, 197–199

location near the market versus lowest cost, 201

mass customization, 197

mitigating strategies, 201

product and service flexibility (mass customization), 196–197

replacement of inventory with information, 202

shielding strategies, 201

General Electric, 210, 217–250

adaptability, 249

aggressive advertising and promotion, 219

background, 217–218

benign cycle strategy, 219

big data, 243

business computers, 224

commercial jet engines, 224

community development and housing, 226

conclusions, 249–250

conclusions about Growth Council ventures, 226–227
consignment selling, 219
develop a market for the product, 218–219
early financing, 244
early years, 217–218
education, 226
electronics, 221
enter a more profitable business, 220–221
entertainment, 225
evolution of products and services over stages of GE history, 222–223
evolution of services, 248
examples of segment products and services, 241–243
extend known technology into a new product area, 219–220
GE Credit, 219
General Electric Supply Company, 219
GE progress over time, 221–227
GE revenues as percent of total, 234
GE sales and income, 229–230, 231–232, 235–236
Growth Council product proposals (1970s), 224–225
Growth Council service proposals (1970s), 225–226
growth during World War II, 221–224
help sell products by financing sales, 220
Immelt years, 234–237
industrial automation and productivity systems, 224
influence, 249
LATIN, 249
leadership, 249
materials, 224
medicine, 226
model illustrating GE's strategy, 244–249
more recent product profiles, 227–244
networks, 249
nuclear energy, 224
pattern, 244, 245, 246
personal and financial services, 226
polymer chemicals, 224–225
portfolio strategy, 240
product life cycle, 245, 246
product line (2006), 238
product-to-service progression, 247
propulsion systems, 221

provide infrastructure in emerging countries, 220
provide post-sales support, 220
recent changes, 243–244
relationship selling (supporting the retailers), 219
retailer franchising, 219
segment revenues and profits, 239
seven dwarfs, 224
strategies in developing services businesses, 246
talent, 249
total revenues by products and services, 233
ways for manufacturers to add services, 218–221
Welch years, 227–234
why services are important to a manufacturing company, 218
Hewlett-Packard, 251–275
Adaptive Enterprise strategy, 263
Apotheker as CEO (September 30, 2010, to September 22, 2011), 265
distribution of revenues between products and services, 272
financial results, 267, 270–271
Fiorina as CEO (July 19, 1999, to February 9, 2005), 263–264
hierarchy of services provided to businesses, 275
Hurd as CEO (April 1, 2005, to August 6, 2010), 264–265
lawsuit, 265
product groups, transformation in, 256
product innovations, 251
product segments revenues and income, 257
revenues and income, 259, 262
segment results before Agilent divestiture, 260
services since 2000, 267–273
stages of evolution, 251
summary of revenues and earnings by major product segment, 268–269
timeline of product line evolution at HP, 252
transformation one (1939–1959), 253–254
transformation two (1959–1968), 254
transformation three (1968–1976), 254–255

transformation four (1976–1986), 255–256

transformation five (1986–1999), 256–260

transformation six (1999–present), 260–267

Whitman as CEO (September 22, 2011, to present), 265–267

increasing complexity of marketplace, 173

inter-organizational communications, 202–205

  company culture, 204

  Internet EDI, 203–205

  order processing, 203

  traditional EDI, 202–203

  trust between participants, 204

  value-added networks, 203

  virtual private network, 204

knowledge transfer from services to manufacturing, 184–185

  areas of manufacturing expertise, 184

  areas of service expertise, 184–185

  customer service, 185

  demand/supply relationship, 184

  open-system environment, 184

managing the customer encounter, 206

maturity of customer as shopper, 173

movement from agriculture to manufacturing, 172

movement from make-to-stock to make-to-order, 178

movement from product-centric to customer-centric, 174–177

  comparison of strategies, 175

  customer-engaged organization, 176–177

  customer value guide, 176

  "outside-in" way of looking at business, 175

movement toward mass customization, 178–179

  approach description, 179

  assemble-to-order strategy, 179

  complexity, 179

  growth in services, 179

  postponement, 179

necessary by-product, 171

need for manufacturing companies to add services, 173–174

  aggregator, 174

  capital-intensive facilities, 173

  core competencies, 173

  embedded innovator, 174

  solutionist, 174

  synergists, 174

nonquantitative performance measurement, 206

personal services, 183

producer (business) services, 183

product development as a result of customer inputs, 205

quality as customers' perceptions, not just conformance to specification, 206

response time reduction, 191–196

  bullwhip effect, 191

  Collaborative Planning Forecasting and Replenishment, 191, 193–194

  continuous replenishment programs, 192

  Efficient Consumer Response, 191, 192

  electronic data interchange, 192

  future of, 196

  global online registry, 195

  inventory challenge, 196

  offshore outsourcing movement, 195

  point-of-sales terminals, 192

  present status, 194–196

  Quick Response Systems, 191, 192

  sales and operations planning, 193

  supply chain management, 194

  value-added networks, 192

  vendor-managed inventory, 193

  voluntary inter-industry commerce standards, 193

rise of services as part of the economy, 172

self-service, 183

services as separate new business segment, 177–178

social services, 183

summary, 207–210

  ambiguity, 209

  augmented product, 208

  collaborative efforts, 210

  defensible differences, 208

  differences between manufacturing and services, 208

  from services to manufacturing, 208–209

  intangibility of quality, 209

  niche markets, 208

  service firms, 209

swing of power from manufacturing to retail, 172–173

  focus, 172

  improvement programs, 172

online retailers, 172
retail clothing stores, 172
working in open-system environment, 205
Market
culture, 367
research, 429
Marketing
CRM, 189
factors, 32
lean supply chain, 401
pushing of newer products, 69
Mass customization
alternatives, 197
future of, 459
movement toward, 178–179
approach description, 179
assemble-to-order strategy, 179
complexity, 179
growth in services, 179
postponement, 179
operational effectiveness, 198
requirement of, 12–13
response times, 459
Materials requirements planning (MRP), 117,
131–132, 304
Mathematical models, 51
Matrix organization, 330
Maxwell House coffee, 362
Mayo Clinic, 135, 358, 371
McDonald's, 322, 338, 360
Microsoft, 172, 354
Mobipocket, 146
Morale-building environment, 302
Motorola, 85, 96, 106
MRP, *see* Materials requirements planning
MRP II, *see* Manufacturing resources planning
MTO strategy, *see* Make-to-order strategy
MTS strategy, *see* Make-to-stock strategy
Multi-skilling, 337

N

NAICS, *see* North American Industry
Classification System
NBC, 245
Niche, 208, 24
Nike, 17
Nokia, 323
Nonprofit organizations, 438–439
Nordstrom's, 324, 359
North American Industry Classification System
(NAICS), 420

O

Oakley, 364, 365
OE, *see* Organizational effectiveness
Operational planning, role of CSFs in, 40–43
accounting, 41–42
assemble-to-order strategy, 43
critical success chains, 42
human resources, 41
make-to-order strategy, 43
make-to-stock strategy, 43
management information systems, 42
manufacturing flows, 43
manufacturing strategies, 43
operations, 42
organizational effectiveness, 41
positioning strategy, 42
product focus, 42
purchasing, 41
selection of processing strategies, 43
Optimization paradox, 410
Oracle, 265
Order qualifiers, 21
Organizational culture, 351–380
changing organizational culture, 368–377
change the physical cultural artifacts, 375
codes of ethics, 370
hire the kind of people you want to *see*
perpetuate your desired culture, 371–373
implement the strategy proposed, 373–374
key focus areas, 369
old-timers, 374
orientation sessions, 375–376
partially implemented changes, 373
potential scenarios when hiring a professor, 372
promote an ethical culture, 370–371
recognize employees formally, 376 377
state what you want the culture to be, 369–370
status quo, 374
strategy–culture linkage, 374
take care of your employees, and they will take care of your customers, 373
tell stories, 375–376
components, 359–367
artifacts (display of organizational culture), 360–367
bootlegging, 361

components of culture, 359–360
disastrous reward systems, 366–367
Golden Rule, 360
identifiable value systems and behavioral
    norms, 363–364
innovation, 361, 365
"jerk-free" atmosphere, 360
language, 361–362
open door policy 363
organizational rewards and reward
    systems, 365–367
physical surroundings characterizing a
    culture, 364–365
status symbols, 365
stories, 361
symbols, ceremonies, and rituals,
    362–363
values, 360
vocabulary, 362
description, 352–353
    belief system, 352
    business and nonprofit organizations,
        353
    gag announcements, 353
importance, 353–359
    behavioral quirk, 354
    discrimination lawsuits, 359
    employee morale, 359
    fostering creativity, 357
    functions of culture, 353
    organizational culture enables employees
        to be committed to the company,
        356–358
    organizational culture gives the company
        an identity, 354–355
    organizational culture helps add stability
        to the company, 358–359
    organizational culture helps employees
        make sense of things, 355–356
    reasons, 353
    value statements, 355
    "zero tolerance" approach, 359
ITO Model, 351
restaurant, 352
state agencies, 352
technology and, 283–285
    corporate culture, 283
    enterprise resource planning, 284
    globalization, 285
    implicit knowledge, 285
    knowledge management, 284
    software development, 285

types, 367–368
    adhocracy, 367
    aligning culture and strategy, 368
    clan, 368
    hierarchy, 368
    market, 367
    mismatches in culture, 368
Organizational effectiveness (OE), 41
Organization structure, 327–329
    alternate, 329–333
        core competencies, 331
        cross-functional process teams, 331
        flatter organization, 331
        horizontal organization, 330–331
        inter-organizational virtual organization,
            331
        matrix organization, 330
        re-engineering, 330
        virtual organization, 331–333
    functional, 327–328
    geographic, 329
    product, 328
    trends in, 333–340
        employee empowerment, 335
        job specialization, 335
        lateral communication, 338
        mechanistic versus organic
            organizations, 340
        micro-management, 334
        moving from autocratic managers
            to more empowered employees,
            334–335
        moving from centralization to
            decentralization, 333
        moving from job specialization to higher
            skill variety, 335–336
        moving from line managers to self-
            directed work teams, 336–337
        moving from mechanistic structures to
            organic structures, 339–340
        moving from rigid policies and
            procedures to more flexibility,
            338–339
        moving from specialized departments to
            cross-functional teams, 337
        moving from top-down to
            multidirectional communications,
            337–338
        moving from vertical structures to
            horizontal structures, 334
        multi-skilling, 337
        policies, 339

practices, 340
processes, 339
scientific management principles, 336
self-managed teams, 337
skunk works, 337
"top-down" approach, 337
visual representation, 327
Outsourcing
becoming a more focused activity, 453
creation of goods and services, 17
focus of, 10
future of, 457
post-sales service, 7
product design process, 121
role of, 426–427
Overshooting, 5

P

Palm Inc., 265
Penguin, 143
Pepsi, 337
Person–machine relationship, 11
PERT, *see* Project evaluation and review
technique
Physical cultural artifacts, 375
Physical models, 50, 51
Point-of-sales (POS) terminals, 192, 307
POS terminals, *see* Point-of-sales terminals
Pricewaterhouse Coopers, 142
Proctor & Gamble, 323, 387
Producer services, 424–437
classes, 427
consumer services versus, 420–424
customer categories, 420
distributive services, 422
extended sectoral classification,
421–422
extractive industries, 422
industry groups along supply chain, 420
model of services provided to businesses,
423
social services, 422
determinants of demand, 424–425
functional classification, 426
future of, 436
core idea, 436
depersonalization of communications,
436
outsourcing, 436
services sciences, 436
importance, 425

reasons to acquire, 427–434
acquire expertise not available internally,
431–432
advertising, 429
architectural or facilities design, 431
assist in strategic planning, 432
auditing, 432–433
banking, 429
comply with regulatory or otherwise
mandatory requirements, 432–433
construction, 431
continuous improvement programs, 434
crisis management, 433–434
education providers, 432
employee benefits, 429–430
employee screening, 429
employee training, 429
engineering testing, 429
environmental compliance, 433
facilities location, 431
gain a cost advantage, 427–428
global guidance consultants, 432
import/export services, 433
inventory-taking service, 430
investment counseling, 431
manage major programs, 433–434
management consulting, 432
market research, 429
merger and acquisition specialists, 432
provide flexible capacity or avoid
overload on key departments, 430
registration at colleges and universities,
430
remove noncore types of activities, 428
reservation systems, 430
risk and insurance, 431
security services, 431
supplement internal staffs with added
expertise, 428–430
technology compliance, 433
traceability of product flow, 433
wellness programs, 430
role of outsourcing, 426
role of producer services, 427
steps to interfacing business with, 434–436
develop a strategy for each producer
services area, 434
enterprise resources planning system, 434
implement the service (fit it to the
situation), 435
integrated supply chain, 434
measure the progress, 435–436

prepare a plan that interfaces with the supply chain plan, 435
recognize the role of producer services in the company, 434
revise and assimilate, 436
select suppliers to provide a continuing relationship, 435
troublesome functions, 424
Project evaluation and review technique (PERT), 311
PRTM Group, 389, 393

Q

QFD, *see* Quality function deployment
QRS, *see* Quick Response Systems
Quality
factors, 32
function deployment (QFD), 121
improvement of, 465
intangibility of, 209
measurement, 132
Quasi-private firms, 439
Quick Response Systems (QRS), 191, 192

R

Radiofrequency identification (RFID), 64, 287, 433
Re-engineering movement, 287
REI (Recreation Equipment), 354, 371
Reservation systems, 430
Response time reduction, 191–196
bullwhip effect, 191
Collaborative Planning Forecasting and Replenishment, 191, 193–194
continuous replenishment programs, 192
Efficient Consumer Response, 191, 192
electronic data interchange, 192
future of, 196
global online registry, 195
inventory challenge, 196
offshore outsourcing movement, 195
point-of-sales terminals, 192
present status, 194–196
Quick Response Systems, 191, 192
sales and operations planning, 193
supply chain management, 194
value-added networks, 192
vendor-managed inventory, 193
voluntary inter-industry commerce standards, 193

Retrenchment, 323
Reverse logistics, 70–75
barriers to implementation, 73
benefits of, 72
closed-loop supply chain, 71
forward and reverse supply chains, 74
green supply chains and, 70
information and communications technology support, 73
intent, 71
interest in, 71–72
laws, 72
public awareness of environmental concerns, 71
recommendations, 74
system design and implementation, 73–74
update, 74–75
RFID, *see* Radiofrequency identification
Risk
management, improvement of, 468
paradox, 411
Ritz Carlton Hotels, 339
Robot, 292, 346
Rockwell Automation, 460

S

Samsung, 267
Sarbanes–Oxley, 130
SAS, 357, 358
Satisficing, 27, 66, 452, 454
SBU, *see* Strategic business unit
Schmenner model, 114
Scientific management, 91
Scientific Management philosophy (Taylor), 26
SCM, *see* Supply chain management
SCOR framework, *see* Supply Chain Council framework
Sears, 333
Self-publishing, 146
Service-oriented architecture (SOA), 306, 460
Services, adapting manufacturing techniques to, 109–169, *see also* Supply chain, role of services to complement
Amazon, 138–149
agency pricing, 143
Amazon Cloud, 145
Amazon Web Services, 139, 143
antitrust lawsuit, 143
cloud computing services, 148
as competitor, 149
distribution (fulfillment centers), 143–145

e-books, 145, 148
employees, 144
expansion into new businesses, 138
fulfillment centers, 139
incorporation, 138
inventory management, 145
major business segments, 148
manufacturing (Kindle), 145–146
origination (book authors), 146
overview, 138–140
primary source of revenue, 138
retailing (electronic products), 142–143
retailing (physical products), 140–142
sales, income, and cash flow, 140
self-publishing, 146
services, 138
shipping companies, 144
Simple Storage Service, 140
summary, 146–149
supply chain, 146, 147
variable costs, 139
virtual manufacturers, 145
comparison of manufacturing and services,
    114–116
assumption, 116
change, 116
process characteristics, 114–115
product–process matrix, 115
description of manufacturing process types,
    110–113
amount of labor content, 110
assemble-to-order process, 113
batch sizes, 112
employee skill requirements, mental, 110
employee skill requirements, physical,
    111
engineer-to-order producers, 113
level of employee flexibility through
    cross-training, 112
level of equipment intensity, 112
level of product standardization, 112
make-to-stock process, 113
process classification, 110
product orientation, 113
regularity of process flow, 112
type of facilities required, 112
types of manufacturing processes, 111
volume of products, 112–113
JIT and lean, 124–127
association with Six Sigma, 12
bid quotation process, 125
customer service, 126

door-to-balloon time, 126
kanban, 125
Lean Sigma, 126
local government, 125–126
mass services, 125
need for integrated system, 124
service characteristics, 124–125
state government, 125
keys to extending manufacturing techniques
    to services, 133–134
breaking free from product-based roots,
    134
ITO Model, 133
return to roots, 134
service awakening, 134
service management era, 134
manufacturing objectives, 116–123
"almost finished" products, 121
batch flow, 122
batch sizes, 117
concurrent engineering, 121
demand forecasting, 120
employee empowerment, 120
equipment utilization, 118
improve quality, 119–120
in-control operation, 118
increase resource utilization, 118–119
initiatives, 122
materials requirements planning, 117
outsourcing, 121
product standardization, 118
quality function deployment, 121
reduce inventories, 117–118
reduce product costs, 116–117
reduce product development time,
    121–123
reduce response time, 120–121
repetitive industry businesses, 119
specialization of labor, 116
statistical techniques, 119–120
product–process relationship, 113
demands of marketplace, 113
job-shop approach, 113
life-cycle curve, 113
programs more difficult to adapt to service
    operations, 130–133
activity-based costing, 131
automation, 132–133
enterprise resources planning, 132
materials requirements planning,
    131–132
performance measures, 132

product costing, 131
resource utilization, 133
programs that work in services, 123–130
   JIT and lean, 124–127
   services science, 123
   TQM and Six Sigma, 127–130
   understudied field, 123
quality programs, 110
service industry classifications, 113–114
   Chase model, 114
   Schmenner model, 114
service objectives, 123
Total Quality Management, 109
TQM and Six Sigma, 127–130
   decision support systems, 130
   health care, 127–128
   hospital processes, 129
   Human Sigma, 128
   intensive care, 129
   ISO certification, 129
   local government, 128–129
   need for committed employees, 128
   retail, 130
   Sarbanes–Oxley, 130
   similarity of manufacturing and service
      applications, 130
   TQM and change management, 129
   universities, 127
types of manufacturing processes, 111
United Parcel Service, 150–168
   acquisitions, 151
   business segments, 157–162
   company history, 150–152
   compressed natural gas vehicles, 151
   concern for employee welfare, 155
   constructive dissatisfaction, 153
   crusade for continual improvement,
      152–154
   Delivery Information Acquisition
      Device, 151
   embedded culture of efficiency, 152
   emphasis on drivers as key customer
      interface, 154–155
   evolution along supply chain, 156–157
   hierarchy of services provided to
      businesses, 164
   industrial engineering techniques,
      152–153
   key financial results, 162
   profitability ratios, 163
   progressive use of technology, 153–154

   relationship with unions (Teamsters),
      155
   revenues and income, 163
   stability at CEO level (promote from
      within), 154
   summary, 162–165
   sustainability, 155–156
   timeline of services, 166–168
   UPS Supply Chain Solutions, 151
Service sciences, future of, 461
Seven dwarfs, 224
SIFE, *see* Students in Free Enterprise
Silos, 337
Simon & Schuster, 143
Single minute exchange of dies (SMED), 118
Six Sigma, 5, 106, 451
Six Sigma, TQM and, 127–130
   decision support systems, 130
   health care, 127–128
   hospital processes, 129
   Human Sigma, 128
   intensive care, 129
   ISO certification, 129
   local government, 128–129
   need for committed employees, 128
   retail, 130
   Sarbanes–Oxley, 130
   similarity of manufacturing and service
      applications, 130
   TQM and change management, 129
   universities, 127
SKU, *see* Stock keeping unit
Skunk works, 337
SMED, *see* Single minute exchange of dies
SOA, *see* Service-oriented architecture
Social services, 422, 437–443
   classes, 438–439
      government, 438
      nonprofit organizations, 438–439
      quasi-private firms, 439
   future of, 442–443
      emphasis on business-related topics, 443
      level of effectiveness and efficiency, 443
      level of privatization, 442–443
      level of regulation, 442
   importance, 437
   role of, 437–438
   specific, 439–441
      communications systems, 439
      education and training, 440
      health care, 440–441
      information, 439

location incentives, 440
protection, 440
standards, 439
support infrastructure, 440
steps to interfacing business with, 441–442
businesses may have less choice in selecting the supplier, 441
consequences may be more than economic, 441
political climates affect social supplier actions, 441
relationships with the supplier may be more permanent, 441
social services may not be as good as private enterprises, 441–442
Socio-technologies, 280
Sodexo, 333
Soft technologies, 280, 289
Southwest Airlines, 134, 353, 356, 360
SPC, *see* Statistical process control
Statistical process control (SPC), 34, 35, 106
Stock keeping unit (SKU), 191–192
Strategic business unit (SBU), 34
Strategic planning, *see* Critical success factors and strategic planning
Straying acronym syndrome, 187
Students in Free Enterprise (SIFE), 376
Sun Microsystems, 354
Supply chain
closed-loop, 71
configurations, extension of ITO Model into, 62–70
building relationships, 68–70
classical management thinkers, 65–66
closed and open systems model, 62–68
closed systems, 62–63
conflicts of interest, 69
core competency, 67
determinate system, 65
effect of variation on process, 67
entropy, 63
feedback, 67–68
making decisions in bounded rationality, 66
marketing, 69
multiple supply chains, 70
open systems, 63–64
original conditions in relationship building, 68
progression to seamless flow, 69
radiofrequency identification tags, 64
realignment, 70

relationship building, 68
satisficing, 66
supply chain, 62
forward, 74
green, 70
management (SCM), 188, 194, 381, 386
need to integrate companies into, 12
reverse, 74
Supply chain, integrated, 381–417
background of supply chains, 382–386
cross-functional teams, 384
division of labor concept, 384
evolution of the supply chain, 385
"functional foreman" recommendations, 384
functional silos, 384
Input–Transformation–Output Model, 386
inter-organizational systems, 386
material flows, 382–383
organization hierarchy, 384
role of the supply chain, 386
systems theory, 384–386
expectations, 386–391
collaboration, 390
competition, 387, 388
complexity of supply chains, 388
corporate strategy, 388
definitions, 386
demand, 386
e-commerce, 389
Extended Mark-up Language technology, 390
full network connectivity, 387
global network, 386
information sharing, 389
lean supply chain, 386
recursive interplay, 389
scenario, 390
surveys, 387
variability, 388
lean production, 396–398
collaboration with customers, 398
delivery process, 398
demand variation, 398
key values, 396
limitations, 396–397
product, 397
production process, 397–398
purchasing process, 397
third-party logistics providers, 398

looking ahead, 410–413
  clairvoyance, 411–412
  complex adaptive system, 411
  complexity, 410–411
  customer collaboration paradox, 411
  flexibility paradox, 411
  innovation paradox, 411
  innovation skills, 410
  left-brain thinking, 412–413
  optimization paradox, 410
  risk paradox, 411
models, 391–398
  action steps, 395
  comprehensive supply chain model,
    393–395
  cross-enterprise collaboration, 393
  demand management, 392
  evolution of supply chain models,
    391–393
  external integration, 393
  functional focus, 393
  Global Supply Chain Forum framework,
    391
  industrial organization, 391
  integrated process redesign, 391
  internal integration, 393
  intersectionist perspective, 392
  lean production, 396–398
  modern logistics, 391
  order fulfillment, 392
  processes, 391–392
  purchasing, 392
  relabeling perspective, 392
  requirement for interactive movement,
    395
  supply chain awareness, 391
  Supply Chain Council framework,
    objectives of, 392
  traditional logistics, 391
  transformation, 393
  transition from batch flow to integrated
    and agile supply chain, 394
  unionist perspective, 392
present status of supply chains, 381–382
  benefits, 381
  difficulty of building, 382
  global strategies, 381
  progress in developing lean supply
    chains, 384
  status of lean supply chain
    implementation, 383
  study, 382

steps to achieve lean and agile supply chain,
  398–408
  collaboration, 407–408
  commitment, 399–400
  communication, 404–405
  concept, 400
  configuration, 400–404
  coordination, 407
  culture, 405
  customization, 406
  distribution, 402
  finance and accounting, 403
  integrated supply chain, 406–407
  manufacturing, 402
  marketing, 401
  purchasing, 401
  top management, 403–404
  trust, 407–408
steps in the change process, 408–409
  implement, 409
  initiate, 408
  innovate, 409
  institutionalize, 409
  integrate, 409
  investigate, 408
  invigorate, 409
Supply chain, role of services to complement,
  419–448
  consumer services, 443–444
    data analytics, 444
    explicit knowledge, 444
    future of, 444
    hospitality industries, 444
    roles of, 443–444
    tacit knowledge, 444
    ways of collecting customer data, 444
  future of producer services, 436
    core idea, 436
    depersonalization of communications,
      436
    outsourcing, 436
    services sciences, 436
  integrated service package, 444–447
    develop objectives for each component,
      446–447
    develop objectives for the service
      package, 446
    develop strategies for each component,
      447
    develop strategies for the service
      package, 446
    evaluate the results, 447

identify the service package components, 445–446
implement the strategic plan, 447
need for, 445
recognize the need to change, 445
revise as needed, 447
producer services, 424–437
classes, 427
determinants of demand, 424–425
functional classification, 426
importance, 425
reasons to acquire, 427–434
role of outsourcing, 426
role of producer services, 427
troublesome functions, 424
producer services firms buy producer services, 437
producer services versus consumer services, 420–424
customer categories, 420
distributive services, 422
extended sectoral classification, 421–422
extractive industries, 422
industry groups along supply chain, 420
model of services provided to businesses, 423
social services, 422
social services, 437–443
businesses may have less choice in selecting the supplier, 441
classes, 438–439
communications systems, 439
consequences may be more than economic, 441
education and training, 440
emphasis on business-related topics, 443
future of, 442–443
government, 438
health care, 440–441
importance, 437
information, 439
level of effectiveness and efficiency, 443
level of privatization, 442–443
level of regulation, 442
location incentives, 440
nonprofit organizations, 438–439
political climates affect social supplier actions, 441
protection, 440
quasi-private firms, 439
relationships with the supplier may be more permanent, 441

role of, 437–438
social services may not be as good as private enterprises, 441–442
specific, 439–441
standards, 439
steps to interfacing business with, 441–442
support infrastructure, 440
steps to interfacing business with producer services, 434–436
develop a strategy for each producer services area, 434
enterprise resources planning system, 434
implement the service (fit it to the situation), 435
integrated supply chain, 434
measure the progress, 435–436
prepare a plan that interfaces with the supply chain plan, 435
recognize the role of producer services in the company, 434
revise and assimilate, 436
select suppliers to provide a continuing relationship, 435
summary, 447–448
Supply Chain Council (SCOR) framework, 392
Systems theory, 95, 384

T

Tacit knowledge, 286, 444, 475
Teamsters, 155
Technology, role of in continuous improvement, 277–320
agriculture, mining, construction, and manufacturing (goods producers), 287–290
flexible manufacturing systems, 287
lean production, 287
process technologies in manufacturing, 288
re-engineering movement, 287
RFID, 287
simulation, 290
soft or conceptual technologies, 289
decision making, criteria used in, 306–309
behavioral versus scientific management issues, 307–308
break-even analysis, 308
costs versus benefits of added technology, 308–309
decision tree technique, 308

job content, 307
preference matrix approach, 308
strategic needs versus short-term needs,
  306–307
trade-offs, 307
vision statement, 306
definitions, 280
e-business, 293–295
  characteristics of successful
    implementations, 295
  customer-focused processes, 294
  definitions and concepts, 293–295
  infrastructure, 294
  management-focused processes, 294
  processes, 294
  production-focused processes, 294
  transactions, 295
enterprise resource planning, 278
equipment, 300
  enhance performance of equipment, 300
  provide source of performance
    information, 300
  sensing devices, 300
facilities, 300–303
  design, 301–302
  diversity of services, 301
  facilitating the flow of goods and
    information, 302
  hard results, 301
  improving human and machine
    productivity, 302
  layout, 303
  location, 302–303
  optimizing the customer encounter, 302
  optimizing facility flexibility, 302
  safe and morale-building environment,
    302
future considerations for technology,
  312–316
  Bill Gates's view, 315–316
  collaboration, 315
  customer acceptance, 313–314
  economic feasibility, 314–315
  electronic data interchange, 314
  older generations, 313
  technical feasibility, 315
  workforce acceptance, 314
hard technologies, 280
humanness, 277
human resources, 296–300
  as aid to employee, 296–297
  automated guided vehicles, 297

automated materials-handling systems,
  296
balance, 299
computer-integrated manufacturing, 297
demand for customization, 299
digital document systems, 296
displacement of people, 298
examples of helpful technology, 297
flexibility, 299
flexible manufacturing systems, 297
as integral part of process, 298–299
interdisciplinary design methodology,
  300
lights-out factory, 298
as substitute for employee, 297–298
information technology, 303–304, 305
  description of, 303
  function, 303
  issues matching IT to business, 305
integrated systems, 304–306
  compatibility problems, 304
  enterprise resource planning systems,
    304
  extensible markup language, 304
  inter-organizational systems, 304
  manufacturing resources planning
    systems, 304
  materials requirements planning, 304
  service-oriented architecture, 306
  structure, trust, and collaboration, 306
inter-company communication, 278
ITO Model, 279, 280
resistance to new technologies, 279
resource enhancement, technology for,
  296–306
  equipment, 300
  facilities, 300–303
  human resources, 296–300
  information technology, 303–305
  integrated systems, 304–306
self-directed teams, 278
self-service, 292–293
services, 290–292
  back-room operation, 290
  bank tellers, 290
  categories of robots, 292
  challenge, 292
  examples of automation in services, 291
  personalized service, 291
  robots, categories of, 292
socio-technologies, 280
soft technologies, 280, 289

steps in adding technology to the process, 309–312
    step 1 (communicate), 309
    step 2 (identify needs and opportunities), 309–310
    step 3 (evaluate alternatives and select the optimum alternative), 310
    step 4 (educate and orient), 310
    step 5 (develop the implementation plan), 311
    step 6 (implement the technological changes), 311–312
    step 7 (evaluate results, redefine needs, and redefine additional increments), 312
summary, 316–318
    competitiveness, 316
    decision making, 317
    degree of customer contact, 316
    employee acceptance of new technology, 317
    flatter organizations, 316
    humanness, 318
    relationship of customers and technology, 316
    service delivery system, 317
    technology in our future, 318
technology categories, 281
technology and the infrastructure, 281–283
    decentralization of management responsibilities, 282
    focused factory approach, 282
    implications, 282
    supply chain, 282
    unwelcome changes, 281
technology and organizational culture, 283–285
    corporate culture, 283
    enterprise resource planning, 284
    globalization, 285
    implicit knowledge, 285
    knowledge management, 284
    software development, 285
technology for process improvement, 286–295
    agriculture, mining, construction, and manufacturing (goods producers), 287–290
    e-business, 293–295
    self-service, 292–293
    services, 290–292

technology transfer, 286
    explicit knowledge, 286
    scenarios, 286
    tacit knowledge, 286
    ways of learning new technology, 286
    typical improvement program, 278
Third-party logistics providers, 398
3M, 106, 361
Timberland, 366
Time Warner Book Group, 146
Top-down communication formats, 337
Total Quality Management (TQM), 16, 78, 96, 109
Total Quality Management (TQM), Six Sigma and, 127–130
    decision support systems, 130
    health care, 127–128
    hospital processes, 129
    Human Sigma, 128
    intensive care, 129
    ISO certification, 129
    local government, 128–129
    need for committed employees, 128
    retail, 130
    Sarbanes–Oxley, 130
    similarity of manufacturing and service applications, 130
    TQM and change management, 129
    universities, 127
Toyota, 85, 346
Toyota Production System (TPS), 78, 124
TPS, *see* Toyota Production System
TQM, *see* Total Quality Management
Triple bottom line, 27
Troublesome functions, 424
Tylenol, 433

U

Unforeseen circumstances, 33
Union Carbide, 323
United Parcel Service (UPS), 109, 135, 150–168, 183, 360, 402, 451
    acquisitions, 151
    business segments, 157–162
        global small package, 157–159
        supply chain capital, 162
        supply chain solutions, 160–162
    company history, 150–152
    compressed natural gas vehicles, 151
    concern for employee welfare, 155
    constructive dissatisfaction, 153

crusade for continual improvement,
152–154
Delivery Information Acquisition Device,
151
embedded culture of efficiency, 152
emphasis on drivers as key customer
interface, 154–155
evolution along supply chain, 156–157
heavy focus on customer, 156
integrated logistics services (domestic
and global), 157
movement downstream in supply chain
(retail stores), 157
movement upstream in supply chain,
156–157
hierarchy of services provided to businesses,
164
industrial engineering techniques, 152–153
key financial results, 162
profitability ratios, 163
progressive use of technology, 153–154
airplane scheduling and operation,
153–154
fleet planning, 153
package tracking, 153
routing trucks, 153
sorting and loading packages onto
trucks, 153
relationship with unions (Teamsters), 155
revenues and income, 163
stability at CEO level (promote from
within), 154
summary, 162–165
sustainability, 155–156
timeline of services, 166–168
UPS Supply Chain Solutions, 151
United States Marine Corps, 360
UPS, *see* United Parcel Service

V

Value-added networks (VANs), 192, 203
Value statements, 355
Vanishing manufacturing/services boundary,
1–18
differences between manufacturing and
service, 1–3
complementary package, 1
knowledgeable employees, 2
line flow management, 2
service organization assessment, 2

forces that are eliminating the boundary,
3–13
automobile manufacturers, 7
banks facing competition, 5
batch processing of transactions, 10
blending of tasks, resources, and
techniques into programs, 4
changing personal and organizational
relationships, 11
chaos theory, 13
continuous improvements blurring the
boundary between manufacturing
and services, 12–13
creditworthiness, 8
customer input to product design, 8
economy going through a natural
evolution, 3
enhanced customer relationship
management, 8
financial services, 3
financing of purchases, 8
health care providers looking for ways to
cut costs and improve quality, 5
health care service, 3
introduction of technology, 3
investment bankers facing competition,
6
Lean Sigma, 5
manufacturing adding services to be
more customer focused, 6–8
mass customization concept, 12–13
movement toward industry focus, 12
movement toward process perspective, 10
need to identify critical success factors, 4
need to integrate companies into supply
chains, 12
online purchasing, 8
outsourcing, 10–11
overshooting, 5
person–machine relationship, 11
post-sales services, 7
product-service continuum, 7
relative participation of change agents, 9
retail industry moving in two strategic
directions, 6
retail industry stalwarts, 6
"sense and respond" approach, 8
service businesses looking for ways to
operate more efficiently, 5–6
services becoming more relevant to all
types of customers, 12
services segregated into industries, 3

supply chains expanding, 4
technology development, 9–10
view of technology, 9
virtual organization, 10
quest to find ways to compete, 1
summary, 17
vanishing manufacturing/services boundary,
    14–17
    bundling of services with products, 16
    make-and-sell strategy, 16
    phase 1 (separated disciplines), 14
    phase 2 (internal improvements in costs
        and quality), 14–16
    phase 3 (customer service
        improvements), 16
    phase 4 (integrated product and service
        functions), 16–17
    total quality management, 16
VANs, *see* Value-added networks
VICS, *see* Voluntary inter-industry commerce
        standards
Virtual manufacturers, 145
Virtual organization, 331–333, 460
    creation of, 17
    future of, 460
    process perspective, 10
Virtual private network (VPN), 204

Visio, 290
Voluntary inter-industry commerce standards
        (VICS), 193
VPN, *see* Virtual private network

W

Walmart, 6, 134, 172, 323, 387
Walt Disney Company, 322, 360
Wellness programs, 430
Wendy's, 338
Western Electric, 96, 127
William Wrigley Jr. Company, 358
Wind tunnels, 50
Wireless Internet service, 375
Wisdom, 345, 476
WorldNow, 362, 375

X

Xerox, 267
XML, *see* Extensible markup language

Y

Yahoo!, 354
Y2K, 188